Applied Artificial Intelligence for Industrial Nondestructive Evaluation NDE4.0

Applied Artificial Intelligence for Industrial Nondestructive Evaluation NDE4.0

Guest Editors

Ahmad Osman
Valérie Kaftandjian

MDPI

Basel • Beijing • Wuhan • Barcelona • Belgrade • Novi Sad • Cluj • Manchester

Guest Editors

Ahmad Osman
School of Engineering
Saarland University of
Applied Sciences
Saarbrücken
Germany

Valérie Kaftandjian
Mechanical Engineering
Department
National Institute of Applied
Sciences of Lyon
Villeurbanne
France

Editorial Office
MDPI AG
Grosspeteranlage 5
4052 Basel, Switzerland

This is a reprint of the Special Issue, published open access by the journal *Applied Sciences* (ISSN 2076-3417), freely accessible at: https://www.mdpi.com/journal/applsci/special_issues/SIJC7946A3.

For citation purposes, cite each article independently as indicated on the article page online and as indicated below:

Lastname, A.A.; Lastname, B.B. Article Title. *Journal Name* **Year**, *Volume Number*, Page Range.

ISBN 978-3-7258-5127-0 (Hbk)
ISBN 978-3-7258-5128-7 (PDF)
https://doi.org/10.3390/books978-3-7258-5128-7

Contents

About the Editors

Ahmad Osman

Ahmad Osman is a professor of signal and image processing with a focus on modern AI at Saarland University of Applied Sciences and academic lead at the Fraunhofer Institute for Nondestructive Testing IZFP in Saarbrücken, Germany. He is also affiliated as an adjunct Professor with Université Laval in Quebec, Canada. His research focuses on the integration of artificial intelligence, sensor systems, and data-driven innovation into industrial applications, particularly in the fields of nondestructive evaluation (NDE), predictive maintenance, and intelligent inspection systems. He has led numerous international R&D projects and industry collaborations, contributing to the development of real-world AI solutions for manufacturing and infrastructure. Professor Osman also serves as director of the Franco-German Institute of Higher Education (DFHI) and plays an active role in mentoring start-ups and advising international technology transfer initiatives. He is a senior member of the IEEE and has held visiting positions at renowned universities while serving as an advisor on AI and Digital Transformation to the private sector and public organizations.

Valérie Kaftandjian

Valérie Kaftandjian is a professor of physics at the National Institute of Applied Sciences of Lyon (INSA Lyon), France, and a senior researcher in the Laboratory of Vibration and Acoustics (LVA). Her scientific work centers on modeling and applying X-ray and ultrasonic techniques in nondestructive evaluation, with a strong focus on structural health monitoring, defect characterization, and inverse methods. She has coordinated multiple collaborative projects at the European and national levels, contributing to advances in simulation-assisted NDE and the deployment of hybrid methods that combine physical models with AI-driven signal processing. Professor Kaftandjian actively supports the development of open-access research and interdisciplinary exchange. As a chair of the working group "AI and NDT" of the COFREND French NDT society, she remains a driving force in the academic and industrial dissemination of NDE technologies.

Preface

This Special Issue reprint, "Applied Artificial Intelligence for Industrial Nondestructive Evaluation" (NDE4.0), brings together a curated selection of research articles that demonstrate the growing role of artificial intelligence in enhancing industrial reliability and safety. Motivated by the increasing complexity of materials, processes, and inspection environments, this reprint addresses how intelligent systems, particularly machine learning and deep learning, can support and advance nondestructive evaluation methods.

This collection aims to provide insight into state-of-the-art techniques and practical applications, where AI enables faster, more accurate, and scalable detection of defects. It is intended for researchers, engineers, industry professionals, and academic audiences interested in the intersection of AI and industrial inspection technologies.

As Guest Editors, we thank all contributing authors for their valuable work and collaboration, and we express our gratitude to the reviewers and editorial team at *Applied Sciences* for their support in bringing this reprint to completion.

Ahmad Osman and Valérie Kaftandjian
Guest Editors

applied sciences MDPI

Editorial

Applied Artificial Intelligence for Industrial Nondestructive Evaluation NDE4.0

Ahmad Osman [1,2,*] and Valerie Kaftandjian [3]

1 Fraunhofer Institute for Nondestructive Testing IZFP, Campus E 3.1, 66123 Saarbruecken, Germany
2 School of Engineering, Saarland University of Applied Sciences, Goebenstrasse 40, 66117 Saarbrücken, Germany
3 Laboratory of Vibration and Acoustics, Mechanical Engineering and Design Department, National Institute of Applied Sciences of Lyon, INSA-Lyon, UR677, CEDEX, 69621 Villeurbanne, France; valerie.kaftandjian@insa-lyon.fr
* Correspondence: ahmad.osman@izfp.fraunhofer.de

1. Introduction

Imagine a world where a single undetected crack in a turbine blade or a hidden flaw in a 3D-printed component could halt production or worse. This is the stakes-driven reality that nondestructive evaluation (NDE) confronts, now revolutionized by artificial intelligence (AI) and deep learning (DL) in the era of NDE 4.0. By fusing advanced sensors, robotics, and intelligent data analysis, NDE 4.0 empowers industries, from aerospace to energy, to detect defects with unprecedented precision. This Special Issue "Applied Artificial Intelligence for Industrial Nondestructive Evaluation NDE4.0" addresses the following challenges: noisy data drowning out signals, complex defects that are difficult to interpret, and the struggle of AI to scale or earn trust, with 11 pioneering papers, illuminating solutions and igniting curiosity for what lies ahead.

2. An Overview of the Published Articles

This collection showcases AI's transformative potential across NDE applications, from factory floors to infrastructure maintenance. These 11 contributions, diverse yet interconnected, address critical gaps with ingenuity and practical relevance.

Lang et al. (contribution 1) harnessed deep learning to monitor additive co-extrusion processes in real time using laser profiler data. Their work strengthened thermoplastic components—vital for industries such as the automotive—by detecting defects as they form in advanced manufacturing settings.

Bai et al. (contribution 2) blended temporal convolutional networks with bidirectional LSTM to predict equipment faults, mastering the complexity of time-series data and addressing limitations in traditional fault prediction methods.

Wang and Xie (contribution 3) pivoted to acoustic signals for rolling bearing diagnostics in noisy environments, leveraging microphones' affordability and sensitivity over vibration sensors to tackle signal interference.

Ko and Lee (contribution 4) introduced a multi-patch time series transformer to detect bearing faults under varying noise, mimicking the chaotic conditions of industrial settings with novel noise techniques.

Jin et al. (contribution 5) used multiple attention mechanisms to segment defects in drainage pipeline sonar images, safeguarding aging infrastructure from collapse by targeting complex defect types.

Li and Chen (contribution 6) deployed machine learning, optimized by particle swarm optimization, to predict lithium-ion battery health—a linchpin for sustainable energy systems—using LSTM, CNN, and SVR models.

Wang et al. (contribution 7) elevated steel production with an improved YOLOv8 model, tackling the complex reality of surface defects amid noise and complex backgrounds.

Val et al. (contribution 8) turned to AI to analyze wear in milling insert tools, optimizing lifespans and slashing costs in modern machining by examining flank wear patterns.

Baumeyer et al. (contribution 9) enhanced magnetic particle inspection with AI assistance, improving defect detection under UV-A lighting—a boon for quality control in harsh industrial environments.

Wei et al. (contribution 10, Appl. Sci. 2023, 13(24), 13093) provided a pulsed thermography dataset for automated defect depth estimation in PVC, addressing the challenge of subsurface defect characterization.

Wei et al. (contribution 11, Appl. Sci. 2023, 13(5), 2901) offered another dataset for training DL models in pulsed thermography, easing the interpretive burden and fueling research with high-quality data.

Together, these papers bridge noisy data, defect complexity, and scalability gaps, blending AI with human expertise to redefine industrial reliability.

3. Conclusions

The journey of NDE 4.0 is just beginning, and this Special Issue paves the way toward new horizons. How can we ensure AI earns the trust of engineers and regulators in life-or-death applications? Developing explainable models and universal validation metrics is a pressing next step. As sensor networks flood us with data, scaling NDE through cloud, edge, and fog computing beckons as an urgent challenge—and opportunity. The partnership between human inspectors and AI assistants remains uncharted; optimizing this relationship could unlock new efficiencies. With industries racing toward composites and 3D-printed parts, adapting these methods to novel materials is also worthwhile.

We invite researchers, engineers, and industry pioneers to address these questions. Imagine an NDE 4.0 that does not just detect flaws but predicts and prevents them—saving time, costs, and lives. This collection is your springboard; dive in, explore, and shape the future of this vibrant field.

Author Contributions: A.O.: Writing—original draft, Writing—review and editing. V.K.: Review and editing. All authors have read and agreed to the published version of the manuscript.

Conflicts of Interest: The authors declare no conflicts of interest.

List of Contributions:

1. Lang, V.; Herrmann, C.; Fuchs, M.; Ihlenfeldt, S. Deep Learning Utilization for In-Line Monitoring of an Additive Co-Extrusion Process Based on Evaluation of Laser Profiler Data. *Appl. Sci.* **2025**, *15*, 1727. https://doi.org/10.3390/app15041727.
2. Bai, J.; Zhu, W.; Liu, S.; Ye, C.; Zheng, P.; Wang, X. A Temporal Convolutional Network–Bidirectional Long Short-Term Memory (TCN-BiLSTM) Prediction Model for Temporal Faults in Industrial Equipment. *Appl. Sci.* **2025**, *15*, 1702. https://doi.org/10.3390/app15041702.
3. Wang, H.; Xie, J. Fault Diagnosis of Rolling Bearings Based on Acoustic Signals in Strong Noise Environments. *Appl. Sci.* **2025**, *15*, 1389. https://doi.org/10.3390/app15031389.
4. Ko, S.; Lee, S. Multi-Patch Time Series Transformer for Robust Bearing Fault Detection with Varying Noise. *Appl. Sci.* **2025**, *15*, 1257. https://doi.org/10.3390/app15031257.

5. Jin, Q.; Han, Q.; Qian, J.; Sun, L.; Ge, K.; Xia, J. Drainage Pipeline Multi-Defect Segmentation Assisted by Multiple Attention for Sonar Images. *Appl. Sci.* **2025**, *15*, 597. https://doi.org/10.3390/app15020597.
6. Li, K.; Chen, X. Machine Learning-Based Lithium Battery State of Health Prediction Research. *Appl. Sci.* **2025**, *15*, 516. https://doi.org/10.3390/app15020516.
7. Wang, J.; Chen, T.; Xu, X.; Zhao, L.; Yuan, D.; Du, Y.; Guo, X.; Chen, N. An Improved YOLOv8 Model for Strip Steel Surface Defect Detection. *Appl. Sci.* **2025**, *15*, 52. https://doi.org/10.3390/app15010052.
8. Val, S.; Lambán, M.; Lucia, J.; Royo, J. Analysis and Prediction of Wear in Interchangeable Milling Insert Tools Using Artificial Intelligence Techniques. *Appl. Sci.* **2024**, *14*, 11840. https://doi.org/10.3390/app142411840.
9. Baumeyer, J.; Chatoux, H.; Pelletier, A.; Marquié, P. Industrial Application of AI-Based Assistive Magnetic Particle Inspection. *Appl. Sci.* **2024**, *14*, 1499. https://doi.org/10.3390/app14041499.
10. Wei, Z.; Osman, A.; Valeske, B.; Maldague, X. A Dataset of Pulsed Thermography for Automated Defect Depth Estimation. *Appl. Sci.* **2023**, *13*, 13093. https://doi.org/10.3390/app132413093.
11. Wei, Z.; Osman, A.; Valeske, B.; Maldague, X. Pulsed Thermography Dataset for Training Deep Learning Models. *Appl. Sci.* **2023**, *13*, 2901. https://doi.org/10.3390/app13052901.

applied sciences

MDPI

Article

Deep Learning Utilization for In-Line Monitoring of an Additive Co-Extrusion Process Based on Evaluation of Laser Profiler Data

Valentin Lang [1,*], Christian Thomas Ernst Herrmann [1,2], Mirco Fuchs [2] and Steffen Ihlenfeldt [1,3]

[1] Institute of Mechatronic Engineering, Faculty of Mechanical Engineering, Technische Universität Dresden (TUD), 01062 Dresden, Germany; steffen.ihlenfeldt@tu-dresden.de (S.I.)

[2] Institute of Electronics and Biomedical Information Technology, Faculty of Engineering, Hochschule für Technik, Wirtschaft und Kultur (HTWK Leipzig), 04277 Leipzig, Germany; mirco.fuchs@htwk-leipzig.de

[3] Fraunhofer Institute for Machine Tools and Forming Technology, 01187 Dresden, Germany

* Correspondence: valentin.lang@tu-dresden.de; Tel.: +49-351-463-34338

Abstract: Additive manufacturing is gaining importance in a number of application areas, and there is an increased demand for mechanically resilient components. A way to improve the mechanical properties of parts made of thermoplastics is by using reinforcing material. The study demonstrates the development of a monitoring procedure for a fused filament fabrication-based co-extrusion process for manufacturing wire-reinforced thermoplastic components. Test components in two variants are produced, and data acquisition is carried out with a laser line scanner. The collected data are employed to train deep neural networks to classify the printed layers, aiming for the deep neural networks to be able to classify four different classes and identify layers with insufficient quality. A dedicated convolutional neural network is designed taking into account various factors such as layer architecture, data pre-processing and optimization methods. Several network architectures, including transfer learning (based on VGG16 and ResNet50), with and without fine-tuning, are compared in terms of their performance based on the F1 score. Both the transfer learning model with ResNet50 and the fine-tuning model achieve an F1 score of 84% and 83%, respectively, for the decisive class 'wire bad' classifying inadequate reinforcement.

Keywords: artificial intelligence; computer vision; deep learning; artificial neural networks; convolutional neural networks; additive manufacturing; fused filament fabrication; co-extrusion

1. Introduction

In the future vision of industry, intelligent machines coordinate manufacturing processes independently, and autonomous devices perform most of the human operators' routine tasks. Networking will take place both within factories and across company boundaries. Digital networking combined with additive manufacturing technologies enables new production processes, business models and products [1–3]. Thanks to continuous monitoring of machine statuses, potential defects can be detected quickly to prevent breakdowns and costly downtimes [4–6].

Additive manufacturing processes excel by combining maximum design freedom with the best efficiency for the production of small series [7–9]. On the other hand, improvements in dimensional accuracy, load-bearing capacity of processable materials and reproducibility of production quality still need to be achieved in order to make additive manufacturing processes robust for industrial production [10–12]. Reinforced plastic composite parts are recognized by structural characteristics tailored to the load profile and therefore offer particularly high lightweight construction capabilities [13–15]. Compared to established manufacturing processes with the necessary toolmaking and the comparatively complex and correspondingly costly production of plastic composite components, additive manufacturing of reinforced plastic components can be expected to yield major gains in manufacturing efficiency [16–18]. Multi-material 3D printing can combine the advantages of additive manufacturing with the benefits of fiber-reinforced plastic composite components. Quality assurance constitutes a key factor in the use of wire-reinforced 3D printing processes, as the quality of the printed parts significantly influences the component properties and reliability in various applications [19–21]. However, traditional methods of quality control, such as visual inspection, face limitations, especially when it comes to detecting and evaluating complex structures and subtle defects inside a component. In this study, routines are developed and evaluated facilitating in-line monitoring for a thermoplastic co-extrusion printing process utilizing data from a laser profiler, scanning each printed layer and recording corresponding height profiles.

Computer vision is an area of artificial intelligence (AI) that enables computers to extract relevant information from digital images, videos and other visual inputs and, as a result, take action or make recommendations [22]. When it comes to large volumes of data, as typically can arise in ongoing production processes, deep learning (DL) methods are a particularly promising way of leveraging conventional domain-specific process knowledge for automated production monitoring solutions [23–25]. Artificial neural networks (ANNs) can significantly enhance process monitoring in industrial processes through their ability to learn patterns, adapt to complex data, and make predictions [26,27]. By leveraging these capabilities, ANNs improve decision-making, efficiency, and reliability in industrial process monitoring [28,29]. Therefore, this study investigates and evaluates the application of artificial neural networks for quality control of wire-reinforced 3D printing processes. The focus is on the development and implementation of deep learning models that can accurately identify and characterize defects and quality anomalies in printed components. Test components in two variants are produced, and data acquisition is carried out with a laser profiler. The collected data are used to train deep neural networks to classify the printed layers into four different classes and identify those layers with insufficient quality. The stages of data acquisition, data preparation, modeling and evaluation are documented in detail. Data acquisition essentially describes the design and printing of test components, scanning with the laser profiler and importing the recorded data. Data pre-processing involves selecting the relevant areas in the raw data and subsequently splitting the matrices into sections and thus increasing the total number of instances. Grayscale images are converted such that they can be used with neural networks designed for RGB data. A dedicated convolutional neural network (CNN) is designed taking into account various factors such as layer architecture, data pre-processing and optimization methods. The various network architectures, including the custom CNN, as well as additional transfer learning models (based on VGG16 and ResNet50), and transfer learning with fine-tuning, are compared in terms of their performance based on the F1 score.

2. Literature Review

With the rising availability of large data storage and increased computing power, it becomes further feasible to record and utilize large amounts of data. Deep learning has emerged as a particularly promising method for applications with complex correlations.

Deep learning is also increasingly applied in computer vision to analyze images and videos and recognize specific objects. This technology is used in various areas such as autonomous vehicle technology, security technology and medical technology [30–33].

Impressing with excellent recognition of correlations in space and time, e.g., in images or signal sequences, a frequently used algorithm is CNN. This is achieved by distributing weights to neuron connections and successively reducing the number of parameters in the forward progression through the artificial network architecture. A typical CNN consists of different layers, including convolutional layers, pooling layers and at least one dense layer. The algorithm is based on discrete convolution, where the input in two- or multi-dimensional matrices passes through one or more convolution matrices with filter kernels. It is worth noting that, depending on the problem, no standard solution exists; rather, the optimal architecture for a given problem must be found by trying out different approaches and structures [34].

A major challenge in training CNNs is the availability of labeled data. To ensure robust feature recognition, CNNs usually consist of a large number of deep layers. In order to develop a model, datasets with several thousand images per class are required. In practice, this is not always feasible if such large datasets are not available. The concept of transfer learning (TL) was therefore developed. TL involves using a model that has already been trained on an extensive dataset with millions of images. These images are typically not tailored to a specific application but represent a wide range of examples [35].

In its first few layers, which are close to the input, a CNN is designed to detect simple general features such as edges and corners. In the following layers, these features are combined to form high-level feature detectors. These are responsible for recognizing more complex and domain-specific features. Therefore, a model trained on general data can be used as a starting point for specific tasks to be trained. In this respect, robust models can be trained with limited datasets by truncating the fully connected (FC) layers of the base model and adding an application-specific head [36].

LeNet, also known as LeNet-5, is regarded as the starting point for the use of CNN for image recognition. In its basic form, the model consists of five layers, two of which are designed as convolution, followed by three FC layers. This model was initially used to recognize handwritten numbers [34]. AlexNet is an evolution of the LeNet architecture. It is considered a milestone in the development of CNN, in which it demonstrated the potential of deep networks. It consists of a total of eight layers, five of which are designed as convolution and three as FC layers [37]. With VGG, a distinction can be made between VGG-16 and VGG-19 depending on the type of implementation of this network. The suffix gives an overview of the number of layers used. VGG-16 consists of 13 convolution layers, 2 FC layers and 1 softmax output layer [38]. As the depth of CNN increases, a problem of convolution becomes apparent. With the layer depth, the amount of information still available from input at deep layers drops excessively. The ResNet architecture therefore does not exclusively consist of simple convolution layers but is also capable of forwarding full-size input information to deep layers of the network via residual connections [39].

A study by Dion and Pattinson [40] deals with the application of CNN in the monitoring of material extrusion as an additive manufacturing method. The authors trained a multi-headed neural network that utilizes automatically labeled images to detect deviations from optimal printing parameters. Automated data acquisition and labeling enabled the creation of an extensive and diverse 3D printing dataset with 1.2 million images from 192 different parts, which are tagged with printing parameters. The trained neural network, in conjunction with a control loop, enables real-time detection and rapid correction of various errors. The authors also create visualizations of the network predictions to provide insights into decision making [40].

Banadaki et al. [41] present an automated quality assessment system for an FFF process using a proprietary real-time CNN model. The CNN model is trained and tested on images showing internal and surface defects. The performance of the model in detecting and evaluating defects in the AM process at different extruder speeds and temperatures is examined. The model achieves a precision of 94%, and the F-score, sensitivity and precision in real-time classification of the AM process quality are all above 75% for five different classes. The proposed model adds an automated, non-contact quality monitoring system to the AM process with the aim of stopping the AM process and eliminating time-consuming inspection of printed parts for internal defects [41].

Lin et al. [42] developed an online quality monitoring system for material extrusion additive manufacturing by integrating laser scanning technology. This system captures real-time geometric data of each printed layer, enabling the detection of dimensional deviations and surface defects during the printing process [42].

In a layerwise monitoring methodology for material extrusion AM processes Lafirenza et al. [43] utilized a high-resolution blue laser line profilometer. This approach involves capturing point cloud data for each printed layer, enabling computation of layer height deviations and assessment of surface quality by detecting defects. The authors derived specific quality indexes, like average deviation of layer height, relative area of defects, and process stability slopes, to quantify layer height accuracy, defect occurrence and distribution, and overall process stability, respectively. The methodology's effectiveness is demonstrated in identifying process anomalies in real-time, thereby enhancing the reliability and repeatability of the AM process for producing functional parts [43].

Reckert et al. [44] explored the use of computer vision and deep learning to classify and predict the quality of multi-material jetting additive manufacturing processes. The study focuses on defects arising from improper droplet overlap settings. An automated classification system using custom CNNs trained on in-line surface measurements of printed layers is developed. Notably, models trained solely on first-layer images attained validation accuracies of 90%, indicating that analyzing just the initial layer suffices to predict the quality of subsequent layers [44].

Hasib Zubayer et al. [45] focus on developing DL-based image processing techniques for real-time defect detection in additively manufactured continuous carbon fiber-reinforced polymers. While the paper does not specifically address co-extrusion processes or the use of laser profiler data, its emphasis on real-time defect detection through AI and deep learning methodologies offers valuable insights. The approaches discussed could potentially be adapted for in-line monitoring and quality assurance in various AM processes, including those involving co-extrusion and laser profiling techniques.

Although a large amount of the available literature addresses the application of DL for quality monitoring in additive manufacturing AM, reviewing the available literature, it appears that there is a limited number of studies specifically addressing the application of DL for in-line monitoring of additive co-extrusion processes using laser profiler data. Co-extrusion involves the simultaneous deposition of multiple materials, adding complexity to process monitoring and defect detection, which existing studies have not addressed in particular. While a strong base of existing work exists, co-extrusion processes specifically remain less explored compared to other AM processes, opening opportunities for targeted research in this niche and leveraging broader findings in DL-enabled AM quality monitoring.

3. Materials and Methods

The machine system applied (see Figure 1) is designed as a portal machine. The basic structure of the machine frame is realized using aluminium profiles, which enables flexibility in terms of modifications. As is common with FDM printers, the print bed is not mounted in a fixed position but is traversed in the Y-direction. There are two control cabinets in the lower section of the frame. The rear control cabinet contains the power supply and the motor drivers required for the axes. The front control cabinet contains an industrial PC from Beckhoff as well as various input and output terminals with which interaction with the test bed is possible. A monitor with a control panel is located on the left-hand side. The processing area is located above the two control cabinets. During operation, the work area is protected from unauthorized access by a light barrier. There are four motors in the machining area that move the three axes. The Y-axis has a rack and pinion drive in a tandem motor configuration, which counteracts slippage of the rack and pinion pairing. The machining area has an effective construction space of 400 mm × 350 mm × 180 mm (x, y, z). Attached to the Z-axis is a mounting plate with various attachments. Both a multi-material jetting print head and an FFF co-extrusion print head are installed. The machine bench is equipped with a spring-mounted heated bed that enables a wide range of materials to be printed.

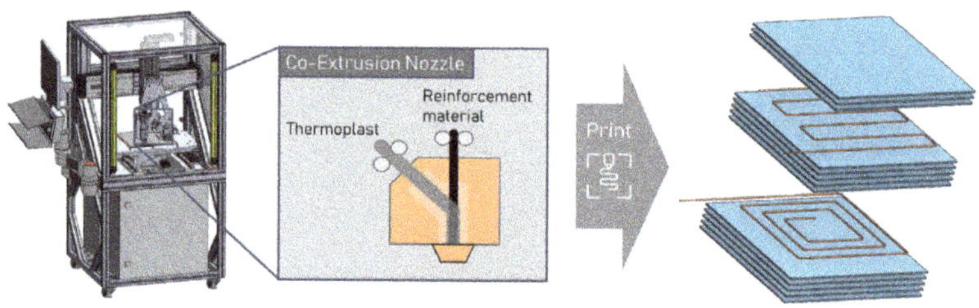

Figure 1. Setup of the experimental equipment consisting of a 3-axis machine platform with a co-extrusion head for wire-reinforced thermoplastics strand deposition.

In the experiments, polylactic acid (PLA) is used as the primary printing material (matrix material). PLA provides several advantages that make it suitable for the intended type of experiments. In general, PLA is a cost-effective material that is particularly popular in the 3D printing community [46,47]. In particular, the PLA printing process is convenient due to its properties. This is mainly due to the comparatively low printing temperature of around 190 degrees Celsius. In addition, the PLA printing process does not require active ventilation and it is also not necessary to heat the entire build chamber. These properties make PLA suitable for prototyping and predestine it for the production of a large number of printed samples [48–50]. Copper is used as a secondary material. The material is in the form of a 0.4 mm thick wire, which is additionally coated with a plastic coating. The choice of copper as a material is based on its superior material properties for the experiments conducted, including its ductility and electrical conductivity. In principle, it is possible to use a stainless steel wire, but this option does not apply in practice as stainless steel requires the use of harder printing nozzles. In addition to 0.4 mm thick wires, it is also possible to embed thinner wires into the thermoplastic. Due to the lower stiffness, these tend to slip off the propulsion unit, blocking the wire feed and clogging the propulsion unit, which is why these wires were found to be unsuitable and are not considered in this study. PrusaSlicer (Version 2.5.2+MacOS-x64) is used to generate the machine code. This software tool enables the definition of basic parameters such as printer parameters, material parameters, geometric setting parameters and other relevant settings.

Consistent component quality is ensured by using a uniform set of print settings. The most important settings are listed in Table 1. Preliminary tests involving a grid search led specifically to printing process parameters with which the technically demanding printing process, carried out using innovative experimental co-extrusion technology, runs as robustly as possible so that the quantity of profilometer data required for process analysis can be obtained. The layer height had to be aligned within certain limits to the material thickness of the wire reinforcement material. The nozzle temperature is also significantly a factor in the functionality of the intended reinforcing wire deposition, in order to ensure that the polymer matrix material is liquid enough, but not excessively liquid and can be replasticized sufficiently quickly.

Table 1. Summary of the basic Slicer settings applied.

Parameter	Value	Unit
Nozzle Diameter	0.7	mm
Layer Height	0.7	mm
Nozzle Temperature	190	°C
Bed Temperature	30	°C
Infill	10	%
Print Velocity	60	$mm{\cdot}s^{-1}$
Num. of Solid Bottom Layers	3	[-]
Num. of Solid Top Layers	3	[-]

In order to monitor the quality of the printed part, a Gocator 2330 laser line profile sensor (LPS) from LMI Technologies is integrated on the Z-axis of the experimental setup. The LPS makes it possible to obtain height information of the printed component and is moved over the component after each printed layer. The scanner works on the principle of optical triangulation, whereby a laser beam is projected onto the component. The light beam is reflected by the surface of the component and recorded by an internal camera. The height of the component relative to the scanner can be determined by the time span between the emission and reception of the laser beam. The scanner records a sequence of neighboring points, which is optically visible as a line on the component. The entire component can be captured by additionally positioning the scanner on the X-axis. Table 2 lists the key characteristics of the laser profilometer applied.

Table 2. Key characteristics of the laser profilometer Gocator 2330 applied.

Parameter	Value	Unit
Data points per profile	1280	[-]
Wavelength	660	nm
Resolution in Z direction	6–14	µm
Resolution in X direction	44–75	µm
Sample rate	170–5000	s^{-1}
Measuring distance	90	mm
Measuring distance	47–85	mm

Statistical metrics are used to evaluate the predictive efficiency of the trained models. The confusion matrix has proven particularly useful for evaluating classification problems, as it provides a range of relevant metrics. Accuracy takes into account the overall correctness of the model. Recall indicates the percentage of correctly classified positive samples.

Precision indicates the percentage of positive samples out of the sum of positive observations. The F1 score provides a harmonic mean of precision and recall, providing a balanced assessment of the predictive quality of a model. The F1 score is a metric for evaluating the model's ability to classify with best value equal to 1. These metrics are typically considered more appropriate than accuracy [51,52]. Due to the equal weighting of the precision and recall metrics, the F1 score ensures a balanced assessment of the prediction performance. Depending on the requirements of a specific application, either a high precision or a high recall rate may be of crucial importance, so in such cases it is advisable to evaluate the corresponding metric separately. $F1_i$, the F1 score for class i, is calculated as:

$$F1_i = 2 \cdot \frac{\text{Precision}_i \cdot \text{Recall}_i}{\text{Precision}_i + \text{Recall}_i}, \tag{1}$$

4. Experiments

4.1. Concept

To fulfill the generic goal of process monitoring, a suitable pipeline is to be demonstrated and, central to the research, a classification model for distinguishing relevant classes of the condition of the printing process results is to be developed, evaluated and tested. Figure 2 illustrates a conceivable structure for the in situ monitoring routine. The concept envisages utilizing the laser line profile sensor to measure the surface topography of the printed parts during the printing process in the phases between the printing of individual layers and feeding this data into an analysis routine incorporating a DL classifier. In order to successfully develop and test the method, certain auxiliary functions of data preparation must still be carried out manually in the course of this study in order to close the entire pipeline required for the overall objective for demonstrative purposes. The routine is able to combine the following sub-functions:

- Initialize the scanning process;
- Prepare raw data (read in and pre-process);
- Prepare data for classification;
- Classify scanned layer in separate patches and evaluate results;
- Transmit feedback to the printer.

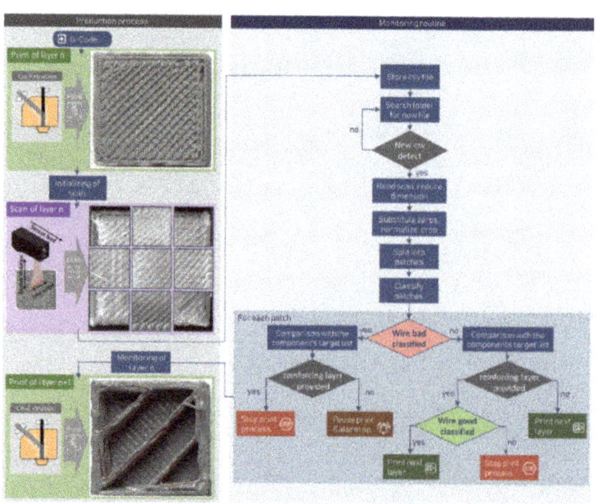

Figure 2. Flowchart of an in situ monitoring routine based on classification of height information of printed layers using convolutional neural networks.

To ensure that the model makes accurate decisions, it is necessary to create a target plan for the component to be printed. This specification plan specifies how the part should be built up layer by layer. In this way, the classification model can compare its predictions with the expected results. If there are deviations between the prediction and the expected layer, the system interrupts the printing process and issues a warning.

In order to generate the required database, various settings for printing thermoplastic with integrated wire must be tested in preliminary trials. With the help of this preliminary knowledge, it is necessary to construct a demo component. The next step is to develop a suitable network architecture for classifying the components. This involves designing a CNN, taking into account various factors such as layer architecture, data pre-processing and optimization methods. The CNN model is trained and evaluated with image data collected in order to assess the classification performance. Attention is paid to ensuring that the model has the desired accuracy and reliability to correctly classify the component layers.

4.2. Data Generation and Data Aquisition

The basis for the data for the classification experiments is formed by printing and scanning the test components. In order to obtain at least 100 unique scans for each class, first the classes must be defined. For each printed component, solid layers and infill structures are generated, which is why these two layers are considered separate classes. However, their classification is not a decisive aspect of the problem under consideration. Nevertheless, they are included in the data collection, as they are suitable for a complete component description in further analyses.

Successful material reinforcement with continuous fiber or wire ensures an elevated load-bearing capacity of the component. In the event of faulty reinforcement, the component may fail unexpectedly. It is therefore important to identify poor reinforcement layers. This should be possible without the need for human control during the printing process. Therefore, the classes "wire good" and "wire bad" are defined as separate classes. This enables printing to be aborted in a production environment if a poor reinforcement layer is printed since the required mechanical properties can no longer be guaranteed. The situation is rather different for solid and infill layers. If an error occurs with solid and infill layers, this can be corrected by overprinting and re-melting during subsequent print layers. Following four classes are therefore distinguished:

- 'Solid';
- 'Infill';
- 'Wire good';
- 'Wire bad'.

As illustrated in Figure 3, the four classes are made up of a total of seven variants. Two different patterns are printed for the reinforcement layers with wire and the infill structure. The test component is a combination of both variants. As is usual in the application, solid layers are printed first. This is followed by infill layers, which can be reinforced with wire. If co-extrusion does not occur as intended, bad layers are created. These are layers in which the wire does not remain in the desired position. There are many causes behind these failures:

- Insufficient bond between thermoplastic and wire;
- Clogging of the nozzle;
- Lack of wire feed due to a malfunction in the feed unit (e.g., too little friction between wire and feed wheel);
- Wire breakage due to excessively small bending radii;

- High feed speed, causing the wire to detach from the hot thermoplastic.

All these factors lead to an unsatisfactory print result. It is possible to correct geometric or design errors. However, process-related errors such as wire breakage can not be completely ruled out.

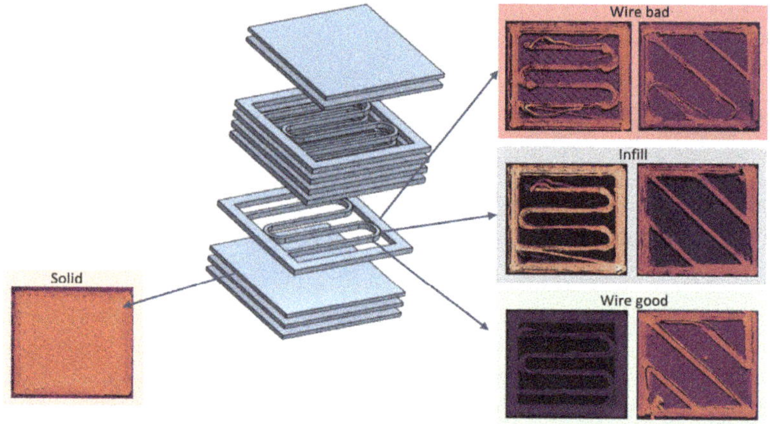

Figure 3. Display of all part configurations of the components in the four distinguished classes 'solid', 'infill', 'wire good' and 'wire bad' (visualization based on scanned, unprocessed data).

During data generation through printing experiments, attention was to be taken to the repeatability of the experiments. Care was taken in creating the machine code to ensure that no unnecessary systemic influences from the extruder trajectory planning had a negative impact on the quality of the wire reinforcement under investigation. In addition, the experiments were consistently used with consistent materials to ensure maximum repeatability with regard to the materials. This is limited by the influence of the ambient climate, as the experiments were carried out in a laboratory without air conditioning, meaning that room temperature and humidity had at least a minor influence on the processes.

Using a dedicated command in G-code, the laser line scanner can be traversed linearly over the component in parallel with the extruder in order to capture the height profile of the printed component. The height profile is saved as a CSV file using a proprietary program. The process for creating the data is divided into three main sub-processes:

- First, a layer is printed;
- Subsequently, the scanner is traversed over the printed layer;
- Finally the captured data are saved in a CSV file.

In this way, a specific CSV file is created for each printed layer. The data can be transferred to other programs for further processing. The CSV file is structured as outlined in Table 2, and contains following information:

- Date and time;
- Package width (based on height measurements);
- Package length (based on height measurements);
- Number of all points (based on height measurements);
- Array of volumetric information acquired through the scanning (x, y, z).

Table 3 provides an example of the structure of such a CSV file. The first five lines are dedicated to descriptive data about the file. Starting with the sixth line, the scanned volumetric measurement values are saved in the format given.

Table 3. Basic structure of a CSV file issued subsequent to the laser line scan.

Line	Content (Example)						
1	*** 11.10.2023 13:58:21						
2	*** Buffer Width: 1272						
3	*** Buffer Length: 1058						
4	*** Amount of Points: 1,345,776						
5	*** x, y, z						
//	x	y	z	x	y	z	...
6	−38.4120	−34.9800	29.9725	−38.3112	−34.9800	29.8765	...
7	−38.4120	−34.9400	29.8525	−38.316	−34.9118	29.8525	...
...	...	...	...	...	...	...	...
1058	−38.4104	−31.8692	29.9785	−38.356	−31.8692	29.9485	...

Sequence '***' serves as indicator for header information: Date and time, the packet width, the packet length and the number of all points (provided by scan software and carried through as such).

In adverse cases, strong scattering reflections of the laser beam from the measuring scanner can occur. This is not uncommon given the glazed surface of the FFF-manufactured components. At these points, no data are stored. At such a measuring point, the X, Y and Z values remain empty. This results in a lack of information at the respective point; however, the dimension of the measurement remains unaffected. The CSV format in its original form is not suitable for further processing. The main problem is the dimensional disparity between the first five lines and the subsequent measured values. It is therefore necessary to adapt the CSV file in order to convert the data into a useful format. To extract useful information from the raw data, performing the following steps is required:

- Reading CSV with exception of lines 1 to 5;
- Localization of the component on the printing plate;
- Cropping to the relevant area;
- Data pre-processing;
- Identifying and removing outliers;
- Imputation of missing values;
- Normalization of data;
- Convert data into a usable dimension;
- Save processed CSV data as PNG.

For cropping the relevant areas of the images, a semi-automated procedure is developed (see Figure 4). This involves localizing the component using a graphical user interface (GUI) and saving it separately. Although this process is time-consuming, it ensures a minimum of the required data quality. The top left corner of the component is targeted via the input windows and a fixed size of 600 × 600 pixels is defined in the source code, assuring that the data generated are always the same size. The output CSV is cropped using the row and column values specified and the fixed dimension. The storage path is specified in the source code and can be changed. After this processing, the output data are available in a reduced format of 600 × 600 pixels. In addition, the GUI saves only the Z values and replaces all missing values with zeros. As a result, self-written scripts for data processing can be used effectively and the required memory space is reduced. The scan program assigns unique names to the scans, consisting of date, time and scan number (layer number), which are subsequently saved unaltered. For further processing of the data, a script is used that performs the following steps:

- Importing a CSV as Pandas Dataframe;
- Replace all zero values with the mean value of the non-zero values minus 0.7 (layer height of the print);
- Calculate the 99the percentile and cut off all values above it;
- Normalize the data using MinMaxScaler function of Scikit-learn.

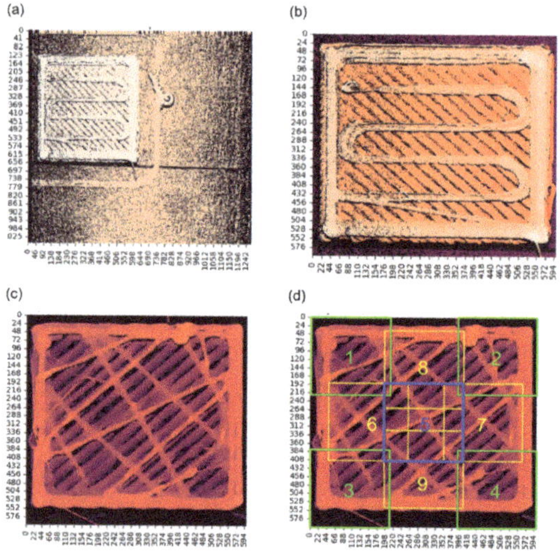

Figure 4. (**a**) Raw data file: all height measurement values recorded with laser line scanner in a CSV. (**b**) Area of interest: area specified via the interface. (**c**) Pre-processed data: output file before splitting into patches. (**d**) Processed output: split patches (numbers 1–9) in format 224 × 224 pixels appropriate for model learning.

In order to extract a proper amount of information from the image files, the output files are divided into nine different patches (see Figure 4d). The splitting is convenient as common transfer learning models do not operate with 600 × 600 pixel data, as given by the images prior to splitting, and in addition, a higher number of image instances are made available for the learning approach, which can always be beneficial in training ANN models. A frequently used image resolution in ANN applications is 224 × 224 pixels. Splitting allows to increase the total number of samples and therefore also the number of samples per class. However, it is important to verify and adjust the class labels of the cropped image tiles according to the visual information in the particular image patch.

The output dataset consists of the processed scans. Not all scans can be used, as some contain a failure or the quality is insufficient. After processing and splitting the individual scans, the dataset has the extent depicted in Table 4. It can be seen that the dataset is unbalanced, as the classes 'solid' and 'infill' are comparatively overrepresented, which is due to the fact that 'solid' and 'infill' layers are consistently printed during printing. The class 'infill' is overrepresented in particular with a share of 44.6% of the total dataset. It is technically not feasible to reinforce each layer with wire, so that regular infill layers are constantly printed in between. If a print fails, this is typically attributable to the reinforcement layers, but this does not apply to the 'infill' and 'solid' layers, where issues occur less frequently. The target of 100 images for the class 'wire good' before data processing was not fulfilled, which is a result of technical difficulties.

In order to examine the influence of the dataset size on the performance of the models, the dataset is extended using augmentation. This involves using the rotation technique for augmentation; the initial data are rotated by 90°, 180° and 270°. These angles are beneficial as there is no loss of information that would otherwise occur with other rotation angles. After the procedure, the dataset volume is four times the size as before (see Table 4). For the sake of unambiguity, an extension is added to each file, consisting of the original file name suffixed by the rotation angle. In order not to falsify the results of the dataset, the data used in the initial dataset (No. 1) are augmented in their respective folders (training, validation, and testing, in a ratio of 0.7/0.2/0.1, respectively). This preserves the comparability of the results of the models among each other. If data are randomly distributed to folders after augmentation, it is possible that the models learn differing parameters from the data and their performance differs as a result.

Table 4. Allocation of the data of the initial dataset and of derived datasets containing RGB images, each of the derived RGB datasets being both without and with augmentation.

Dataset	Description	Color Coding	Augm.	Data Per Class			
				Solid	Infill	Wire Good	Wire Bad
1	Initial	1 Ch., Grayscale	no	1422	2574	811	964
2	Initial	1 Ch., Grayscale	yes	5688	10,296	3244	3856
3	RGB, seriell	3 Ch., RGB	no	1422	2574	811	964
4	RGB, seriell	3 Ch., RGB	yes	5688	10,296	3244	3856
5	RGB, parallel	3 Ch., RGB	no	1422	2574	811	964
6	RGB, parallel	3 Ch., RGB	yes	5688	10,296	3244	3856

Since transfer learning models usually expect three-channel inputs because their base models have been trained on RGB data, a method of generating color images from the gray value images has been tested [53,54], as is illustrated in Figure 5. In principle, the height information is coded as a single intensity value, whereas RGB images encode three values between 0 and 255 for each pixel.

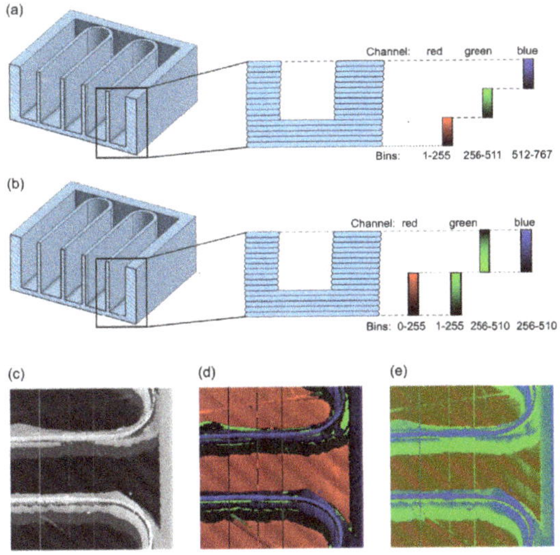

Figure 5. (**a**) Serial color coding of the normalized CSV data using the height information. (**b**) Coding of the height information in two channels parallel to each other. (**c**) Example image output dataset, (**d**) Example image serial coding. (**e**) Example image parallel coding.

With the serial coding method applied, the different channels (red, green and blue) are coded one after the other with ascending values. This means that the lowest Z value in a file has the coding [1,0,0]. When the color coding of a channel reaches the maximum value of 255, this channel is set to 0 and the consecutive channel starts at 1. This means that there are not only grey levels in the range [0–255] but three channels with 256 bins each (see Figure 5a). As the start of each channel begins with the value 1 and the other channels set to 0, it should be noted that black coloration occurs at the transitions from one channel to another. The images obtained are saved as PNGs and can be loaded into a pretrained model without further processing.

The advantage of serial coding is the high number of bins. This makes it possible to divide height information into smaller bins, giving pretrained models more precise information for learning. The disadvantage, however, is the strong differentiation between the color channels. If an important information sector for layer determination is encoded along the channel boundaries, the important information may fall into different channels depending on the image. In the worst case, this can lead to indeterminability of the feature.

Parallel coding solves this shortcoming. The information is encoded in parallel in two channels to enable CNN to recognize features across multiple channels. The number of bins is then 511, whereby the missing bin is due to the transition from the first green color gradient to the following one (see Figure 5b).

Figure 5c–e shows example images of different color coding variants. The result of parallel encoding is shown in Figure 5c. It can be seen that the resulting image is brighter than the same initial image with serial encoding (Figure 5d). This is due to the fact that with serial encoding, each color channel starts with black and the predominant color only gradually appears. With parallel coding, two color channels are always combined so that no black areas are created in the result and the combined colors are lighter on average.

5. Modeling

5.1. Custom Convolutional Neural Network

The experiment with a custom CNN serves to provide insight into the depth of a user-defined CNN and the hyperparameters involved. A custom CNN is advantageous as a dedicated CNN can process the gray value images without further processing steps, whereas pretrained models used for transfer learning are trained on RGB data and therefore require a three-channel input. The aim is to first determine the possible performance with an individually constructed CNN prior to comparing it to results achieved with pretrained models.

The adaptation of an ANN to a classification task involves both the curation of meaningful and, in particular, task-specific data and the optimization of hyperparameters. The latter includes parameters such as the learning rate, the batch size, the number of epochs, the kernel size, the number of filters and the activation function. Some of the hyperparameters are defined in Table 5. The specification of these parameters is based on common standards on the one hand and iterative trials on the other. For further improvement of the CNN performance, it is necessary to conduct additional dedicated hyperparameter studies. However, since the purpose of these experiments is to demonstrate the basic performance expected from a custom CNN, no further in-depth hyperparameter studies are carried out.

The early stopping function is used to avoid unnecessary training of the models. The experiments show that an early stopping value of five is sufficient to ensure appropriate training progress of the model, with the validation loss being monitored. If the model does not improve any further, the best model value, i.e., the epoch with the lowest validation loss, is reset via the callback function and the model performance is verified using the test dataset. A learning rate of 1×10^{-5} demonstrates suitable convergence in the experiments, matching recommended values from the literature.

Table 5. Hyperparameters of the custom CNN models.

	Hyperparameter	Specification
General	batch size epochs loss metrics	32 90 sparse categorical crossentropy accuracy
Early Stopping	monitor patience mode restore best weights	validation loss 5 auto True
Model Checkpoint	monitor save weights only save best only	val loss False True
Optimizer	Optimizer Learning Rate	adam 1×10^{-5}

Table 6 summarizes the examined architectures. As shown in Table 6, the CNNs used in the study consist of four to six convolutional layers. Each network consists of an input part, which processes the input data, a convolutional part, in which the convolutional layers extract features from the data using filter kernels, and a classification part, which is preceded by flattening. The classification part consists of two dense layers with a fixed number of neurons.

Table 6. Overview of the structure and parameters of the custom CNNs examined.

ID	Architecture	Activation	Filter	Kernel	Dropout	Metrices	
						Accuracy	F1 (Wire Bad)
1	[C-B-A-M-O][C-B-A][C-B-A][C-B-A]-F-D-D	ReLU	32-64-80-128	2×2	0.25	0.81 + 0.01	0.67 ± 0.02
2	[C-B-A-M][C-B-A-M][C-B-A][C-B-A][C-B-A-M]-F-D-D	ReLU	32-64-80-128-140	2×2	0.25	0.85	0.73
3	[C-B-A-M-O][C-B-A-O][C-B-A-O][C-B-A-M] [C-B-A-M]-F-D-D	ReLU	32-64-80-128-64	2×2	0.25	0.66 − 0.01	0.27 ± 0.06
4	[C-B-A-M][C-B-A-M][C-B-A-O][C-B-A-O][C-B-A-M-O][C-B-A-M-O]-F-D-D	ReLu	32-64-80-128-128-140	2×2	0.25	0.78 ± 0.03	0.54 ± 0.11
5	[C-B-A-M][C-B-A-M][C-B-A][C-B-A-M][C-B-A][C-B-A-M]-F-D-D	ReLU	$16 \times 32 \times 64 \times 80 \times 128 \times 80$	2×2	-	0.80 ± 0.02	0.60 ± 0.07
6	[C-B-A-M-O][C-B-A-M-O][C-B-A-O][C-B-A][C-B-A][C-B-A-M]-F-D-D	ReLU	$64 \times 88 \times 126 \times 150 \times 128 \times 64$	2×2	0.10	0.84 ± 0.01	0.73 + 0.01

Layer: C—Convolution, B—Batch normalization, A—Activation, M—MaxPooling, O—Dorpout, F—Flatten, D—Dense.

For the experiments with model IDs 1–4, dense layers with 128 and four nodes are used, with an output layer that uses the "softmax" function of Keras. The models with IDs 5–6 also use four nodes for the output layer, but 64 nodes for the penultimate layer. This allows the influence of different density layers on the classification to be investigated. All models use a different number of filters to investigate the extent to which the number of filters influences the F1 score. A kernel size of 2×2 is intended to provide the model with a maximum number of features extracted from the image. The drop-out values are based on iterative experimental conclusions and can be further elaborated in future studies. The evaluation metrics serve to provide an initial insight into the quality of the models and facilitate the comparison of different model approaches.

5.2. Transfer Learning Utilizing Custom Fully Connected Layers

The transfer learning approach is used to examine whether pre-trained models are more accurate in determining the data. For this purpose, ResNet50 and VGG16 models are used, both of which are integrated via Keras and used with their respective Imagenet [55] weights. As both models in the original configuration are not compatible with the classification of the four classes, the pre-trained classification head of the model is truncated and replaced by a custom head (see Figure 6). The architecture of the new classification head is based on experimental iterative experience. The size of the dense layers is gradually reduced in order to minimize the resulting space without using too large dense layers. Dense layers influence the computing time of a model considerably due to their high number of connections. In addition, two drop-out layers are used to prevent overfitting of the model.

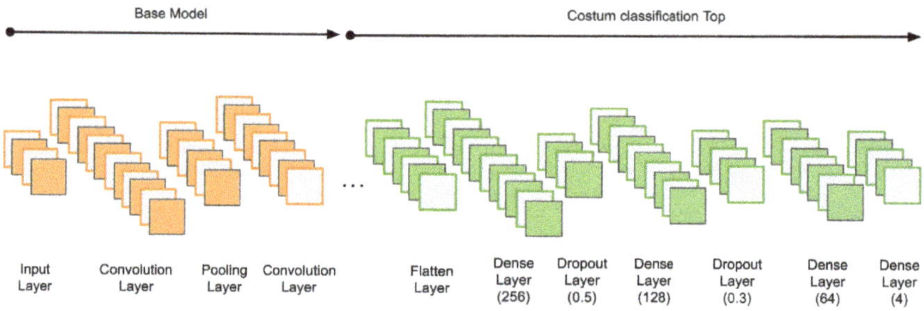

Figure 6. Top-layer architecture of the employed transfer learning models.

It is necessary to determine a sufficient learning rate for the model. Figure 7 shows the accuracy for three different learning rates across the course of the epochs. The number of epochs depends crucially on the initial weights and can therefore vary from seed to seed. Seed refers to a pre-defined value used to initialize the random number generator involved in the modeling process, ensuring that operations involving randomness are reproducible. The graphs of Figure 7 show the progression of accuracy on both training and validation datasets. Figure 7a displays the trend of the accuracy over the course of the epochs for dataset 1 (see Table 4) and Figure 7b displays the trend of the accuracy over the course of the epochs for dataset 2 (see Table 4). It can be seen that the three learning rates investigated achieve almost identical accuracy. In the course of the experiments, the learning rate 1×10^{-5} is utilized, as it offers a good balance between not too rapid convergence and not too prolonged training times. The results of the model with both datasets show that neither the non-augmented nor the augmented dataset yields a validation accuracy of more than 0.9.

The layer architecture settings are already specified by the base network architecture in transfer learning. Settings such as the batch size or the number of epochs provide scope through predefinition. Table 7 lists the relevant hyperparameters for the ResNet50 and VGG16 models. The number of epochs is set to 100. The use of 'early stopping' causes the learning process to be suspended after a phase of 10 epochs without further improvement. The best model weights are found and the performance of the model is checked with the test dataset.

Table 7. Hyperparameters of the pre-trained models (transfer learning).

	Hyperparameter	Specification
General	batch size epochs loss metrics	32 100 sparse categorical crossentropy accuracy
EarlyStopping	monitor patience mode restore best weights	validation loss 10 auto True
Model Checkpoint	monitor save weights only save best only	val loss False True
Optimizer	Optimizer Learning Rate	adam 1×10^{-5}

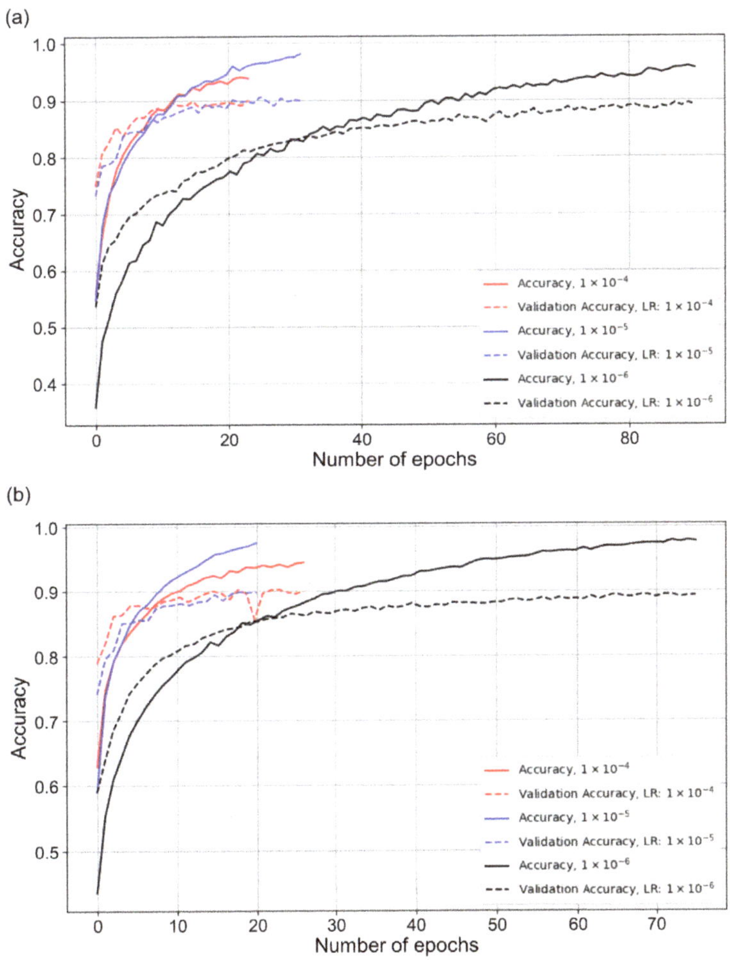

Figure 7. Training progress for learning rates 1×10^{-4}, 1×10^{-5}, and 1×10^{-6} for the ResNet50 model with (**a**) dataset 1 (seed 456) and (**b**) with dataset 2 (seed 456).

The results of the F1 score in Table 8 show that both the ResNet50 model and the VGG16 model achieve almost identical performance. Even the type and nature of the input files are not of significant importance. The use of augmented datasets does not significantly improve the performance of the models either.

The class 'solid' is classified almost optimally in all models. This can be attributed to the large amount of data examples in this class and the homogeneous data in this class, which only differ in a few features. It should be noted that the class 'solid' is the only one with a completely distinct layer geometric structure during printing. The classes 'wire bad' and 'wire good' are merely a modification of the 'infill' layer. The two classes 'wire bad' and 'wire good' in particular are classified significantly worse than the others, with the class 'wire bad' still being classified a few percent better.

Table 8. Overview of the F1 scores achieved by the various TL models.

Model	Dataset	F1 Score Averaged			
		Infill	**Solid**	**Wire Bad**	**Wire Good**
ResNet50	Dataset 1	0.89 ± 0.01	0.99	0.82 ± 0.05	0.77 ± 0.04
ResNet50	Dataset 2	0.90 ± 0.01	0.99	0.81 ± 0.01	0.79 ± 0.04
ResNet50	Dataset 3	0.88 ± 0.01	0.98 + 0.01	0.82 ± 0.03	0.77 ± 0.04
ResNet50	Dataset 4	0.89+0.01	0.99 − 0.01	0.84	0.80 ± 0.02
ResNet50	Dataset 5	0.89 ± 0.01	0.99 − 0.01	0.80 ± 0.04	0.79 ± 0.02
ResNet50	Dataset 6	0.89 ± 0.01	0.99	0.82 ± 0.02	0.81 ± 0.02
VGG16	Dataset 2	0.89 ± 0.01	0.99	0.82 ± 0.05	0.77 ± 0.04
VGG16	Dataset 4	0.90 ± 0.01	0.99	0.81 ± 0.01	0.79 ± 0.04
VGG16	Dataset 6	0.88 ± 0.01	0.98 + 0.01	0.82 ± 0.03	0.77 ± 0.04

The F1 score is a suitable metric for a classification task, as it takes into account both the correct identification of the 'wire bad' class and the avoidance of misclassifications of other classes. In order to further improve the F1 score, it will be investigated whether fine-tuning can yield an improved result. For this purpose, the ResNet50 model is used together with dataset 4 (see Table 4), as this combination has the highest F1 score for the 'wire bad' class.

5.3. Fine-Tuning

During fine-tuning, the classification head is trained as before with the locked weight of the pre-trained model. Once an optimum is reached, the layers of the base model are unlocked in order to be trained more finely. In this way, models can be better adapted to the data. The ResNet50 model with dataset 4 (see Table 4) is used to investigate fine-tuning, as this model achieves the highest F1 score on average for the 'wire bad' class. The influence of the fine-tuning process on the results is investigated, taking into account various aspects such as the number of unlocked layers, the learning rate and the batch size. It is expected that a larger batch size leads to an improvement in the F1 score [56].

Table 9 shows the hyperparameter settings of the fine-tuning models. The settings relate exclusively to the fine-tuning part. All other model settings are identical to those in Table 7. In contrast to the investigations of the transfer learning models, training is not aborted after early stopping. Instead, a certain number of the last layers (layer base model) are unlocked and retrained with a defined learning rate (LR fine-tuning) over a limited number of epochs (epoch fine-tuning). The learning rate of the initial model is multiplied by the LR fine-tuning in order to achieve a finer tuning.

Table 9. Hyperparameters of the fine-tuning models.

	FT1	FT2	FT3
Layer Basemodel	20	25	25
LR Fine-Tuning	0.10	0.10	0.05
Epoch Fine-Tuning	15	20	20
Batch Size	32	32	48

Figure 8a shows the progression of the validation accuracy of the fine-tuning models along the training progress (number of epochs) for seed 42. It can be seen that after unlocking the base model layers, a decrease in validation accuracy occurs. Subsequently, the accuracy continuously increases towards the initial level, but no significant further improvement can be observed for any of the models. Figure 8b shows the progression of the validation loss of the fine-tuning models along the training progress (number of epochs) for seed 42. It can be seen that an increase in validation accuracy occurs after unlocking the base model layers. Subsequently, the loss continuously decreases and returns to the initial level, but no significant further improvement can be observed for any of the models.

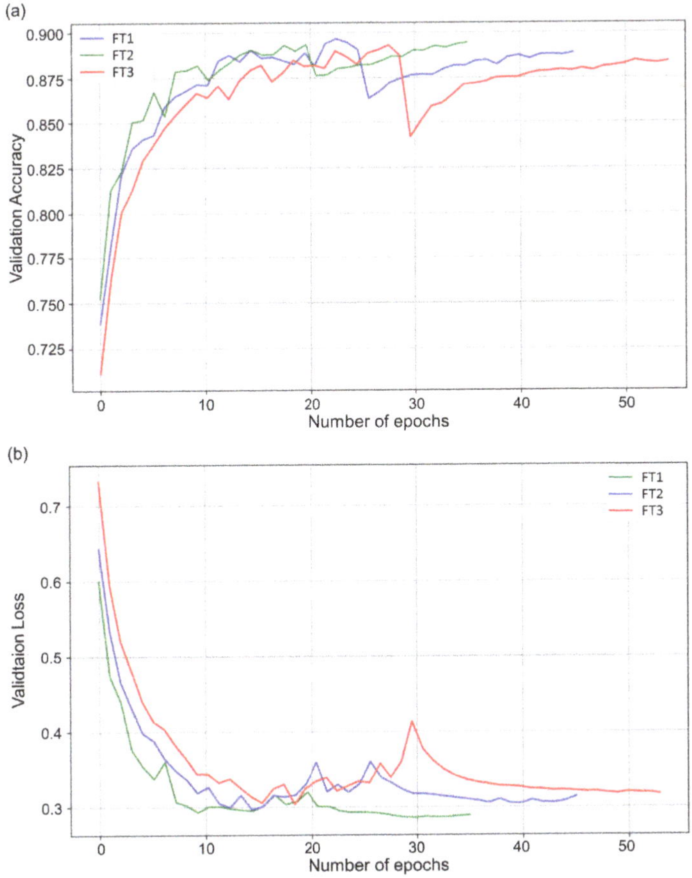

Figure 8. (**a**) Validation accuracy and (**b**) validation loss as metrics for the training progress of the fine-tuning models.

Figure 8a,b indicate that the model tuning from an epoch prior to fine-tuning is best suited for testing models FT2 and FT3, whereas for testing model 1, the model tuning after fine-tuning is best suited. The results of testing the three models using the test dataset are shown in Table 10. The table shows that both models FT1 and FT3 deliver almost identical solutions. A possible further increase in the performance of model FT1 is expected by increasing the batch size or increasing the fine-tuning epochs.

Table 10. Overview of the F1 scores achieved by the pre-trained models utilizing fine-tuning.

Model	Dataset	F1 Score Averaged			
		Infill	Solid	Wire Bad	Wire Good
FT1	Dataset 4	0.89	0.99	0.83 ± 0.01	0.81 ± 0.03
FT2	Dataset 4	0.89 ± 0.01	0.99	0.81 ± 0.02	0.82 ± 0.01
FT3	Dataset 4	0.90 ± 0.01	0.99	0.83 ± 0.02	0.83 ± 0.02

The confusion matrix in Figure 9 indicates the performance of a model by illustration of a comparison of the labels of the predictions to the true labels. The matrix contains counts of correct and incorrect classifications; in the case of multi-class classification, information on which incorrect classes the incorrect prediction was assigned to by the model. This helps calculate and interpret metrics like accuracy, precision, recall, and F1 score to understand a model's strengths and weaknesses, identifying where the model is making errors, particularly in different classes.

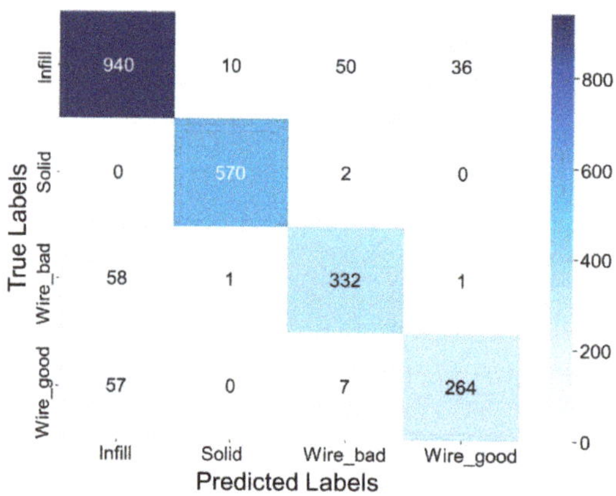

Figure 9. Confusion matrix of the FT3 model for seed 42.

The confusion matrix in Figure 9 shows that the model provides the best prediction for the 'solid' class, with a precision ('solid') of 0.981 and a recall ('solid') of 0.997. This is plausible as, of all the classes considered, the geometric characteristics of the printed surface for the 'solid' class differ most clearly from those of the other classes. The prediction of the 'infill' class is less favorable for the model, with a precision ('infill') of 0.890 and a recall ('infill') of 0.907. This can be explained by the fact that the geometric characteristics of the printed surface for the 'infill' class occasionally hardly differ from the surface geometries of the 'wire good' and 'wire bad' classes. The model's predictions are worst for the wire-reinforced instances, with a precision ('wire good') of 0.877 and a recall ('wire good') of 0.805 for the 'wire good' class, and with a precision ('wire bad') of 0.849 and a recall ('wire bad') of 0.847 for the 'wire bad' class. As the confusion matrix in Figure 9 shows that the majority of misclassified images of the classes 'wire bad' and 'wire good' are classified as 'infill', it is indicated that the model is able to distinguish between the two classes ('wire good' and 'wire bad'), but is not able to distinguish sufficiently between these and the 'infill' class.

6. Discussion

6.1. Network Architecture

The decision on the best model structure is determined by the reliability of the model with regard to the 'wire bad' class, as the 'wire bad' class is the most critical process step for an FDM coextrusion process. Although there is a possibility of errors occurring in both the 'infill' and 'solid' layers, the probability of this is comparatively low. In addition, with both infill and solid layers, it is possible to correct errors by overprinting with filament or by re-melting using the hot nozzle. However, if problems occur when printing the wire layer, this can not only affect the mechanical properties of the component but can also cause severe damage to the manufacturing machine. It is therefore necessary for the monitoring routine to be particularly adapted to this task.

Table 11 lists the models with the best F1 score from each of the experimental approaches (custom CNN, transfer learning, fine-tuning). It can be seen that there is no significant difference between the models in their ability to correctly classify the decisive class (custom CNN: F1 = 0.73; transfer learning: F1 = 0.84; fine-tuning: F1 = 0.83). The experiments with transfer learning and with additional fine-tuning achieve the best results with results being almost identical. The custom CNN performed nearly 10% worse than the other two models but has not been optimized in a similar manner.

Table 11. Overview of the averaged F1 scores achieved by the best models of each approach conducted.

Model with ID	Dataset	F1 Score Averaged			
		Infill	Solid	Wire Bad	Wire Good
Custom CNN 2	Dataset 1	0.86 + 0.01	0.98 − 0.01	0.73	0.69
Transfer Learning 4	Dataset 4	0.89 + 0.01	0.99 − 0.01	0.84	0.80 ± 0.02
Fine-Tuning 3	Dataset 4	0.90 ± 0.01	0.99	0.83 ± 0.02	0.81 ± 0.02

All models can accurately classify the class 'solid'. This is primarily due to the consistent quality of the data and to the fact that the other three classes deviate considerably from 'solid' in their appearance and merely represent variations from one another. For both the classes 'wire good' (transfer learning: $F1_{wire\,good} = 0.8$; fine-tuning: $F1_{wire\,good} = 0.81$) and 'infill' (transfer learning: $F1_{infill} = 0.89$; fine-tuning: $F1_{Infill} = 0.9$), the F1 scores of both the transfer learning and fine-tuning models are similar, with the fine-tuning model classifying slightly more accurately.

6.2. In Situ Classification Routine

As illustrated in Figure 10, several misclassifications are demonstrated. The predicted and true labels are displayed adjacent to the images to serve as a reference. The images emphasise the challenges associated with the conversion of height data to image data. In Figure 10a, the phenomenon known as 'stringing' is evident, where thermoplastic strings form the printed part as a consequence of the extruder nozzle oozing on its travel trajectory. This phenomenon can potentially lead to misclassification, as the hardened thermoplastic strings may result in the model erroneously identifying an out-of-place copper wire. Another salient issue visible in this image is the absence of height data in the shadow of the printed part, attributable to the angle at which the light sensor is mounted.

However, in this particular instance, the absence of height data does not constitute a significant problem, as only the height data within the printed plane is relevant.

As illustrated in Figure 10c–h, sensor artifacts are evident and can originate from various sources, including fluctuations in ambient light, reflections, or mechanical oscillations of the machine. Upon examination of the data, it is observed that these artifacts frequently result in a localized offset relative to the surrounding data, manifesting as intensity variations in the corresponding regions of the image.

Furthermore, Figure 10c–g illustrate the difficulty of distinguishing wire-related data from infill data. When a copper wire is fully embedded within the thermoplastic, the model struggles to make accurate predictions. Similarly, if the nozzle or a partially retracted copper wire leaves a slight groove in the infill structure, the model may confuse it with the groove that forms around the wire when it is embedded.

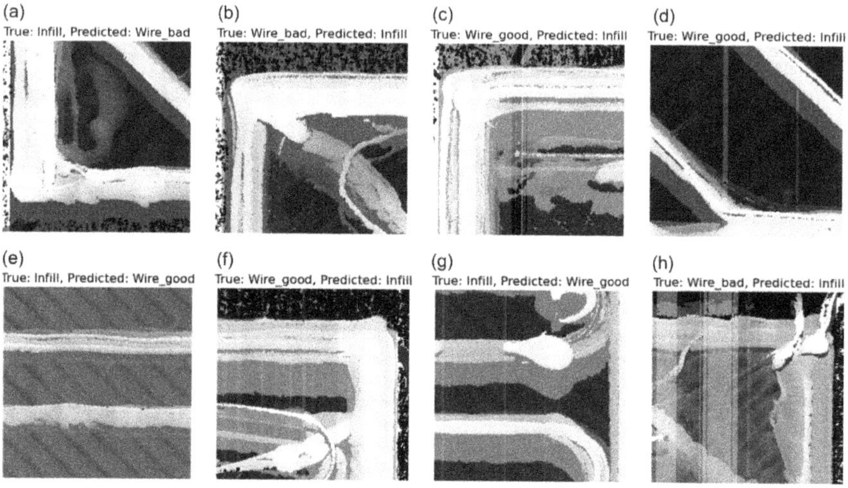

Figure 10. Misclassifications regarding the corresponding images from the dataset illustrating deficiencies in effectiveness of proposed method: (**a**) true: infill, prediction: wire bad, (**b**) true: wire bad, prediction: infill, (**c**) true: wire good, prediction: infill, (**d**) true: wire good, prediction: infill, (**e**) true: infill, prediction: wire good, (**f**) true: wire good, prediction: infill, (**g**) true: infill, prediction: wire good, and (**h**) true: wire bad, prediction: infill.

An isolated classification model is not an adequate solution for in situ monitoring of the printing process. After each printed layer, the controller must initiate a scan, process the recorded data and analyze all the resulting patches. The first obstacle in this procedure is the format of the scan. Each scan is manually prepared for training the CNN using a purpose-built GUI. However, this is not practical during a continuous printing process.

It is therefore necessary to develop a way of detecting the component in a scan. The overall objective of the study is to enable a DL classifier that can be incorporated into an automatic in-line monitoring method. The top-level objective of fully automated in-line monitoring remains impaired by the manual image cropping still present in the developed method. Consequently, automated solutions for cropping the raw data must be integrated into the overall solution in the future. This can be achieved, for example, by using predefined target image sizes and a simple cutting routine based on edge detection. Methods for the automated cropping of images are available in the literature, so it can be considered technically feasible to automate this task. Examples include filter-based methods [57,58], attention-based methods [59], and aesthetic-based methods [60]. The use of an upstream ML model for object recognition is also a good option for this task. Alternatively, any separate program can also be used for localization; therefore, this step is omitted in this study. However, it is necessary to develop a comparable solution in order to exploit the full potential of the laser line scanner for monitoring 3D printing in situ.

6.3. Findings

In particular, in the most technically interesting and probably most application-critical areas of the 'wire good' and 'wire bad' classes, the prediction results of the developed models showed weaknesses, with F1 scores below 85% even for the best models in the study. The probable main reasons are to be found in the following categories:

- Data input;
- Model structure;
- Hyperparameter settings.

The experiments illustrate that the maximum F1 score is achieved by using pre-trained models. However, such models are not only more complex but also not inherently compatible with the initial data from the source. In particular, the performance of custom CNNs is highly dependent on the architecture and requires detailed investigation. No model–dataset combination achieves an F1 score of over 85% for the 'wire bad' class. This is particularly problematic for components with multiple reinforcement layers, as an incorrect classification of just one layer is bound to result in the printing process being aborted.

The results in Table 11 indicate that fine-tuning does not yield a significant improvement. It is possible to optimize the hyperparameters of the fine-tuning, but a significant increase in the results is not expected. For example, the different number of enabled base model layers as well as the learning rate and batch size did not have any influence on the classification quality.

A comparison with the results of all models in Table 8 suggests that a significant improvement cannot be expected. Various causes are conceivable:

- Insufficient quality of the dataset;
- Insufficient scope of the data;
- Differences among the classes are too small to distinguish significantly from each other.

The lack of quality of the dataset provides a plausible explanation for the performance of the models, which in part requires improvement. It is striking that all models exhibit a similar maximum F1 score for the 'wire bad' class, each of which ranges closely to the other. This suggests that the models are poor at correctly classifying some images. It is conceivable that some images are incorrectly labeled or contain excessive noise. The quality of the images captured by the laser scanner depends heavily on ambient conditions. Excessive ambient light or surfaces with excessive reflection can severely distort the images.

Prior to the start of the experiments, the dataset was checked for poor records and totally invalid records were excluded. However, measurement artifacts are also to be expected in a prospective real application scenario; therefore, the recorded noisy data does not necessarily need to be an unsuitable basis for model development. As a consequence, a selection of noise recordings characterized by inferior quality were retained in the dataset. This was carried out deliberately to give the models a stronger generalization capability. It can be assumed that these records are responsible for the stagnation of the F1 scores. This also shows the importance of the correct setting for the laser line scanner. In subsequent experiments, it is conceivable to darken the print chamber in order to eliminate ambient light components. A revision of the dataset used in this work and the use of explicit images only for the training of the models are also an alternative.

The reason for the distribution of the data in the dataset of the study can be attributed to the characteristics of the printing process utilized in the experiments. As a matter of procedure, more images are recorded from the 'solid' and 'infill' areas as opposed to the 'wire good' and 'wire bad' areas. Consideration can be given to removing data from the 'solid' and 'infill' classes from the dataset in order to restore parity. Because a disparity between the classes equivalent to the experimental results is to be expected in a real application scenario, it was decided to retain the data distribution resulting from the printing experiments'.

The effect of the investigated data augmentation techniques is minimal, although it is theoretically expected and cannot be completely dismissed in the results. For example, the model trained with the augmented dataset 4 in Table 8 shows the highest F1 score for the critical classes 'wire good' and 'wire bad', with F1 ('wire good') = 0.84 and F1 ('wire bad') = 0.80. For this reason, the augmented dataset 4 was used for the in-depth investigations with the fine-tuning models. Although no significant improvement was observed with the augmentation methods investigated, it does not necessarily imply that the increased training effort due to augmentation is useless. It is possible that variations learned through the augmented training images are not sufficiently included in the validation data used. If validation data is used that was printed with other materials or with other parameterizations or contains image sections of different sizes, the variations learned through augmentation may certainly prove useful. In any case, data augmentation is known as an appropriate tool for improving the generalization capability of models [61]; therefore, data augmentation techniques were included in the investigations. Since the augmentation techniques used did not have a strong improvement effect on the observed model classification performance, it can be assumed that the most appropriate augmentation techniques are not being used in the investigations. In general, augmentations must be well adapted to the task at hand in order to have a positive effect on model performance. In the described study, only rotations of 90° increments are realized. Rotations with smaller increments, for example, 45° or 30°, may be better suited to improve the generalization capability of the trained model in the application under consideration. Mirroring (or flipping, respectively), random cropping, or random erasing of the images can also be considered as rather non-complex but well established and thus promising augmentation methods [37,62–64]. In case more sophisticated augmentation methods than simple geometric operations such as rotation and flipping are needed, more elaborated augmentation methods such as style transfer [65,66] and generative adversarial networks [61,67], or combinations of standard and advanced augmentation methods [68], are available.

It is also plausible to assume that the models are unable to learn the data effectively, resulting in inadequate classification. It is conceivable that the data requires deeper models. Although the difference between the VGG16 and ResNet50 models is not pronounced, a deeper model can lead to higher results. Examples of this are the EfficientNet ans ResNet models with different deeper architectures. However, the observation that both the VGG16-based pre-trained network and the deeper REsNet50-based pre-trained network exhibit barely any performance differences also allows the opposite assumption that for the given problem with comparatively low geometric complexity and low heterogeneity of the input dataset, the model architecture is already too deep to be optimally adapted to the given use case. In this respect, the limitation of the input dataset in terms of quantity, heterogeneity and balancedness of classes can probably be attributed to the highest weight in causing deficits in observed model performance. To fulfill the overall goal of process monitoring, the successfully demonstrated method needs to be trained through future experiments upon a larger and more heterogeneous dataset.

In this study, a systematic hyperparameter optimization has deliberately been omitted, which implies the possibility of a potential improvement of the F1 score by such an operation. In particular, the number and size of filters and the stride should be systematically investigated, as these parameters have a significant impact on the performance of convolutional neural networks. A comprehensive hyperparameter optimization can help to find optimal settings that improve the performance of the model. Frameworks such as Optuna, Hyper-opt, Optunity, and sequential model-based algorithm configuration can be used to automatize the optimization process and make the search for optimum hyperparameters more efficient [69]. Through the utilization of hyperparameter optimization libraries, potentially untapped opportunities to improve the performance of the model can be identified and leveraged to further increase classification accuracy [70].

An alternative strategy involves the use of a more comprehensive custom CNN, which is able to process images in an extended format, thus eliminating the need for splitting into batches and reducing the data preparation effort. However, this would require a significantly larger dataset, as splitting the data into patches has already increased the size of the dataset ninefold. A custom CNN can be trained specifically for a particular use case and the associated data input format. It is to be expected that this advantage will have a strong effect with a more abundant database. The aim of this study was to determine the maximum model performance with the given dataset based on the experiments conducted with prototype printing technology. The use of pre-trained models specifically addresses the issue of limited datasets and is well established for this type of problem. Although the pre-trained models are not tailored to the input data format of the use case under investigation, the better performance of pre-trained models in the results obtained is therefore plausible.

The challenge of the simultaneous occurrence of good and bad patches in one layer can be circumvented by training the model to recognize that, even in a bad layer, there may be a part with a successfully placed wire. However, such a model will require a longer development time due to the higher dimension of the input data, yet it could overcome certain challenges and provide potential improvements.

7. Conclusions

This study examines the process of printing wire-reinforced thermoplastic test parts and the subsequent classification of image recordings from each printed layer using deep learning. The main objective is to distinguish layers with sufficient mechanical reinforcement from those with insufficient mechanical reinforcement. As part of the investigations, test components are produced in two variants, and the data acquisition is carried out with a laser profiler. Data collected from the laser profiler is used to classify the printed layers sensed into classes 'solid', 'infill', 'wire good' and 'wire bad'. Various machine learning methods are used for this purpose, in particular deep learning techniques using convolutional neural networks. Alongside dedicated CNNs with different architectures, transfer learning models based on pre-trained ResNet50 and VGG16 models, with the model's classification head truncated and replaced by a custom head, were investigated. The models are primarily evaluated by means of averaged F1 scores for the class 'wire bad'. As a result, a transfer learning model with fine-tuning achieves the highest F1 score by correctly classifying the decisive class 'wire bad' with an F1 score of 0.83%.

In view of the limited contribution of current augmentation methods to improving model accuracy, alternatives are worth investigating. Methods other than those used in this work are conceivable, such as zooming in and out of images or loading a differently augmented image as input concurrently. Particular potential for improving model performance is seen in conducting a hyperparameter study. This work can be considered a framework for further work using the concept presented to develop advanced models.

Reliable CNN models have the potential to be utilized for in situ monitoring systems that improve the quality of additively manufactured parts. This requires the function of automated scanning to be combined with a trained CNN model and a part recognition algorithm, as well as feedback to the 3D printer's controllers. In addition, the capabilities of automated scanning and subsequent segmentation can be used to build up a dedicated database of 2D relief images from the wire-reinforced FFF manufacturing process.

Author Contributions: Conceptualization, V.L. and C.T.E.H.; methodology, V.L. and C.T.E.H.; software, C.T.E.H.; validation, V.L. and C.T.E.H.; formal analysis, V.L. and C.T.E.H.; investigation, C.T.E.H.; resources, M.F. and S.I.; data curation, C.T.E.H.; writing—original draft preparation, V.L. and C.T.E.H.; writing—review and editing, V.L.; visualization, V.L. and C.T.E.H.; supervision, M.F. and S.I.; project administration, S.I.; funding acquisition, V.L. and S.I. All authors have read and agreed to the published version of the manuscript.

Funding: This research was funded by the Federal Republic of Germany's Ministry Bundesministerium für Wirtschaft und Klimaschutz (BMWK)/Federal Ministry for Economic Affairs and Climate Action, respectively, grant number 16KN 112726.

Institutional Review Board Statement: Not applicable.

Informed Consent Statement: Not applicable.

Data Availability Statement: The raw data supporting the conclusions of this article will be made available by the authors on request.

Conflicts of Interest: The authors declare no conflicts of interest.

Abbreviations

The following abbreviations are used in this manuscript:

AI	Artificial Intelligence
ANN	Artificial Neural Network
CNN	Convolutional Neural Network
DL	Deep Learning
FFF	Fused Filament Fabrication
FT	Fine-Tuning
LPS	Laser Line Profile Sensor
ML	Machine Learning
PLA	Polyactic acid
TL	Transfer Learning

References

1. Araújo, N.; Pacheco, V.; Costa, L. Smart additive manufacturing: The path to the digital value chain. *Technologies* **2021**, *9*, 88. [CrossRef]
2. Bogers, M.; Hadar, R.; Bilberg, A. Business models for additive manufacturing: Exploring digital technologies, consumer roles, and supply chains. *Technol. Forecast. Soc. Change* **2015**. Available online: https://ssrn.com/abstract=2638054 (accessed on 16 October 2024).
3. Weller, C.; Kleer, R.; Piller, F.T. Economic implications of 3D printing: Market structure models in light of additive manufacturing revisited. *Int. J. Prod. Econ.* **2015**, *164*, 43–56. [CrossRef]
4. Mohan, T.R.; Roselyn, J.P.; Uthra, R.A.; Devaraj, D.; Umachandran, K. Intelligent machine learning based total productive maintenance approach for achieving zero downtime in industrial machinery. *Comput. Ind. Eng.* **2021**, *157*, 107267. [CrossRef]

5. Shahin, M.; Chen, F.F.; Hosseinzadeh, A.; Zand, N. Using machine learning and deep learning algorithms for downtime minimization in manufacturing systems: An early failure detection diagnostic service. *Int. J. Adv. Manuf. Technol.* **2023**, *128*, 3857–3883. [CrossRef]
6. Fang, Q.; Xiong, G.; Zhou, M.; Tamir, T.S.; Yan, C.B.; Wu, H.; Shen, Z.; Wang, F.Y. Process monitoring, diagnosis and control of additive manufacturing. *IEEE Trans. Autom. Sci. Eng.* **2022**, *21*, 1041–1067. [CrossRef]
7. Thomas, D.S.; Gilbert, S.W. Costs and cost effectiveness of additive manufacturing. *NIST Spec. Publ.* **2014**, *1176*, 12.
8. Verboeket, V.; Krikke, H. The disruptive impact of additive manufacturing on supply chains: A literature study, conceptual framework and research agenda. *Comput. Ind.* **2019**, *111*, 91–107. [CrossRef]
9. Peng, T.; Kellens, K.; Tang, R.; Chen, C.; Chen, G. Sustainability of additive manufacturing: An overview on its energy demand and environmental impact. *Addit. Manuf.* **2018**, *21*, 694–704. [CrossRef]
10. Gao, W.; Zhang, Y.; Ramanujan, D.; Ramani, K.; Chen, Y.; Williams, C.B.; Wang, C.C.; Shin, Y.C.; Zhang, S.; Zavattieri, P.D. The status, challenges, and future of additive manufacturing in engineering. *Comput.-Aided Des.* **2015**, *69*, 65–89. [CrossRef]
11. Tofail, S.A.; Koumoulos, E.P.; Bandyopadhyay, A.; Bose, S.; O'Donoghue, L.; Charitidis, C. Additive manufacturing: Scientific and technological challenges, market uptake and opportunities. *Mater. Today* **2018**, *21*, 22–37. [CrossRef]
12. Kanishka, K.; Acherjee, B. Revolutionizing manufacturing: A comprehensive overview of additive manufacturing processes, materials, developments, and challenges. *J. Manuf. Process.* **2023**, *107*, 574–619. [CrossRef]
13. Diniță, A.; Ripeanu, R.G.; Ilincă, C.N.; Cursaru, D.; Matei, D.; Naim, R.I.; Tănase, M.; Portoacă, A.I. Advancements in Fiber-Reinforced Polymer Composites: A Comprehensive Analysis. *Polymers* **2024**, *16*, 2. [CrossRef]
14. Hassan, H.Z.; Saeed, N.M. Advancements and applications of lightweight structures: A comprehensive review. *Discov. Civ. Eng.* **2024**, *1*, 47. [CrossRef]
15. Chauhan, V.; Kärki, T.; Varis, J. Review of natural fiber-reinforced engineering plastic composites, their applications in the transportation sector and processing techniques. *J. Thermoplast. Compos. Mater.* **2022**, *35*, 1169–1209. [CrossRef]
16. Cheng, P.; Li, S.; Peng, Y.; Duigou, A.L.; Wang, K.; Ahzi, S. 3D/4D Printed Functional Continuous Fiber-reinforced Polymer Composites: Progress and Perspectives. *Chin. J. Mech. Eng. Addit. Manuf. Front.* **2023**, *2*, 100090. [CrossRef]
17. Cheng, P.; Peng, Y.; Li, S.; Rao, Y.; Le Duigou, A.; Wang, K.; Ahzi, S. 3D printed continuous fiber reinforced composite lightweight structures: A review and outlook. *Compos. Part B Eng.* **2023**, *250*, 110450. [CrossRef]
18. Tian, X.; Todoroki, A.; Liu, T.; Wu, L.; Hou, Z.; Ueda, M.; Hirano, Y.; Matsuzaki, R.; Mizukami, K.; Iizuka, K.; et al. 3D Printing of Continuous Fiber Reinforced Polymer Composites: Development, Application, and Prospective. *Chin. J. Mech. Eng. Addit. Manuf. Front.* **2022**, *1*, 100016. [CrossRef]
19. Tamir, T.S.; Xiong, G.; Fang, Q.; Dong, X.; Shen, Z.; Wang, F.Y. A feedback-based print quality improving strategy for FDM 3D printing: An optimal design approach. *Int. J. Adv. Manuf. Technol.* **2022**, *120*, 2777–2791. [CrossRef]
20. Charalampous, P.; Kostavelis, I.; Tzovaras, D. Non-destructive quality control methods in additive manufacturing: A survey. *Rapid Prototyp. J.* **2020**, *26*, 777–790. [CrossRef]
21. Lu, Q.Y.; Wong, C.H. Additive manufacturing process monitoring and control by non-destructive testing techniques: Challenges and in-process monitoring. *Virtual Phys. Prototyp.* **2018**, *13*, 39–48. [CrossRef]
22. Klette, R. *Concise Computer Vision*; Springer: London, UK, 2014; Volume 233.
23. Sahoo, S.; Kumar, S.; Abedin, M.Z.; Lim, W.M.; Jakhar, S.K. Deep learning applications in manufacturing operations: A review of trends and ways forward. *J. Enterp. Inf. Manag.* **2023**, *36*, 221–251. [CrossRef]
24. Sharp, M.; Ak, R.; Hedberg, T., Jr. A survey of the advancing use and development of machine learning in smart manufacturing. *J. Manuf. Syst.* **2018**, *48*, 170–179. [CrossRef]
25. Trakadas, P.; Simoens, P.; Gkonis, P.; Sarakis, L.; Angelopoulos, A.; Ramallo-González, A.P.; Skarmeta, A.; Trochoutsos, C.; Calvo, D.; Pariente, T.; et al. An artificial intelligence-based collaboration approach in industrial iot manufacturing: Key concepts, architectural extensions and potential applications. *Sensors* **2020**, *20*, 5480. [CrossRef] [PubMed]
26. Meireles, M.R.; Almeida, P.E.; Simões, M.G. A comprehensive review for industrial applicability of artificial neural networks. *IEEE Trans. Ind. Electron.* **2003**, *50*, 585–601. [CrossRef]
27. Mumali, F. Artificial neural network-based decision support systems in manufacturing processes: A systematic literature review. *Comput. Ind. Eng.* **2022**, *165*, 107964. [CrossRef]
28. De Filippis, L.A.C.; Serio, L.M.; Facchini, F.; Mummolo, G. ANN modelling to optimize manufacturing process. In *Advanced Applications for Artificial Neural Networks*; InTech: London, UK, 2018; pp. 201–226.
29. Liu, H.; Zhang, G. Enhancing Efficiency and Energy Optimization: Data-Driven Solutions in Process Industrial Manufacturing. *EAI Endorsed Trans. Energy Web* **2024**, *11*. [CrossRef]
30. Akhtar, N.; Mian, A. Threat of adversarial attacks on deep learning in computer vision: A survey. *IEEE Access* **2018**, *6*, 14410–14430. [CrossRef]

31. Uçar, A.; Demir, Y.; Güzeliş, C. Object recognition and detection with deep learning for autonomous driving applications. *Simulation* **2017**, *93*, 759–769. [CrossRef]
32. Brunetti, A.; Buongiorno, D.; Trotta, G.F.; Bevilacqua, V. Computer vision and deep learning techniques for pedestrian detection and tracking: A survey. *Neurocomputing* **2018**, *300*, 17–33. [CrossRef]
33. Elyan, E.; Vuttipittayamongkol, P.; Johnston, P.; Martin, K.; McPherson, K.; Moreno-García, C.F.; Jayne, C.; Sarker, M.M.K. Computer vision and machine learning for medical image analysis: recent advances, challenges, and way forward. *Artif. Intell. Surg.* **2022**, *2*, 24–45. [CrossRef]
34. LeCun, Y. Generalization and network design strategies. In *Connections in Perspective*; Elsevier: Toronto, ON, Canada, 1989.
35. Shaha, M.; Pawar, M. Transfer learning for image classification. In Proceedings of the 2018 Second International Conference on Electronics, Communication and Aerospace Technology (ICECA), Coimbatore, India, 29–31 March 2018; pp. 656–660.
36. Toprak, T.; Belenlioglu, B.; Aydın, B.; Guzelis, C.; Selver, M.A. Conditional weighted ensemble of transferred models for camera based onboard pedestrian detection in railway driver support systems. *IEEE Trans. Veh. Technol.* **2020**, *69*, 5041–5054. [CrossRef]
37. Krizhevsky, A.; Sutskever, I.; Hinton, G.E. Imagenet classification with deep convolutional neural networks. In Proceedings of the Advances in Neural Information Processing Systems 25 (NIPS 2012), Lake Tahoe, NA, USA, 3–6 December 2012.
38. Tammina, S. Transfer learning using vgg-16 with deep convolutional neural network for classifying images. *Int. J. Sci. Res. Publ. (IJSRP)* **2019**, *9*, 143–150. [CrossRef]
39. Shafiq, M.; Gu, Z. Deep residual learning for image recognition: A survey. *Appl. Sci.* **2022**, *12*, 8972. [CrossRef]
40. Brion, D.A.; Pattinson, S.W. Generalisable 3D printing error detection and correction via multi-head neural networks. *Nat. Commun.* **2022**, *13*, 4654. [CrossRef]
41. Banadaki, Y.; Razaviarab, N.; Fekrmandi, H.; Li, G.; Mensah, P.; Bai, S.; Sharifi, S. Automated quality and process control for additive manufacturing using deep convolutional neural networks. *Recent Prog. Mater.* **2021**, *4*. [CrossRef]
42. Lin, W.; Shen, H.; Fu, J.; Wu, S. Online quality monitoring in material extrusion additive manufacturing processes based on laser scanning technology. *Precis. Eng.* **2019**, *60*, 76–84. [CrossRef]
43. Lafirenza, M.; Guerra, M.G.; Lavecchia, F. A layerwise monitoring methodology based on blue laser line profilometer for Material Extrusion processes. *Int. J. Adv. Manuf. Technol.* **2023**, *127*, 5421–5436. [CrossRef]
44. Reckert, A.; Lang, V.; Weingarten, S.; Johne, R.; Klein, J.H.; Ihlenfeldt, S. Quality Prediction and Classification of Process Parameterization for Multi-Material Jetting by Means of Computer Vision and Machine Learning. *J. Manuf. Mater. Process.* **2024**, *8*, 8. [CrossRef]
45. Zubayer, M.H.; Xiong, Y.; Wang, Y.; Imdadul, H.M. Enhancing additive manufacturing precision: Intelligent inspection and optimization for defect-free continuous carbon fiber-reinforced polymer. *Compos. Part C Open Access* **2024**, *14*, 100451. [CrossRef]
46. Iftekar, S.F.; Aabid, A.; Amir, A.; Baig, M. Advancements and limitations in 3D printing materials and technologies: A critical review. *Polymers* **2023**, *15*, 2519. [CrossRef]
47. Hasan, M.R.; Davies, I.J.; Pramanik, A.; John, M.; Biswas, W.K. Potential of Recycled PLA in 3D Printing: A Review. *Sustain. Manuf. Serv. Econ.* **2024**, *3*, 100020. [CrossRef]
48. Gregor, A.; Filová, E.; Novák, M.; Kronek, J.; Chlup, H.; Buzgo, M.; Blahnová, V.; Lukášová, V.; Bartoš, M.; Nečas, A.; et al. Designing of PLA scaffolds for bone tissue replacement fabricated by ordinary commercial 3D printer. *J. Biol. Eng.* **2017**, *11*, 1–21. [CrossRef]
49. Mofokeng, J.P.; Luyt, A.; Tábi, T.; Kovács, J. Comparison of injection moulded, natural fibre-reinforced composites with PP and PLA as matrices. *J. Thermoplast. Compos. Mater.* **2012**, *25*, 927–948. [CrossRef]
50. Molnár, I.; Morovič, L.; Delgado Sobrino, D.R.; Lecký, Š.; Michal, D. Medical applications of biomaterials: The case of design and manufacture of orthopedic corsets made of polylactic acid by additive manufacturing. In *Materials Science Forum*; Trans Tech Publ.: Baech, Switzerland, 2019; Volume 952, pp. 223–232.
51. Gupta, B.B.; Sheng, M. Machine Learning for Computer and Cyber Security; CRC Press: Boca Raton, FL, USA, 2019.
52. Konstantopoulos, G.; Koumoulos, E.P.; Charitidis, C.A. Classification of mechanism of reinforcement in the fiber-matrix interface: Application of Machine Learning on nanoindentation data. *Mater. Des.* **2020**, *192*, 108705. [CrossRef]
53. Salehi, A.W.; Khan, S.; Gupta, G.; Alabduallah, B.I.; Almjally, A.; Alsolai, H.; Siddiqui, T.; Mellit, A. A study of CNN and transfer learning in medical imaging: Advantages, challenges, future scope. *Sustainability* **2023**, *15*, 5930. [CrossRef]
54. Deng, J.; Dong, W.; Socher, R.; Li, L.J.; Li, K.; Fei-Fei, L. Imagenet: A large-scale hierarchical image database. In Proceedings of the 2009 IEEE Conference on Computer Vision and Pattern Recognition, Miami, FL, USA, 20–25 June 2009; pp. 248–255.
55. Li, F.F. The ImageNet Project. 2009. Available online: https://www.image-net.org/ (accessed on 23 December 2024).
56. Smith, L.N. A disciplined approach to neural network hyper-parameters: Part 1–learning rate, batch size, momentum, and weight decay. *arXiv* **2018**, arXiv:1803.09820.

57. Al-Mahadeen, B.; AlTarawneh, M.S.; AlTarawneh, I.H. Signature region of interest using auto cropping. *arXiv* **2010**, arXiv:1004.3549.
58. Gautam, C.M.; Sharma, S.; Verma, J.S. A GUI for automatic extraction of signature from image document. *Int. J. Comput. Appl.* **2012**, *54*, 13–19.
59. Chen, J.; Bai, G.; Liang, S.; Li, Z. Automatic image cropping: A computational complexity study. In Proceedings of the IEEE Conference on Computer Vision and Pattern Recognition, Las Vegas, NV, USA, 27–30 June 2016; pp. 507–515.
60. Yan, J.; Lin, S.; Bing Kang, S.; Tang, X. Learning the change for automatic image cropping. In Proceedings of the IEEE Conference on Computer Vision and Pattern Recognition, Portland, OR, USA, 23–28 June 2013; pp. 971–978.
61. Shorten, C.; Khoshgoftaar, T.M. A survey on image data augmentation for deep learning. *J. Big Data* **2019**, *6*, 1–48. [CrossRef]
62. Takahashi, R.; Matsubara, T.; Uehara, K. Data augmentation using random image cropping and patching for deep CNNs. *IEEE Trans. Circuits Syst. Video Technol.* **2019**, *30*, 2917–2931. [CrossRef]
63. Zhong, Z.; Zheng, L.; Kang, G.; Li, S.; Yang, Y. Random erasing data augmentation. *Proc. AAAI Conf. Artif. Intell.* **2020**, *34*, 13001–13008. [CrossRef]
64. Maharana, K.; Mondal, S.; Nemade, B. A review: Data pre-processing and data augmentation techniques. *Glob. Transit. Proc.* **2022**, *3*, 91–99. [CrossRef]
65. Elad, M.; Milanfar, P. Style transfer via texture synthesis. *IEEE Trans. Image Process.* **2017**, *26*, 2338–2351. [CrossRef]
66. Frigo, O.; Sabater, N.; Delon, J.; Hellier, P. Split and match: Example-based adaptive patch sampling for unsupervised style transfer. In Proceedings of the IEEE Conference on Computer Vision and Pattern Recognition, Las Vegas, NV, USA, 27–30 June 2016; pp. 553–561.
67. Bowles, C.; Chen, L.; Guerrero, R.; Bentley, P.; Gunn, R.; Hammers, A.; Dickie, D.A.; Hernández, M.V.; Wardlaw, J.; Rueckert, D. Gan augmentation: Augmenting training data using generative adversarial networks. *arXiv* **2018**, arXiv:1810.10863.
68. Perez, L. The effectiveness of data augmentation in image classification using deep learning. *arXiv* **2017**, arXiv:1712.04621.
69. Shekhar, S.; Bansode, A.; Salim, A. A comparative study of hyper-parameter optimization tools. In Proceedings of the 2021 IEEE Asia-Pacific Conference on Computer Science and Data Engineering (CSDE), Brisbane, Australia, 8–10 December 2021; pp. 1–6.
70. Hanifi, S.; Cammarono, A.; Zare-Behtash, H. Advanced hyperparameter optimization of deep learning models for wind power prediction. *Renew. Energy* **2024**, *221*, 119700. [CrossRef]

applied sciences

MDPI

Article

A Temporal Convolutional Network–Bidirectional Long Short-Term Memory (TCN-BiLSTM) Prediction Model for Temporal Faults in Industrial Equipment

Jinyin Bai, Wei Zhu *, Shuhong Liu, Chenhao Ye, Peng Zheng and Xiangchen Wang

School of Information and Communication, National University of Defense Technology, Wuhan 430035, China; bbaibai0612@163.com (J.B.); shuhongliu0905@163.com (S.L.); a379763802@163.com (C.Y.); zheng2013@foxmail.com (P.Z.); 13188789626@163.com (X.W.)

* Correspondence: zhuwei929@hotmail.com

Abstract: Traditional algorithms and single predictive models often face challenges such as limited prediction accuracy and insufficient modeling capabilities for complex time-series data in fault prediction tasks. To address these issues, this paper proposes a combined prediction model based on an improved temporal convolutional network (TCN) and bidirectional long short-term memory (BiLSTM), referred to as the TCN-BiLSTM model. This model aims to enhance the reliability and accuracy of time-series fault prediction. It is designed to handle continuous processes but can also be applied to batch and hybrid processes due to its flexible architecture. First, preprocessed industrial operation data are fed into the model, and hyperparameter optimization is conducted using the Optuna framework to improve training efficiency and generalization capability. Then, the model employs an improved TCN layer and a BiLSTM layer for feature extraction and learning. The TCN layer incorporates batch normalization, an optimized activation function (Leaky ReLU), and a dropout mechanism to enhance its ability to capture multi-scale temporal features. The BiLSTM layer further leverages its bidirectional learning mechanism to model the long-term dependencies in the data, enabling effective predictions of complex fault patterns. Finally, the model outputs the prediction results after iterative optimization. To evaluate the performance of the proposed model, simulation experiments were conducted to compare the TCN-BiLSTM model with mainstream prediction methods such as CNN, RNN, BiLSTM, and A-BiLSTM. The experimental results indicate that the TCN-BiLSTM model outperforms the comparison models in terms of prediction accuracy during both the modeling and forecasting stages, providing a feasible solution for time-series fault prediction.

Keywords: time-series fault; TCN; BiLSTM; fault prediction; hyperparameter optimization

1. Introduction

With the rapid development of industrialization and intelligent technologies, modern industrial equipment and mechanical systems are increasingly evolving towards greater complexity, precision, and integration. These systems are widely applied in fields such as manufacturing, transportation, energy production, and intelligent logistics, playing an indispensable role in ensuring the efficient operation of the national economy and social activities [1]. However, as system complexity and operational loads continue to increase, the probability of faults and their unpredictability also rise. This not only poses significant challenges to the safe operation of equipment but may also lead to substantial economic losses and safety risks [2].

Unexpected failures of industrial equipment are phenomena that are difficult to completely avoid, and their impacts often go beyond production downtime, potentially causing cascading system damage. For instance, in the field of energy production, issues such as turbine blade wear, cracks, or excessive vibrations can significantly reduce power generation efficiency and even lead to catastrophic shutdown accidents. In the transportation sector, abnormal temperature rises or fatigue cracks in train bearings may become critical triggers for operational failures, thereby endangering passenger safety [3]. Consequently, ensuring the safe and efficient operation of mechanical systems has become a core requirement for modern industrial development. To achieve this, the timely prediction of potential equipment failures and the implementation of preventive maintenance measures have become critical technological approaches to mitigate operational risks and control maintenance costs [4].

The core of fault prediction technology lies in real-time monitoring of the operating states of industrial equipment to establish a mapping relationship between equipment characteristics and health conditions. This enables the prediction of potential fault modes and the estimation of fault occurrence times. In recent years, this field has garnered significant attention from both industry and academia, achieving rapid advancements in areas such as intelligent manufacturing, the Industrial Internet of Things (IIoT), and Prognostics and Health Management (PHM). However, traditional fault diagnosis methods—such as rule-based periodic maintenance or diagnostic analysis relying on human expertise—often exhibit considerable limitations when dealing with complex and diverse industrial equipment. These limitations arise mainly due to the following distinctive challenges associated with modern industrial equipment:

- High Complexity: Industrial equipment typically consists of multiple highly coupled subsystems, with operating states characterized by strong nonlinearity and dynamic coupling. This makes it challenging to accurately describe the overall system behavior using simple physical models.
- High Dynamism: Industrial equipment operates in complex and ever-changing environments, and its operating parameters (e.g., vibration, temperature, and current) fluctuate significantly over time. Traditional static analysis methods struggle to capture these dynamic variations effectively.
- Fault Concealment: Certain potential faults (e.g., early-stage cracks and minor wear) have minimal impact on the operational state in their initial stages, making them difficult to detect through conventional monitoring techniques.

To address the aforementioned challenges, machine learning and deep learning technologies offer novel research avenues for fault prediction in complex equipment. These techniques can automatically extract critical features from massive sensor data and construct nonlinear, multi-layered fault prediction models. Monitoring data from industrial equipment are typically presented as time-series data, encompassing various physical quantities collected by multiple sensors, such as vibration signals, current, voltage, and temperature. These data are characterized by high dimensionality, nonlinearity, and noise interference, making it difficult for traditional methods to model them comprehensively. The complexity of equipment structures and operating conditions results in fault patterns with multiple mechanisms (e.g., mechanical wear, fatigue, overloading, and nonlinear vibrations) and interactions between fault modes, complicating their analysis and prediction. In practical industrial scenarios, fault data are typically scarce and unevenly distributed. Healthy state data dominate the datasets, while fault state data are often limited and insufficient. This imbalance poses challenges for training models that can effectively recognize and predict faults.

Current research methods for time-series fault prediction can be broadly divided into two categories: statistical model-based methods and artificial intelligence (AI)-based methods.

- Statistical Model-Based Methods. These methods use linear or nonlinear regression models to analyze the relationships between equipment operating parameters and potential faults, building predictive models based on historical data. One of the main advantages of these methods is their simplicity, ease of implementation, and, importantly, their transparency and interpretability. This makes them not only easier to implement but also more likely to be accepted as effective prediction tools. However, their performance can be limited when handling large datasets with complex relationships [5]. For example, traditional time-series statistical models can effectively capture the temporal dependencies of faults. However, they generally assume linearity in the data, which limits their ability to address nonlinear issues [6,7].
- AI-Based Methods. Machine learning methods can extract features from large historical datasets, enabling the recognition of complex, nonlinear fault patterns and improving prediction accuracy. However, these approaches often require extensive manual feature selection and extraction, which can be time-consuming and labor-intensive [8]. Expert systems, on the other hand, can make quick predictions and decisions for specific fault patterns, but they rely heavily on domain experts' knowledge and experience, which limits their scalability and generalizability [9]. Long short-term memory (LSTM) networks, specifically designed to handle sequential data, have demonstrated strong capabilities in capturing long-term dependencies in time-series data. By leveraging mechanisms for memory retention and forgetting, LSTMs dynamically adjust the focus on prior information, resulting in more accurate fault predictions [10–12]. However, LSTM models are sensitive to noise, and their learning capacity may be adversely affected when processing data with significant noise levels. Given the inherent limitations of both statistical model-based and AI-based approaches, hybrid predictive models have emerged as a promising solution. These models combine the strengths of different methods, offering more accurate and stable prediction performance compared to standalone models [13].

To enhance the accuracy and stability of industrial equipment fault prediction, this paper proposes a hybrid predictive model based on the temporal convolutional network (TCN) and bidirectional long short-term memory network (BiLSTM). The main contributions of this study are summarized as follows:

- A novel TCN-BiLSTM hybrid model is proposed, which integrates the ability of BiLSTM to capture bidirectional dependencies in time-series data with the efficient multi-scale temporal feature extraction capability of TCN. This combination improves the model's ability to process complex time-series data.
- The proposed model incorporates techniques such as batch normalization, dropout, and an optimized Leaky ReLU activation function into the network structure. These enhancements effectively mitigate the vanishing and exploding gradient problems in deep networks, significantly improving the training efficiency and stability of the model.
- The model leverages the Optuna framework for hyperparameter optimization, enabling the selection of optimal parameters tailored to different datasets. This approach enhances the model's generalization capability across diverse applications.

In addition to the performance during the modeling phase, the predictive phase is equally critical for evaluating the practical value of fault prediction algorithms. During the predictive phase, models often encounter previously unseen data distributions, while observation data in industrial scenarios are frequently accompanied by noise, uncertainty, and unknown error structures. Therefore, a successful predictive model must not only exhibit strong modeling capabilities but also effectively handle noise interference and observational errors during the predictive phase to ensure the robustness and reliability of its predictions.

The proposed TCN-BiLSTM model leverages multi-scale temporal feature extraction and bidirectional sequential dependency modeling, giving it the theoretical and practical ability to address complex observational errors. Furthermore, the network design incorporates techniques such as batch normalization, Leaky ReLU activation functions, and residual connections. These mechanisms significantly mitigate the impact of noise during the predictive phase and enhance the model's adaptability to unseen data distributions.

2. TCN-BiLSTM Predictive Model

2.1. TCN Model Architecture

TCN is a temporal convolutional network specifically designed for time-series data processing. Compared to traditional Convolutional Neural Networks (CNNs), TCN introduces causal convolutions to ensure that the output at any given time step does not depend on future time steps, preserving the temporal order of the data. Furthermore, through the use of dilated convolutions, TCN can effectively expand its receptive field without increasing the number of parameters, allowing it to capture long-range dependencies within sequences. The basic structure of the dilated convolutions model is illustrated in Figure 1. The receptive field regions of the TCN with different dilation factors (d = 4, d = 2, and d = 1) are also depicted.

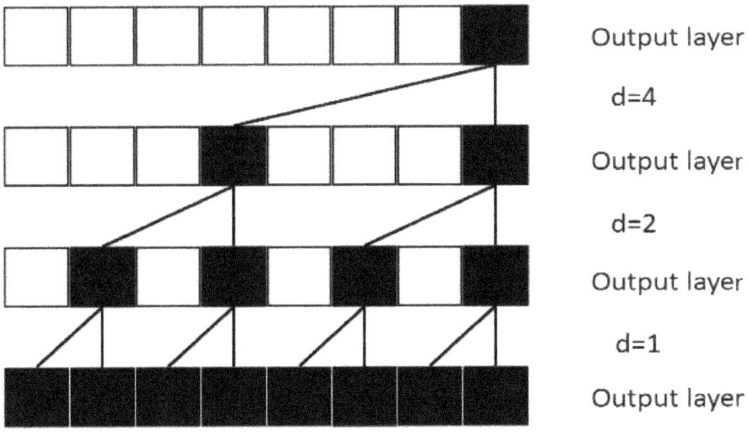

Figure 1. Dilated convolution architecture diagram.

The calculation formulas for the causal convolution in TCN are shown as follows:

$$y(t) = \sum_{k=0}^{K-1} f(k) \cdot x(t - d \cdot k) \tag{1}$$

In this equation, $x(t)$ denotes the input sequence, f represents the convolutional kernel, K is the kernel size, $y(t)$ is the output sequence, and d is the dilation rate.

To mitigate the overfitting issue in the network and enhance its performance, learning efficiency, and accuracy, techniques such as batch normalization, activation functions, and dropout are commonly used. Additionally, to ensure connection stability in deep TCN networks, residual connections are introduced. Combined with other methods, these form the residual module, which further optimizes the performance of TCN.

In this study, Leaky ReLU is employed as the activation function instead of the traditional ReLU. Leaky ReLU retains a small slope on the negative axis, allowing some negative values to pass through. This prevents the issue of completely inactive neurons for negative inputs, which is commonly observed with ReLU. Leaky ReLU is more adaptive to time-series data with uneven feature distributions. The formula is expressed as follows:

$$f(x) = \begin{cases} x, x > 0 \\ ax, x \leq 0 \end{cases} \quad (2)$$

In this formula, x represents the input value, and "a" denotes the slope for the negative input region. The ax part indicates that when the input x is negative, the output will be the product of the input value x and the constant "a". Here, "a" is a small constant, typically set to 0.01, which ensures that there is still a slight output even for negative inputs. This prevents neurons from becoming "dead"—that is, not activating at all—when faced with negative input.

2.2. BiLSTM Model Architecture

LSTM (long short-term memory) is a specialized type of recurrent neural network (RNN) designed for processing and predicting sequential data. Its core lies in the gating mechanism, which regulates the flow of information through the forget gate, input gate, and output gate [14]. These gates work together to manage the storage of short-term and long-term memory via the hidden state and cell state. By addressing the vanishing gradient problem inherent in standard RNNs [15], LSTMs are particularly effective in learning from long sequences. The specific structure of a single-layer LSTM cell is illustrated in Figure 2 [16].

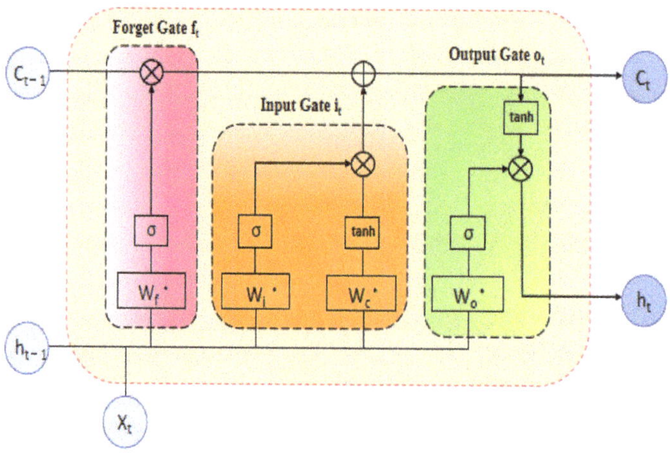

Figure 2. LSTM architecture diagram.

In LSTM, the sigmoid function plays a crucial role in regulating the flow of information by mapping input values to a range of (0, 1). This enables the normalization of outputs for each neuron and helps determine the extent to which different pieces of information are stored or forgotten within the network. Due to these characteristics, the sigmoid function is well suited as the activation function in the gating mechanism.

$$\sigma(x) = \frac{1}{1+e^{-x}} \tag{3}$$

In this equation, σ represents the sigmoid activation function.

The hidden state reflects the "memory" of sequential data at the current time step. By being updated iteratively, it dynamically extracts and retains features relevant to the current prediction task.

$$h_t^l = o_t^l \cdot \tanh(C_t^l) \tag{4}$$

In the equation, l represents the number of LSTM layers, C_t represents the cell state, and *tanh* represents the hyperbolic tangent activation function, which ensures that the generated hidden state is constrained within the range of $[-1, 1]$.

The forget gate often functions to regulate the proportion of past information retained, determining the extent to which information from the cell state C_{t-1} at the previous time step is preserved.

$$f_t^l = \sigma(W_f^l \cdot [h_{t-1}^l, x_t] + b_f^l) \tag{5}$$

In the equation, W_f represents the weight matrix of the forget gate, and b_f represents the bias matrix of the forget gate.

The input gate is responsible for determining the information to be updated in the cell state.

$$i_t^l = \sigma(W_i^l \cdot [h_{t-1}^l, x_t] + b_i^l) \tag{6}$$

In this equation, W_i represents the weight matrix of the input gate, and b_i represents the bias matrix of the input gate.

Meanwhile, at each time step, the candidate cell state is computed based on the current input vector x_t and the hidden state h_{t-1} from the previous time step.

$$\overset{\sim l}{C_t} = \tanh(W_c^l \cdot [h_{t-1}^l, x_t] + b_c^l) \tag{7}$$

In this equation, W_c represents the weight matrix of the cell state, and b_c represents the bias matrix of the cell state.

The cell state is updated through the operations of the forget gate and the input gate.

$$C_t^l = f_t^l \cdot C_{t-1}^l + i_t^l \cdot \overset{\sim l}{C_t} \tag{8}$$

Finally, the output gate is computed.

$$o_t^l = \sigma(W_o^l \cdot [h_{t-1}^l, x_t] + b_o^l) \tag{9}$$

In this equation, W_o represents the weight matrix of the output gate, and b_o represents the bias matrix of the output gate.

In fault prediction, early signs of certain faults may emerge at later time steps, which standard LSTM models may fail to capture. If LSTM considers both forward and backward information at each time step, it can significantly enhance the model's ability to capture contextual information and improve the accuracy of equipment fault prediction. Moreover, considering that fault manifestations may become more complex in certain scenarios, stacking multiple layers of bidirectional LSTM allows each layer to learn features at different levels, enabling deeper feature extraction.

This paper proposes a deep bidirectional long short-term memory network model. On the basis of standard LSTM, the model incorporates a backward LSTM and employs a stacked structure. The architecture of the proposed model is shown in Figure 3.

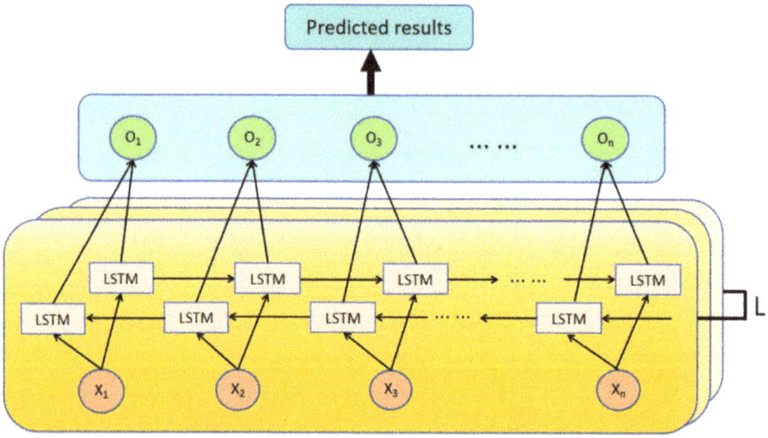

Figure 3. BiLSTM architecture diagram.

2.3. TCN-BiLSTM Prediction Method

The structure of the TCN-BiLSTM model is mainly divided into three components: the data preprocessing module, the TCN layer, and the BiLSTM layer. The detailed architecture of the model is illustrated in Figure 4.

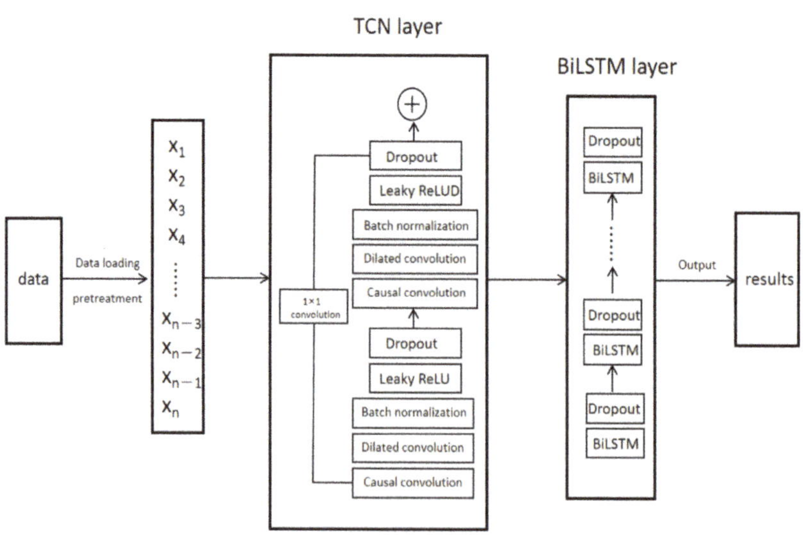

Figure 4. TCN-BiLSTM architecture diagram.

In practical industrial applications, the predictive phase of a model often encounters challenges such as unmodeled noise, observational errors, and uneven data distributions in industrial sensor data. The proposed model, through its design and mechanisms, not only enhances its capability during the modeling phase but also improves its robustness during the predictive phase.

The TCN layer serves as the first core component of the model, specifically designed for feature extraction from time-series data. To enhance the model's performance, the TCN layer adopts the following technical designs:

- Causal Convolution: Causal convolution in TCN is one of its core features, ensuring that the output at each time step only depends on the current and past time steps. This means that future time steps (potential sources of noise) do not influence the prediction of the current time step, thereby preventing future information from interfering with the model's training and reducing the risk of noise propagating to subsequent time steps.
- Dilated Convolution: Dilated convolution increases the receptive field of the convolution, enabling the model to capture longer temporal dependencies without adding more parameters. Dilated convolution helps reduce the impact of noise on short time scales, thereby allowing the model to more stably capture key dynamic changes in long time series. When dealing with industrial time-series data, noise typically manifests as high-frequency signals. Dilated convolution effectively diminishes the interference of high-frequency noise, allowing the model to focus more on the long-term trends of the sequence rather than on local random fluctuations.
- Leaky ReLU Activation Function: Compared to the traditional ReLU activation function, the Leaky ReLU activation function maintains a small slope for negative inputs, effectively preventing the "dying neuron" problem. Noisy data often contain negative values, and Leaky ReLU can effectively mitigate the impact of noise on the network, allowing the neural network to continue training effectively even in the presence of noise. This enhances the model's robustness and its ability to capture nonlinear features.
- Dropout: By randomly dropping out a portion of neurons during training, dropout effectively combats overfitting and enhances the model's generalization capabilities. This technique "turns off" a random subset of neurons during each iteration, reducing dependencies between neurons and improving the model's ability to adapt to unseen data. Dropout is a simple yet powerful regularization method commonly used in deep neural networks, particularly in deeper architectures, to significantly boost performance.
- Batch Normalization: Batch normalization normalizes the inputs to each layer by adjusting their mean to 0 and variance to 1, thus mitigating the internal covariate shift and improving model stability. This technique accelerates convergence during training, alleviates the vanishing gradient problem, and enhances the stability of the training process. Additionally, batch normalization can act as a form of regularization, reducing the risk of overfitting.

Additionally, to facilitate efficient information flow and alleviate the gradient vanishing problem in deep networks, the TCN layer incorporates a 1×1 convolutional residual connection mechanism, ensuring smooth feature propagation between the input and output.

Building on the deep features extracted by the TCN layer, the BiLSTM layer further models the time-series data. Compared to traditional LSTM, BiLSTM captures bidirectional dependencies in the time series, providing a more comprehensive understanding of the dynamic characteristics of fault data. Since minority class faults typically have fewer samples and may occur at specific time points, BiLSTM can capture these critical pieces of information through its bidirectional structure, thereby enhancing the detection capability for minority class faults. By stacking multiple layers, the BiLSTM enhances its capacity to learn complex fault patterns. Moreover, for different application scenarios, the model leverages hyperparameter optimization (based on the Optuna framework) to fine-tune critical parameters such as the number of hidden units, learning rate, and regularization, effectively improving the model's robustness and adaptability.

Both the TCN and BiLSTM layers in this model are designed as multi-layer structures. By conducting hyperparameter optimization tailored to specific problems, the model not only utilizes the TCN layer to capture multi-scale temporal features but also employs the BiLSTM layer to further extract long-term dependencies from the data. This complementary deep learning framework integrates features across different levels, significantly enhancing the model's generalization and overall adaptability.

3. Simulation Experiments

3.1. Simulation Data

To validate the effectiveness and superiority of the proposed hybrid model, simulation experiments were conducted to evaluate its performance. The experiments were implemented based on the TensorFlow deep learning framework. The detailed experimental environment configuration is presented in Table 1.

Table 1. Experimental environment configuration.

Environment	Specifications
Operating System	Windows11 (Version 22H2)
CPU	13th Gen Intel(R) Core(TM) i9-13900HX 2.20 GHz
GPU	NVIDIA GeForce RTX 4080
Memory	32 GB
Software	Visual Studio 2022 (Version 17.5.4)
Language	Python 3.9.12
CUDA	11.7.1

The AI4I 2020 Predictive Maintenance Dataset was used as the test dataset for this experiment. This dataset reflects real-world predictive maintenance scenarios, consisting of a total of 100,000 samples, 8 features, and 6 labels. The 8 features include UID, product ID, air temperature, process temperature, rotational speed, torque, and tool wear. The 6 labels represent different types of failures, which include overall failure, tool wear failure, heat dissipation failure, power failure, overstrain failure, and random failure.

Specific details of the fault types are summarized in Table 2.

Table 2. Fault descriptions.

Fault Label	Description
Machine Failure	This indicates whether the equipment has experienced any failure.
Tool Wear Failure	Failure occurs when the equipment reaches the end of its tool wear lifespan or requires replacement.
Heat Dissipation Failure	Failure is caused by inadequate heat dissipation due to the equipment's condition and temperature.
Power Failure	Failure occurs when the equipment's power exceeds the power supply threshold.
Overstrain Failure	Failure is caused by excessive mechanical strain on the equipment.
Random Failure	Failure occurs due to random, unpredictable factors.

To validate the effectiveness of the proposed model, the dataset was divided into a training set and a testing set in a ratio of 80% to 20%. Specifically, 80,000 samples were randomly selected for model training, while the remaining 20,000 samples were used to evaluate the model's performance. The datasets used for training and testing are contiguous in the time domain. This partitioning strategy ensures that the training data are sufficient for the model to learn effectively, while also leaving enough samples in the testing set to comprehensively assess the model's generalization ability.

During the training phase, the training set was used to optimize the model's parameters, ensuring that the model learns the mapping relationship between input features and output labels. The testing set, on the other hand, was utilized to evaluate the model's performance on unseen data, validating its practical applicability and robustness. The random partitioning of the dataset eliminates potential biases caused by manual selection, thereby ensuring the fairness and reliability of the evaluation process [17,18].

In the experimental design, multiple operational variables of equipment (e.g., temperature, rotational speed, torque, etc.) were monitored to enable fault prediction. Binary fault labels (0 indicating normal operation; 1 indicating fault occurrence) were employed as supervisory signals to guide the training and testing processes of the model, thereby evaluating its predictive capability and real-world performance. Table 3 presents information on a subset of samples from the dataset.

Table 3. Partial real predictive maintenance data.

UID	Product ID	Type	Air Temperature [k]	Process Temperature [k]	Rotational Speed [rpm]	Torque [Nm]	Tool Wear [min]
1	M14860	M	298.1	308.6	1551	42.8	0
2	L47181	L	298.2	308.7	1408	46.3	3
3	L47182	L	298.1	308.5	1498	49.4	5
4	L47183	L	298.2	308.6	1433	39.5	7
5	L47184	L	298.2	308.7	1408	40	9
6	M14865	M	298.1	308.6	1425	41.9	11
7	L47186	L	298.1	308.6	1558	42.4	14
8	L47187	L	298.1	308.6	1527	40.2	16
9	M14868	M	298.3	308.7	1667	28.6	18

3.2. Data Preprocessing

Data preprocessing is a critical step in ensuring the performance of models, particularly when handling complex time-series data and equipment fault prediction tasks. In this study, detailed preprocessing of raw industrial operation data was conducted to enhance the model's adaptability to data features and its generalization ability to unseen data. The preprocessing steps include handling missing values, encoding categorical features, feature scaling, and label transformation, all aimed at optimizing data quality and laying a solid foundation for subsequent model training.

During data loading, the dataset underwent a comprehensive examination to ensure its completeness and consistency. Missing values were addressed using the median imputation method. For intermittent missing data commonly found in industrial datasets, the median imputation method helps mitigate the influence of outliers on data distribution, ensuring data stability. By filling in missing values, the continuity of time-series data was preserved, preventing potential adverse effects on model training caused by missing entries.

The median imputation method demonstrates strong generalizability and can effectively fill in missing information across different industrial datasets, particularly when handling sensor data with missing values [19].

The dataset also contained several categorical features (e.g., product ID and equipment type) that needed to be converted into numerical formats for deep learning models. This study employed label encoding, which maps each categorical feature to an integer value rather than its original string representation. Label encoding is widely applicable in industrial equipment fault prediction and effectively handles various types of categorical data [20].

To ensure comparability across features on the same scale, Min-Max scaling was applied to numerical features, normalizing their values to the [0, 1] range. By normalizing the features, all attributes were optimized on a uniform scale. Feature scaling is a commonly used technique for processing various time-series data, including physical quantities such as machine temperature, rotational speed, current, and voltage. Proper scaling ensures the model can efficiently utilize information from each feature.

For target labels, One-Hot encoding was employed, transforming each category label into a binary vector. One-Hot encoding effectively converts discrete categorical data into a numerical format that can be understood by the model. In industrial equipment fault prediction, different fault types often have distinct characteristics. By using One-Hot encoding, the model can process each fault category independently, thus improving prediction accuracy and classification performance [21].

The preprocessing steps adopted in this study exhibit strong generalizability. The imputation techniques, feature scaling, and encoding methods utilized here are not only suitable for this dataset but can also be applied to other time-series fault datasets. A robust data preprocessing pipeline can significantly improve data quality and, consequently, the model's performance.

3.3. Model Implementation

In this experiment, categorical cross-entropy was adopted as the performance evaluation metric for the model. Categorical cross-entropy is a commonly used loss function for multi-class classification tasks and is designed to measure the divergence between the predicted probability distribution and the actual target distribution. The formulation is expressed as follows:

$$L = -\frac{1}{N}\sum_{i=1}^{N}\sum_{j=1}^{C} y_{ij}\log(\hat{y_{ij}}) \tag{10}$$

where N represents the total number of samples, C is the total number of classes, y_{ij} denotes the true label for sample i belonging to class j (1 for belonging and 0 for not belonging), and $\hat{y_{ij}}$ denotes the normalized predicted probability distribution for the same. By minimizing the categorical cross-entropy loss, the model predictions can be made closer to the true distribution, thereby improving classification accuracy.

The experimental results of the models were evaluated and analyzed using Precision, Recall, and F1 Score as evaluation metrics. These metrics provide a comprehensive assessment of model performance from different perspectives.

Precision measures the proportion of true positive predictions among all samples predicted as positive by the model. It reflects the accuracy of the model, indicating its reliability in correctly identifying positive samples. A higher Precision value demonstrates that the model makes fewer false positive predictions. Recall evaluates the proportion of true positive predictions among all actual positive samples. It reflects the sensitivity or coverage of the model, indicating its ability to correctly identify as many positive samples as possible. A higher Recall value signifies that the model misses fewer actual positive cases. The F1 Score represents the harmonic mean of Precision and Recall, providing a balanced metric that considers both false positives and false negatives. It represents a comprehensive indicator that is particularly useful when there is an imbalance between the number of positive and negative samples. A higher F1 Score represents that the model achieves a better trade-off between Precision and Recall.

$$Precision = \frac{TP}{TP + FP} \tag{11}$$

$$Recall = \frac{TP}{TP + FN} \tag{12}$$

$$F1 = 2 \times \frac{Precision \times Recall}{Precision + Recall} \tag{13}$$

In these equations, TP (true positive) represents the number of samples correctly predicted as positive and FP (false positive) represents the number of samples incorrectly predicted as positive.

To optimize the model's performance, a systematic adjustment and evaluation of the hyperparameters were conducted, which significantly enhances the model's accuracy and generalization ability while avoiding overfitting and underfitting. To ensure improved accuracy and generalization capability for complex time-series data, this study employed an automated hyperparameter tuning approach based on the Optuna optimization framework to systematically optimize the key hyperparameters of the TCN-BiLSTM model [22,23]. The detailed optimization process is as follows:

- Hyperparameter Selection: Critical hyperparameters for both the TCN and BiLSTM layers were identified, including kernel size, dilation factor, number of hidden layer units, learning rate, dropout rate, and activation function type. These hyperparameters significantly influence the learning capacity and performance of the model. For instance, the kernel size and dilation factor determine the receptive field of the TCN, while the number of hidden units and dropout rate affect the BiLSTM's ability to model long-term dependencies and the model's resistance to overfitting.
- Automated Optimization: Using the Optuna framework, automated search and tuning of the aforementioned hyperparameters were conducted. A target function was created (i.e., the loss value on the validation set), and Bayesian optimization was employed to dynamically adjust the search space during each iteration. This approach gradually approximates the global optimum by leveraging feedback from the target function, overcoming the inefficiencies associated with traditional grid search or random search methods.
- Evaluation and Validation: For every set of hyperparameter configurations, the resulting model was evaluated on the validation set to ensure robust performance not only on the training set but also on unseen data. Cross-validation techniques were employed to further mitigate the risks of overfitting or underfitting.

The key hyperparameters for the internal structure of the TCN-BiLSTM model are summarized in Table 4. Batch size refers to the number of training examples processed in a single forward and backward pass. It is a crucial hyperparameter that affects training efficiency, convergence speed, and model generalization. Larger batch sizes may accelerate computation but could reduce model generalization, while smaller batch sizes may improve generalization at the cost of slower training.

Table 4. Hyperparameter optimization range.

No.	Hyperparameter	Search Range
1	Number of TCN Layers	1–10
2	Number of BiLSTM Layers	1–10
3	batch_size	16, 32, 64
4	Dropout Rate	0.1–0.5
5	Number of Hidden Neurons	8–128
6	Dilation Factor	1–16
7	Kernel Size	1–10
8	Learning Rate	0.0001–1

3.4. Analysis of Model Hyperparameters

Taking categorical_crossentropy as the training error metric for the experiments, 20,000 samples were randomly selected from the dataset for hyperparameter tuning to ensure both training efficiency and accuracy. For each combination of hyperparameters, the model was trained for 10 iterations. After conducting 50 rounds of hyperparameter optimization experiments on the training set, the historical optimization curve was created and is shown in Figure 5, while the hyperparameter optimization results are illustrated in Figure 6.

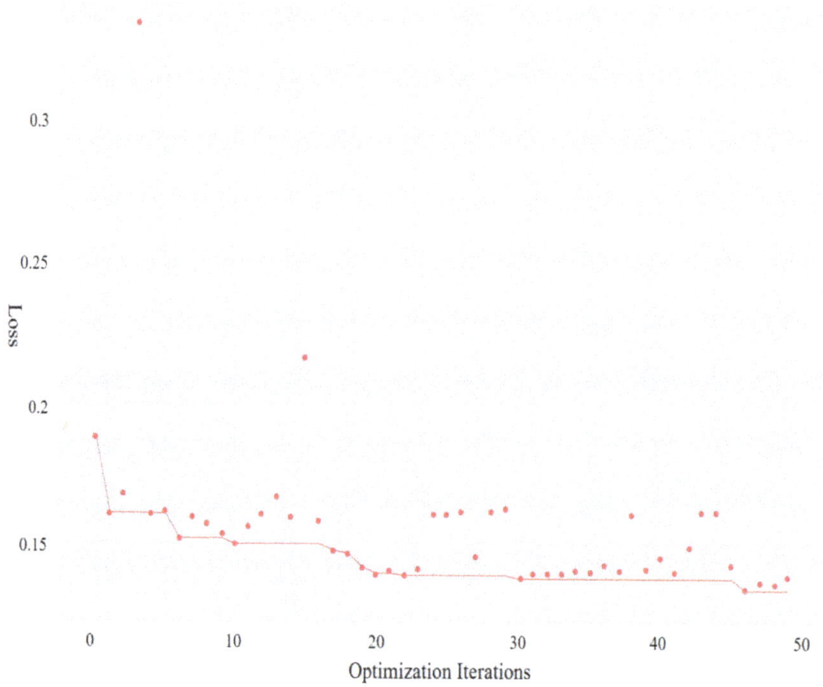

Figure 5. Historical optimization plot.

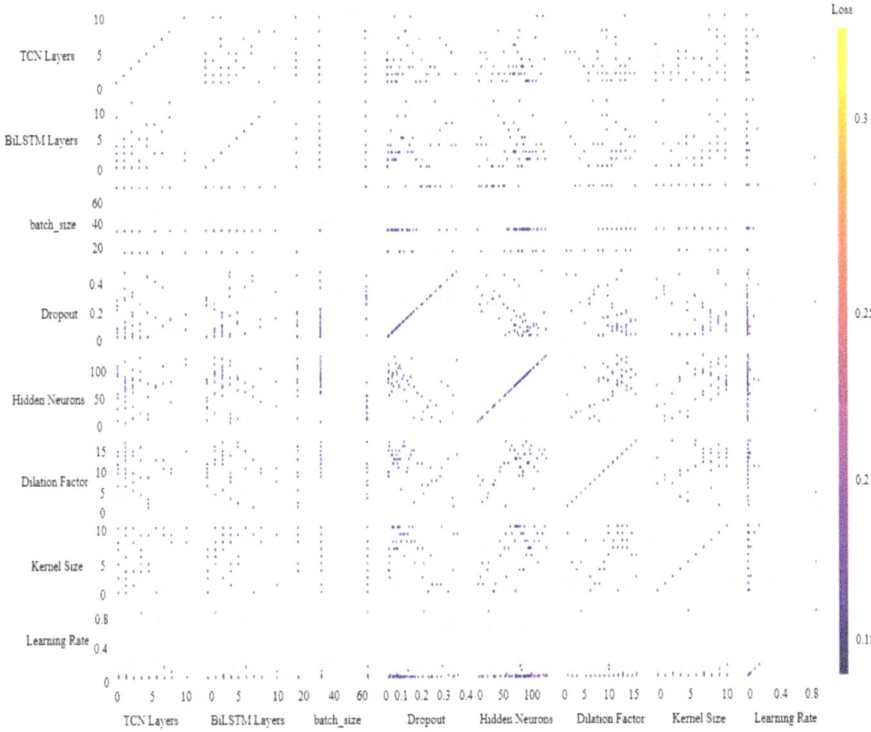

Figure 6. Hyperparameter optimization plot.

The scatter points in Figure 5 represent the results of each optimization attempt, while the red curve indicates the best result achieved at each iteration. As the number of optimization iterations increases, the combination and adjustment of different hyperparameters gradually improve the model's performance. Specifically, the training error is reduced from a maximum of 0.335 to 0.134, achieving a 60% reduction. This demonstrates that the hyperparameter optimization process significantly enhances the model's performance on the given dataset.

In Figure 6, the deeper the color, the smaller the training error, indicating better performance. The horizontal and vertical axes represent the seven hyperparameters used in the model's training process. Based on the hyperparameter optimization results, the optimal hyperparameter combination for this dataset is summarized in Table 5.

Table 5. Hyperparameter optimization results.

No.	Hyperparameter	Search Result
1	Number of TCN Layers	4
2	Number of BiLSTM Layers	3
3	batch_size	32
4	Dropout Rate	0.19
5	Number of Hidden Neurons	98
6	Dilation Factor	13
7	Kernel Size	4
8	Learning Rate	0.00089

3.5. Experimental Results and Analysis

To validate the effectiveness of the proposed TCN-BiLSTM model for time-series fault prediction tasks, comparative simulation experiments with multiple models were conducted. Under the same dataset and experimental conditions, the proposed model was compared with RNN, CNN, BiLSTM, and A-BiLSTM models [24]. The final experimental results of the test set are illustrated in Figures 7 and 8, showcasing the performance and classification outcomes of various models during multiple iterations of the training process.

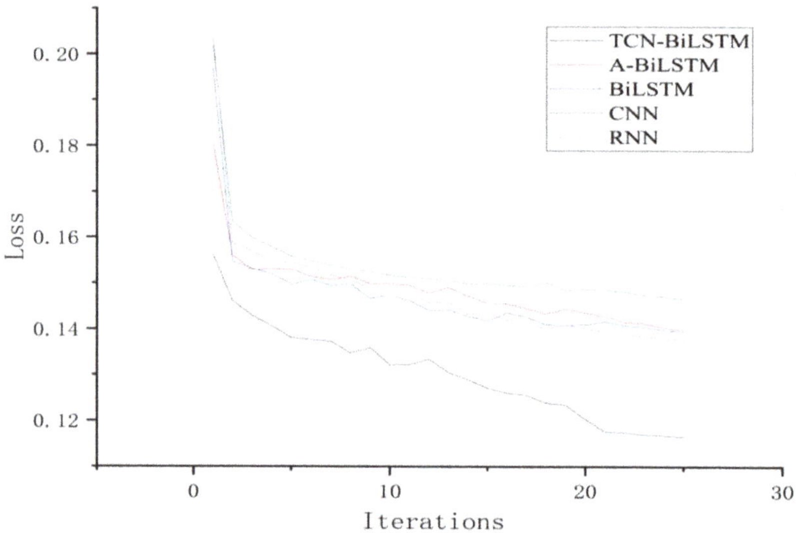

Figure 7. Line chart of experimental results.

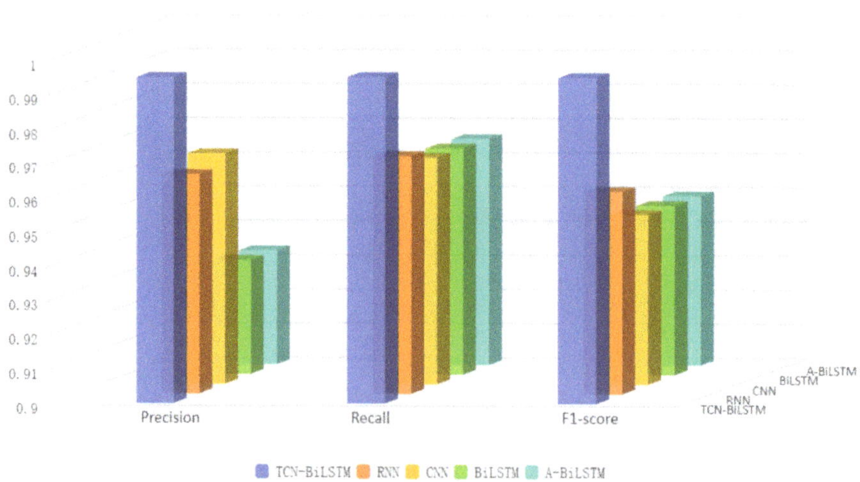

Figure 8. Histogram of experimental results.

An in-depth analysis of the experimental data reveals that all models exhibited a rapid convergence trend during the training phase. Examination of the iterative loss curves indicates that the loss values of the models began to stabilize after approximately

20 iterations, suggesting that these models possess relatively high training efficiency for the given task. Among them, the proposed TCN-BiLSTM model demonstrated superior convergence characteristics, achieving significantly lower loss values within a shorter training duration. This reflects its enhanced capacity for model fitting and generalization. Furthermore, the TCN-BiLSTM model consistently maintained high performance across key evaluation metrics, including Precision, Recall, and F1 Score, underscoring its robustness.

When evaluating the experimental results on the test set, the TCN-BiLSTM model delivered consistently superior classification performance compared to the baseline models. Specifically, it achieved a 20.79% reduction in loss compared to the CNN model, highlighting its remarkable ability to capture temporal dependencies and process sequential data. While CNN excels at extracting localized features, its limited capacity for modeling long-range temporal dependencies constrained its performance in this task. Compared to the BiLSTM model, the TCN-BiLSTM model achieved a 16.67% reduction in loss. Although BiLSTM is proficient at modeling temporal features, its inability to fully leverage long-term dependencies limited its effectiveness, a shortcoming addressed by the introduction of the TCN architecture. Similarly, when compared to the A-BiLSTM model, the TCN-BiLSTM model achieved a 17.09% reduction in loss, and compared to the RNN model, it yielded a 15.54% reduction in loss.

The dilated convolution and batch normalization mechanisms in the TCN-BiLSTM model effectively mitigate the impact of noise in observational data during the predictive phase. In contrast, while the CNN model has certain advantages in extracting local features, its relatively small receptive field makes it more susceptible to being misled by local noise, thereby reducing the accuracy of its predictions. Similarly, although the BiLSTM model is capable of capturing long-term dependencies in time series, it lacks the ability to model multi-scale temporal features.

To further validate the performance of the model, significance testing experiments were conducted. Under the same configuration parameters, each model was independently repeated 20 times. The final results of each experiment on the test set are shown in the Figure 9 below.

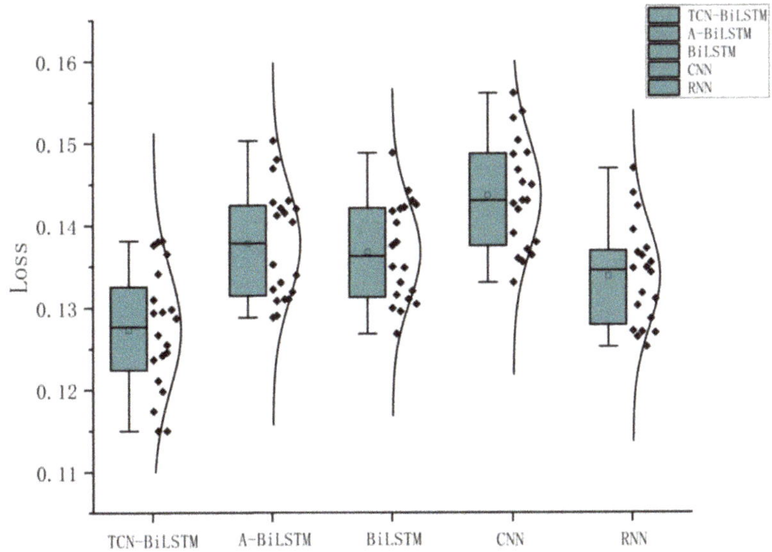

Figure 9. Boxplot of experimental results.

The paired t-test is a commonly used statistical method primarily designed to assess whether the mean difference between paired observations is statistically significant. In this study, paired t-tests were performed on the loss values of different models to evaluate whether the performance differences between the TCN-BiLSTM model and other comparative models are significant. The final loss values of each model, obtained from 20 identical experimental runs, were paired for comparison, and the corresponding p-values were calculated to determine the statistical significance of the differences.The experimental results are shown in Table 6.

Table 6. Results of the paired t-test.

p-Values	TCN-BiLSTM	A-BiLSTM	BiLSTM	CNN	RNN
TCN-BiLSTM	/	0.0006	0.0004	0	0.0011
A-BiLSTM	0.0006	/	0.6273	0.0068	0.0999
BiLSTM	0.0004	0.6273	/	0.0016	0.1772
CNN	0	0.0068	0.0016	/	0.0002
RNN	0.0011	0.0999	0.1772	0.0002	/

The experimental results demonstrate that the p-values for the performance comparisons between the TCN-BiLSTM model and all other comparative models were consistently less than 0.05. According to statistical standards, when the p-value is smaller than the significance level (commonly set at 0.05), it can be concluded that the observed differences between the two groups are unlikely due to random variations and are statistically significant. This indicates that the TCN-BiLSTM model outperforms all other comparative models in terms of performance.

3.6. Error Classification Case Analysis

In the experiments using the TCN-BiLSTM model for fault classification, although the overall performance outperformed other models, there were still some notable instances of misclassification. By conducting an in-depth analysis of specific misclassified cases, we gained further insights into the limitations of the model.

First, the model misclassified a heat dissipation failure as a random failure. Heat dissipation failure is typically caused by uneven temperature distribution or insufficient heat dissipation characterized by a minimal difference between air temperature and process temperature. It is also closely associated with parameters such as rotational speed and torque. In contrast, a random failure is an extremely rare type of fault (with an occurrence probability of approximately 0.1%) that is completely independent of the operational state of the equipment and occurs randomly. The model may have failed to sufficiently recognize the unique physical characteristics of heat dissipation failure, leading to its misclassification as a random failure. The critical distinction between these two failure types is that heat dissipation failure follows a certain physical mechanism, whereas random failure is entirely stochastic. The model likely struggled to capture the essential differences between these two types of failures.

Second, the model misclassified an overstrain failure as a tool wear failure. Overstrain failure typically occurs when the product of tool wear and torque exceeds a certain threshold, leading to sudden mechanical strain or a violent response of the equipment. In contrast, tool wear failure is a gradual process that progresses more slowly over time. The model may not have effectively captured the temporal dependencies and abrupt characteristics that distinguish overstrain failure from tool wear failure. Moreover, due to the imbalanced dataset, overstrain failure had relatively fewer samples compared to tool wear failure. This imbalance may have caused the model to favor predicting tool wear failure, thereby overlooking the distinguishing features of overstrain failure.

In conclusion, while the comparison of experimental results validated the superior performance of the proposed TCN-BiLSTM model in time-series fault prediction tasks, the analysis of misclassification cases revealed areas for further improvement. The model not only demonstrated higher prediction accuracy but also exhibited faster convergence and stronger adaptability to sequential data. These findings establish the TCN-BiLSTM model as a robust and effective solution for tackling complex time-series prediction challenges, offering a significant advantage over traditional approaches.

3.7. Model Scalability

The TCN-BiLSTM model demonstrates scalability and is capable of adapting to a wide range of real-world application scenarios. For instance, in high-dimensional small-sample problems where the data sample size is limited but the feature dimensions are high, traditional algorithms often struggle to handle such complex situations. The TCN-BiLSTM model, through the following mechanisms, exhibits its potential and scalability when applied to high-dimensional small-sample datasets.

When dealing with high-dimensional datasets, the dilated convolution operation in the TCN-BiLSTM model expands the receptive field, enabling the extraction of multi-level temporal dynamic features over a longer time span. This characteristic allows the model to effectively capture useful signals within the data while reducing the interference of redundant features. The ability to extend the receptive field can be generalized to other high-dimensional small-sample problems. For example, in application scenarios involving large volumes of sensor data, while the number of features might be extremely high, the effective sample size could be quite limited. The scalability of the TCN-BiLSTM model facilitates learning significant features from limited data, thereby enhancing its effectiveness in other high-dimensional small-sample problems.

Through its bidirectional learning mechanism, the TCN-BiLSTM model is able to capture dependencies between previous and subsequent time steps simultaneously. This mechanism contributes to its scalability in small-sample datasets. Small-sample problems are prevalent across various industrial domains, particularly in equipment fault prediction, where certain fault patterns may only manifest during a few time points. By enhancing the model's contextual understanding of time-series data, BiLSTM enables the extraction of latent long-term dependencies even when the sample size is limited.

To further enhance the model's adaptability to broader applications, a series of optimization techniques have been incorporated during the network design and training process. These techniques not only improve performance on current tasks but also provide support for future expansions. Batch normalization assists the model in achieving faster and more stable training across diverse tasks, accommodating variations in data distributions. The Leaky ReLU activation function addresses the "dead neuron" problem that may arise when processing high-dimensional data, ensuring model stability in broader application scenarios. The dropout mechanism prevents overfitting while simultaneously enhancing the model's predictive capabilities on unseen data. With the integration of these techniques, the TCN-BiLSTM model exhibits enhanced generalizability, making it expandable to different domains and datasets, thereby boosting its potential for cross-domain applications.

The TCN-BiLSTM model showcases outstanding scalability, making it applicable to a broader range of industrial fields and complex data types. Through continued algorithmic optimization and technological innovation, the application scenarios and processing capabilities of the model are expected to expand further, driving its adoption across more diverse fields.

4. Conclusions

This study addresses the issue of insufficient accuracy in traditional algorithms and single fault prediction models for time-series fault prediction by proposing a combined prediction model based on an improved temporal convolutional network (TCN) and bidirectional long short-term memory network (BiLSTM), referred to as the TCN-BiLSTM model. By incorporating techniques such as hyperparameter optimization, improved activation functions, and dropout, the model achieves enhanced prediction accuracy and generalization capability. Simulation experiments conducted on a publicly available dataset demonstrate that the proposed model exhibits higher accuracy in time-series fault prediction tasks, along with strong applicability and robustness.

This research provides a novel technical perspective and practical value for the field of time-series fault prediction through the design of an improved algorithm and combined model. Future work will explore the adoption of advanced techniques such as model ensemble methods and transfer learning to further improve the adaptability and generalization of the model, enabling its application to a broader range of time-series fault prediction scenarios.

Author Contributions: Conceptualization, C.Y.; Methodology, S.L.; Writing—original draft, J.B.; Writing—review & editing, J.B.; Visualization, P.Z.; Supervision, W.Z.; Project administration, X.W.; Funding acquisition, W.Z. All authors have read and agreed to the published version of the manuscript.

Funding: This research received no external funding.

Institutional Review Board Statement: Not applicable.

Informed Consent Statement: Not applicable.

Data Availability Statement: The original contributions presented in the study are included in the article, further inquiries can be directed to the corresponding author.

Conflicts of Interest: The authors declare that they have no conflict of interest.

References

1. Zheng, X. Research on Fault Prediction Algorithm Based on Multi-Method Fusion with State Estimation. Ph.D. Thesis, Huazhong University of Science and Technology, Wuhan, China, 2016.
2. Li, J. Research on Fault Prediction Algorithms for Complex Equipment Based on Machine Learning. Master's Thesis, Xi'an Technological University, Xi'an, China, 2024. [CrossRef]
3. Huang, R.; Li, J.; Wang, Z.; Xia, J.; Chen, Z.; Li, W. Intelligent Diagnosis and Life Prediction Method for Equipment Based on Multi-Task Learning. *Sci. China Technol. Sci.* **2022**, *52*, 123–137.
4. Yang, J.; Ren, X.; Hong, D.; Zhou, C.; Liu, C. Simulation Training System for Equipment Training Scenarios. *Ordnance Autom.* **2024**, *43*, 21–27.
5. Zhou, J. Fault Prediction of Wind Turbines Based on Multiple Statistical Models. Master's Thesis, Changsha University of Science and Technology, Changsha, China, 2019. [CrossRef]
6. Lu, Z.; Long, F.; Chen, G. Anomaly Analysis of Generator Equipment Based on Time Series. *Mod. Inf. Technol.* **2024**, *8*, 121–124. [CrossRef]
7. Zhao, J.; Lei, Y.; Li, Z. A Review of Kalman Filtering in Equipment Fault Prediction. *Ordnance Autom.* **2024**, *43*, 16–20.
8. Zhang, X.; Sun, G.; Gao, L.; Wang, Y.; Wang, X.; Zhang, K. Research on Substation Fault Prediction Method Based on Machine Learning. *Electron. Devices* **2024**, *47*, 1047–1052.

9. Yang, K. Research on Hydraulic System Fault Diagnosis Technology of Excavators Based on Prediction Models and Expert Systems. Master's Thesis, Jilin University, Changchun, China, 2020. [CrossRef]
10. Du, W. Research on Mechanical Fault Prediction Based on Cointegration Theory and Convolutional LSTM Networks. Master's Thesis, Guilin University of Electronic Technology, Guilin, China, 2020. [CrossRef]
11. Peng, F.; Wang, E.; Gao, X. Disk Fault Prediction Method Based on CNN-LSTM Neural Network. *Comput. Appl. Softw.* **2024**, *41*, 92–100+149.
12. Jiang, F.; Yang, Y. Fault Prediction of Marine LNG Dual-Fuel Engines Based on LSTM-PSO-SVM. *Ship Sci. Technol.* **2024**, *46*, 120–126.
13. Xu, D.; Zhou, C.; Guan, C.; Wang, X. Fault Rate Prediction Method of Equipment Based on ARMA-BP Combined Model. *Fire Control Command Control* **2021**, *46*, 83–87.
14. Zhong, Y.; Wang, J.; Song, G.; Wu, B.; Wang, T. Ultra-Short-Term Power Load Prediction under Extreme Weather Based on Secondary Reconstruction Decomposition Denoising and BiLSTM [J/OL]. *Power Syst. Technol.* **2024**, 1–15. [CrossRef]
15. Cheng, X.; Dong, M. Dynamic Multi-Objective Optimization Algorithm Based on RNN Information Accumulation. *Comput. Sci.* **2024**, *51*, 333–344.
16. Liang, T.; Sun, K.; Fan, J.; Zheng, X.; Liao, R.; Tang, J. Prediction of Mortar Trajectory and Impact Points Based on BiLSTM-AM [J/OL]. *J. Gun Launch Control* **2024**, 1–10. [CrossRef]
17. Li, H.; Chen, F.; Xu, T.; Luo, S.; Mao, Y. Estimation of Lithium Battery SOC Based on EKF-AEKF [J/OL]. *Appl. Sci. Technol.* **2024**, 1–9. Available online: http://kns.cnki.net/kcms/detail/23.1191.U.20240926.1714.004.html (accessed on 3 October 2024).
18. Yang, J.; Li, L.; Qian, Y.; Tang, S.; Yang, P. Fuel Flow Prediction for Domestic Civil Aircraft Based on LSTM-XGBoost [J/OL]. *Prog. Aerosp. Eng.* **2024**, 1–8. Available online: http://kns.cnki.net/kcms/detail/61.1479.V.20240925.1106.002.html (accessed on 5 October 2024).
19. Pan, Z.; Wang, Y.; Wang, K.; Chen, H.; Yang, C.; Gui, W. Imputation of Missing Values in Time Series Using an Adaptive-Learned Median-Filled Deep Autoencoder. *IEEE Trans. Cybern.* **2023**, *53*, 695–706. [CrossRef]
20. Jia, B.-B.; Zhang, M.-L. Multi-Dimensional Classification via Decomposed Label Encoding. *IEEE Trans. Knowl. Data Eng.* **2023**, *35*, 1844–1856. [CrossRef]
21. Wang, Z.; Li, W.; Tang, Z. Enhancing the genomic prediction accuracy of swine agricultural economic traits using an expanded one-hot encoding in CNN models. *J. Integr. Agric.* **2024**. [CrossRef]
22. Zhao, Y.; Huang, W.; Zhang, J. Lithium Battery Life Prediction Based on LSTM-Transformer with Hybrid Scale Health Factors [J/OL]. *Electron. Meas. Technol.* **2024**, *47*, 112–122. Available online: http://kns.cnki.net/kcms/detail/11.2175.tn.20240929.1613.067.html (accessed on 7 October 2024).
23. Akiba, T.; Sano, S.; Yanase, T.; Ohta, T.; Koyama, M. Optuna: A Next-generation Hyperparameter Optimization Framework. *arXiv* **2019**, arXiv:1907.10902.
24. Huang, Y.; Dai, X.; Huang, W.; Yu, J.; Huang, Z. Text Classification Based on A-BiLSTM and CNN. *Comput. Eng. Des.* **2024**, *45*, 1428–1434.

applied sciences

MDPI

Article

Fault Diagnosis of Rolling Bearings Based on Acoustic Signals in Strong Noise Environments

Hengdi Wang * and Jizhan Xie

School of Mechatronics Engineering, Henan University of Science and Technology, Luoyang 471023, China; 13148054890@163.com
* Correspondence: sywhd@sina.com

Abstract: Compared to vibration sensors, microphones offer several advantages, including non-contact detection, high sensitivity, low cost, and ease of installation. To address the challenges posed by the complex components and significant interference in rolling bearing sound signals, we proposed a fault diagnosis method for rolling bearing acoustic signals based on Secretary Bird Optimization Algorithm (SBOA)-optimized Feature Mode Decomposition (FMD). Initially, a microphone is utilized to collect sound data while the bearing operates, followed by the application of S-FMD (Secretary Bird Optimization Algorithm-optimized Feature Mode Decomposition) to decompose the sound signal and extract components that may contain fault information related to the bearing. The SBOA is employed to adaptively optimize four influencing parameters of FMD: mode number n, filter length L, frequency band cutting number K, and cycle period m. By minimizing envelope entropy as the objective function, we achieve FMD of the bearing sound signal with the assistance of the SBOA. Additionally, this paper introduces an Integrated Signal Evaluation Index (ISEI) to extract potential bearing failure characteristics from the filtered components. Simulation experiments and test results indicate that, compared to Empirical Mode Decomposition, Complementary Ensemble Empirical Mode Decomposition, fixed-parameter FMD, and adaptive variational mode decomposition methods, the proposed approach more effectively extracts weak characteristic information related to early faults in bearing sound signals.

Keywords: rolling bearing; acoustic signal; feature mode decomposition; fault diagnosis

1. Introduction

Mechanical failure poses a significant threat to the safe and reliable operation of large mechanical equipment, including wind power generation systems, aerospace engines, and high-end Computer Numerical Control (CNC) machine tools [1]. Rolling bearings are critical components for the effective functioning of rotating machinery [2], and their condition is crucial for the overall performance of the entire mechanical system. In the realm of fault diagnosis, signal decomposition technology is regarded as one of the most established and technically comprehensive methods. This technique can effectively suppress environmental noise and extraneous components in mechanical vibration signals, thereby facilitating the extraction of fault features [3,4]. In contrast to vibration signals, sound signals offer advantages such as non-contact monitoring and convenient remote assessment, making them valuable in specialized environments [5]. By performing time–frequency domain analysis on bearing sound signals, various bearing fault types can be identified. The diagnostic features derived from acoustic signals demonstrate greater accuracy than those obtained from vibration signals, allowing for a more effective differentiation between healthy and faulty conditions [6].

In practical measurements, the sound field is inherently complex and is frequently contaminated by noise signals produced by various sources, including reverberation, machinery, motors, rotating shafts, and human activity. This leads to a notably low signal-to-noise ratio for the acoustic fault signal [7,8]. Consequently, traditional signal processing methods often struggle to effectively separate and extract acoustic fault signals. As a result, the acoustic signal diagnosis methodology has been progressing at a slower pace compared to the vibration signal analysis techniques [9].

In recent years, the empirical mode decomposition (EMD) method proposed by Huang et al. [10] has emerged as an adaptive signal processing technique for decomposing non-stationary signals into multiple intrinsic mode functions (IMFs). This method has been successfully applied to fault feature extraction from bearing vibration signals. However, EMD is associated with several challenges, including modal aliasing, under-envelopment, over-envelopment, and boundary effects. To address these issues, ensemble empirical mode decomposition (EEMD) [11] has been proposed, which offers improved accuracy and efficiency compared to traditional EMD. Nonetheless, the integration process of EEMD can be time-consuming. To mitigate this limitation, the Complementary EEMD (CEEMD) [12] method has been introduced, which enhances decomposition efficiency by adding Gaussian white noise with opposite signs to the signal. CEEMD effectively alleviates modal aliasing and reconstruction errors. Nevertheless, not all IMFs contain useful information, making the selection of the most informative IMF a critical step. For instance, Sahu and Rai employed CEEMD to decompose bearing vibration signals into multiple IMFs and determined the grouping and selection of valuable modes based on correlation coefficients [13,14].

Drag et al. [15] proposed variational mode decomposition (VMD) based on empirical mode decomposition (EMD), which is a signal processing method with variable scaling. Unlike traditional recursive algorithms, VMD exhibits strong robustness and is supported by rigorous mathematical theory. The mode number K and the penalty factor α significantly influence the results. If K is set too high, it can result in over-decomposition (modal loss); conversely, if K is too low, it may lead to under-decomposition (modal aliasing). Additionally, improper configuration of the penalty factor can result in mode aliasing and loss, which adversely affects the extraction of important features. Wang [16] proposed an adaptive VMD method utilizing a bat algorithm search (BAS) optimization technique for the early fault diagnosis of rolling bearings. This method employs the kurtosis value as a metric, selecting the intrinsic mode function (IMF) with the highest kurtosis for analysis. Wang [17] also introduced a rolling bearing acoustic signal fault diagnosis method based on adaptive variational mode decomposition (AVMD) to enhance accuracy in noisy environments. Furthermore, Li [18] proposed a novel method based on frequency band entropy (FBE) for selecting sensitive intrinsic mode functions (IMFs) that contain variational mode decompositions (VMDs) rich in fault information. Subsequently, envelope power spectrum analysis is conducted on the selected IMFs to extract the fault characteristic frequencies of rolling bearings.

Miao [19] proposed Feature Mode Decomposition (FMD), a novel adaptive decomposition method specifically designed for mechanical fault feature extraction. FMD updates the adaptive finite impulse response (FIR) filter bank using correlation kurtosis (CK), thereby overcoming the limitations of the variational mode decomposition (VMD) and Empirical Wavelet Transform (EWT) algorithms regarding filter shape and bandwidth. This enhancement allows for the extraction of more fault information, making FMD more advantageous than VMD in mechanical signal decomposition analysis, particularly in scenarios where prior knowledge of the failure cycle is lacking. Jia [20] introduced a fault feature extraction method that integrates an improved whale optimization algorithm (WOA) to optimize FMD. This approach refines the FMD parameter settings, employs information entropy as the fitness function, and selects the optimal intrinsic mode function (IMF) through kurtosis for fault feature extraction. Sumika [21] developed a bearing fault detection method based on adaptive feature mode decomposition (Adaptive FMD) and the health index (HI). Yan [22] proposed a fault diagnosis method based on adaptive multivariate eigenmode decomposition (AMFMD). The method addresses the challenge of fault information in mechanical equipment being easily obscured by significant noise, and it overcomes the limitation of potentially missing valuable features when relying solely on data from a single sensor. Yan [23] designed the signal period kurtosis-to-noise ratio (SCKNR) as the objective function for the adaptive parameter search of FMD. He selected the primary mode component based on the characteristic energy rate and computed its square envelope spectrum for fault identification. However, he did not consider the influence of the frequency band cutting number K on the decomposition results.

Li [24] proposed an adaptive feature mode decomposition (AFMD) method for analyzing noisy bearing vibration signals. The method optimizes parameters using a cuckoo search algorithm and employs the feature frequency ratio (FFR) to improve fault diagnosis accuracy and efficiency, particularly in the presence of strong noise levels (-15 dB). Meanwhile, Wen [25] introduced a novel method known as Graph Modeling Singular Values (GMSV), which successfully balances sensitivity to early faults with robustness against noise. This method employs a Gaussian distribution for fault detection. Nevertheless, in practical engineering scenarios, the dynamic characteristics of bearings may exhibit greater complexity, necessitating verification of the feasibility of this assumption.

The effectiveness of FMD is highly sensitive to key parameters: mode number n, filter length L, frequency band division K, and cycle period m [19]. A mode number n that is too small may miss critical fault information, while an overly large n can introduce noise and redundancy. Similarly, a short filter length L may cause inaccurate separation, whereas an excessively long L increases computational demands. A higher K enhances frequency resolution but also raises computational complexity. An insufficient iteration period m may hinder algorithm convergence, while an excessive period can lengthen calculation time.

Currently, most research [5,6] on the use of sound signals for bearing fault diagnosis is conducted in controlled, quiet laboratory environments. While sound-based diagnostic methods offer distinct advantages over vibration signal methods—such as non-contact measurement, flexibility in sensor placement, and lower costs for large-scale monitoring—these methods face challenges in noisy real-world environments. Sound signals are particularly susceptible to interference from environmental noise, exhibit complex spectral characteristics, and require extensive data processing, all of which can limit diagnostic accuracy. Furthermore, many FMD-based noise reduction methods [21–23] optimize only two of the four parameters, which may diminish their effectiveness in practical applications. To address this issue, this paper investigates the feasibility of sound-based bearing fault diagnosis in high-noise environments by optimizing all four FMD parameters to enhance the extraction of fault information. The Secretary Bird Optimization Algorithm (SBOA) [26] is employed to adaptively optimize the FMD parameters with the objective of minimizing envelope entropy. Subsequently, the ISEI metric introduced in this study is utilized to select the optimal pattern for extracting fault features. Simulation experiments and tests demonstrate that, compared to the EMD [10], CEEMD [11], fixed-parameter FMD [20], and adaptive VMD [16] methods, this approach is more effective in capturing subtle feature information related to early faults in bearing sound signals, exhibiting improved performance and robustness under noisy conditions.

The structure of this paper is organized as follows: the Section 1 presents a literature review and the relevant research background; the Section 2 discusses the relevant algorithms; the Section 3 elaborates on the proposed fault diagnosis method along with its technical details; the Section 4 details the simulation tests; and the Section 5 focuses on experimental design and result analysis. Finally, the conclusion summarizes the main contributions of this article.

2. Methodologies

2.1. FMD

FMD decomposes a signal into multiple sub-signals by utilizing an FIR filter bank and employs correlation kurtosis (CK) in conjunction with signal impulse and periodicity for deconvolution. The mode is selected based on the correlation coefficient (CC) of the sub-signals, discarding those with lower CK values until the number of retained sub-signals aligns with the expected number of modes. The specific implementation process is detailed as follows [19]:

Step 1: Input the original signal $x(t)$ and the parameters of FMD initialization: mode number n, filter length L, frequency band cutting number K, and cycle period m.

Step 2: Initialize the FIR filter bank through K Hanning windows, usually set in the range of 5 to 10, and initialize iteration i =1.

Step 3: Obtain the filtered signal (i.e., the decomposed modal component) through $u_k^i = x^* f_k^i$; u_k^i represents the decomposed mode obtained from the k-th filter during the i-th iteration of the process, where i = 1, 2, 3, ... represents the iteration index, indicating the current step in the iterative process, k = 1, 2, 3, ... indicates the filter index, denoting the k-th filter in the filter bank, and * is the convolution operation.

Step 4: Update the filter coefficients using the original signal $x(t)$, the decomposed modal component u_k^i, and the estimated fault period T_k^i, where the estimated fault period T_k^i is determined by the autocorrelation spectrum of u_k^i. After the zero point, the moment corresponding to the local maximum value R_k^i is reached and $i = i + 1$ is set.

Step 5: Determine whether the current iteration reaches the maximum number of iterations. If it is not reached, return to step 3; otherwise, continue to step 6.

Step 6: Calculate the correlation coefficient between every two modal components, construct a $K \times K$ correlation matrix CC ($K \times K$), select the two modal components with the largest correlation coefficient, pass the estimated failure period T_k^i, calculate their correlation kurtosis (CK), and then select the modal component with the larger correlation kurtosis (CK) as the decomposed modal component and set $K = K - 1$.

Step 7: Determine whether the current number of modes meets the specified threshold, n. If it does, the decomposition process will conclude, and the results will be presented. If it does not, return to step 2 and continue the previous operations. The specific FMD process is illustrated in Figure 1.

2.2. SBOA

The Secretary Bird Optimization Algorithm (SBOA) [26] is a new meta-heuristic optimization algorithm with the advantages of robustness, high efficiency, and fast convergence. It was inspired by the hunting behavior of the secretary bird. The algorithm steps are as follows:

Firstly, initialize the secretary bird's position using the following formula:

$$X_{i,j} = lb_j + r * (ub_j - lb_j) \quad (1)$$

where $X_{i,j}$ represents the value of the ith secretary bird in the jth dimension, lb_j and ub_j are the lower and upper bounds of the jth dimension, respectively, and r is a random number within the range of [0, 1].

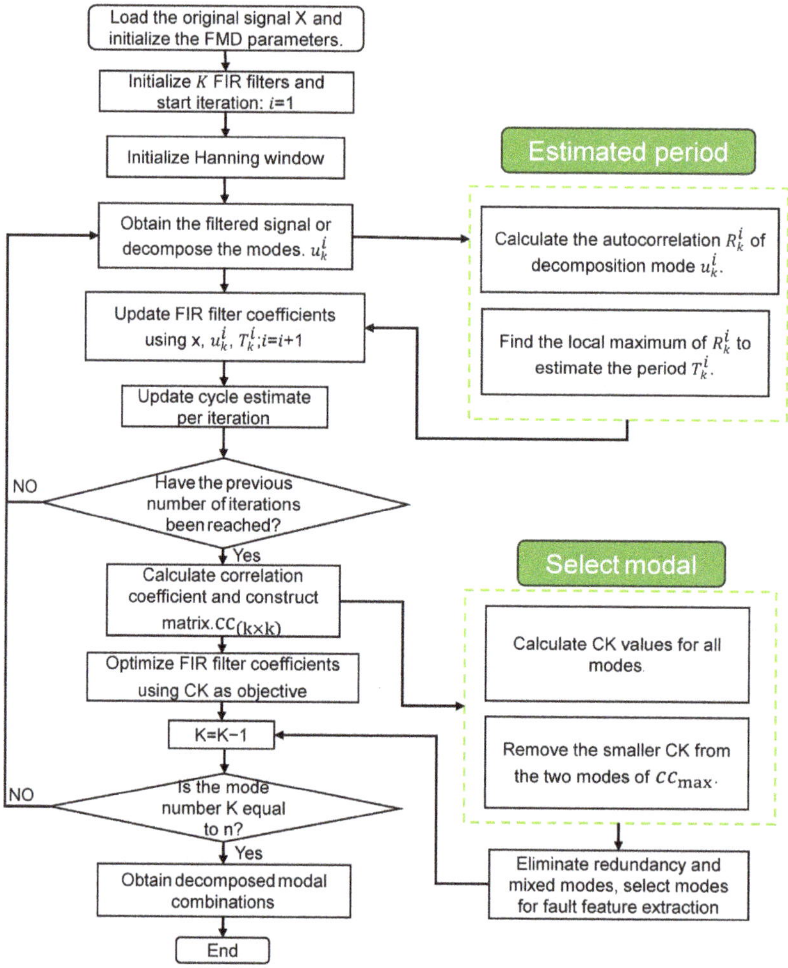

Figure 1. Illustration of the FMD algorithm.

The SBOA represents the secretary bird group by the following formula:

$$X = \begin{bmatrix} x_{1,1} & x_{1,2} & \cdots & x_{1,j} & \cdots & x_{1,Dim} \\ x_{2,1} & x_{2,2} & \cdots & x_{2,j} & \cdots & x_{2,Dim} \\ \vdots & \vdots & \ddots & \vdots & \ddots & \vdots \\ x_{i,1} & x_{i,2} & \cdots & x_{i,j} & \cdots & x_{i,Dim} \\ \vdots & \vdots & \ddots & \vdots & \ddots & \vdots \\ x_{N,1} & x_{N,2} & \cdots & x_{N,j} & \cdots & x_{N,Dim} \end{bmatrix}_{N \times Dim} \tag{2}$$

where N is the number of secretary birds and Dim is the dimension of the problem variable, which is equal to 4 in this paper. The SBOA is used to optimize the FMD parameters: the number of modes n, the filter length L, the number of frequency band cuts K, and the cycle period m. Each secretary bird represents a parameter combination, and the objective function value is given by the following formula:

$$F = \begin{bmatrix} F_1 \\ \vdots \\ F_i \\ \vdots \\ F_N \end{bmatrix}_{N \times 1} = \begin{bmatrix} F(X_1) \\ \vdots \\ F(X_i) \\ \vdots \\ F(X_N) \end{bmatrix}_{N \times 1} \tag{3}$$

where F is the objective function value vector.

In this paper, the envelope entropy value is used as the objective function. By comparing the objective function values, the secretary bird with the smallest envelope entropy value is selected as the best candidate solution. The position and objective function value of the secretary bird are updated in each iteration, so the best candidate solution needs to be re-determined in each iteration.

The test function described by Formulas (4)–(7) is employed to compare the performance of the Snake Optimization Algorithm (SO) [27], Whale Optimization Algorithm (WOA) [28], and Secretary Bird Optimization Algorithm (SBOA). Figure 2 illustrates the convergence curve of the average efficiency value for the selected function obtained from the SBOA. This algorithm demonstrates a rapid convergence speed, strong stability, and high robustness.

$$F_1(x) = \sum_{i=1}^{n} x_i^2 \tag{4}$$

$$F_2(x) = \sum_{i=1}^{n} |x_i| + \lim_{x \to \infty} \prod_{i=1}^{n} |x_i| \tag{5}$$

$$F_3(x) = \sum_{i=1}^{n} i x_i^2 + \text{random}\,(0,1) \tag{6}$$

$$F_4(x) = -20 \exp\left(-\left(\frac{1}{5}\right) \sqrt{\frac{1}{n \sum_{i=1}^{n} x_i^2}} \right) \tag{7}$$

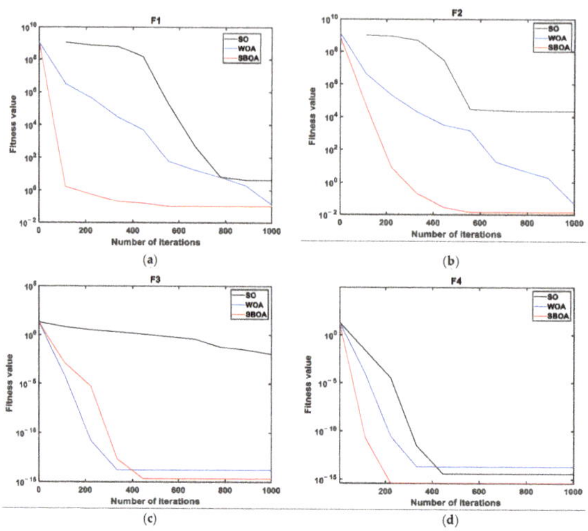

Figure 2. Fitness values convergence curves for selected functions under different optimization methods. (**a**) Convergence curve of the F1 function; (**b**) convergence curve of the F2 function; (**c**) convergence curve of the F3 function; (**d**) convergence curve of the F4 function.

3. SBOA-Optimized FMD

The SBOA uses the envelope entropy value of the reconstructed signal after FMD as the fitness function to optimize the parameters of FMD. At the same time, this paper proposes the Integrated Signal Evaluation Index (ISEI) as the selection criterion for the best mode. The detailed steps are as follows:

(1) Initialize the population: The maximum number of SBOA iterations P is set to 70 and the population size Q is set to 6. P candidate solutions are randomly generated, and each candidate solution represents an eigenmode selection scheme. Combined with the influence of FMD parameters on the noise reduction effect in the literature, this paper sets the range of the mode number n as 3 to 20, the filter length L as 10 to 300, the frequency band cutting number K as 5 to 40, and the cycle period m as 5 to 30, where the frequency band cutting number $K \geq$ the mode number n.

(2) Fitness calculation: The signal is decomposed according to the FMD parameters corresponding to the candidate solution, the envelope entropy value of the reconstructed signal is calculated, and $F(x)$ is constructed as the fitness function as the criterion for stopping the SBOA.

(3) Update location: According to the parameter search and position+n update process outlined in the literature [26], the exploration phase employs a differential evolution strategy to introduce population diversity. This phase combines Brownian motion and Lévy flight strategies [26] to expand the search range, assisting the algorithm in overcoming local optimality. In the utilization phase, the SBOA mimics the secretary bird's behavior of evading natural predators through two strategies: camouflage and rapid movement. The camouflage strategy is implemented using a dynamic perturbation factor $(1 - T_t)^2$, which enables the search agents to disperse more effectively within the solution space, helping to avoid local optima. The rapid movement strategy, on the other hand, adjusts the position and velocity of the search agents to simulate the secretary bird's swift movement, thus improving the efficiency of the search and accelerating convergence. Specifically, the rapid movement strategy incorporates a dynamic perturbation factor $(1 - T_t)^2$, which allows the search agents to move quickly

across the solution space, facilitating a faster approach to the global optimal solution. Additionally, the search direction and step size are dynamically adjusted to guide the algorithm towards efficiently approaching the global optimal solution, thereby ensuring accurate determination of the FMD parameters during the optimization process.

(4) Individuals with better fitness values (FMD combined solutions) are selected according to the criterion of the minimum envelope entropy value, individuals with poor fitness are replaced, and the optimal solution is retained to ensure that the algorithm converges to the global optimum.

(5) Iteration: The fitness calculation, position update, local search and selection, and replacement steps are repeated until the termination condition is met (the maximum number of iterations P is reached).

(6) Output the optimal solution: When the algorithm terminates, the optimal eigenmode combination solution and FMD parameters are output.

Mode Selection Criteria: ISEI

As a common indicator to characterize signal impact characteristics, kurtosis is easily interfered with by a small number of large-value abnormal data or random impacts. In this regard, an Integrated Signal Evaluation Index (ISEI) weighted by kurtosis combined with the harmonic-to-noise ratio is proposed as the Secretary Bird Optimization Algorithm-optimized Feature Mode Decomposition (SFMD) criteria for selecting the best mode after the signal.

(1) Normalization: Kurtosis formula.

$$K = \frac{\frac{1}{N}\sum_{i=1}^{N}(x_i - \mu)^4}{\left(\frac{1}{N}\sum_{i=1}^{N}(x_i - \mu)^2\right)^2} \tag{8}$$

where N is the length of the bearing signal, x_i is the ith data point, and μ is the mean value of the signal.

The formula for the harmonic noise ratio (HNR) is as follows

$$HNR = \frac{\sum_{i=1}^{n} A_{i\times FCO}}{n\times 3\sigma_{\text{noise}}} \tag{9}$$

where $A_{i\times FCO}$ represents the amplitude of the ith harmonic frequency component, FCO represents the fundamental frequency (fault characteristic frequency), n is the number of harmonic frequency components, $\sigma_{\text{noisedddl}}$ is the standard deviation of the noise. The peak search algorithm is used to look for peaks at the bearing's fault frequency. A high HNR value indicates that harmonic components are more significant relative to noise.

The normalized formula of kurtosis K is

$$Z_K = \frac{K-\mu_K}{\sigma_K} \tag{10}$$

where μ_K is the mean value of kurtosis and σ_K is the standard deviation of kurtosis.

The normalized formula of the harmonic noise ratio RFN is

$$Z_{RFN} = \frac{RFN-\mu_{HNR}}{\sigma_{RFN}} \tag{11}$$

where μ_{HFN} is the mean value of the harmonic-to-noise ratio and σ_{RFN} is the standard deviation of the harmonic-to-noise ratio.

(2) Weighted Integrated Signal Evaluation Index (ISEI).

$$ISEI = \delta_1 \cdot |Z_K| + \delta_2 \cdot |Z_{RFN}| \quad (12)$$

where δ_1 and δ_2 are weights. According to the literature [23], the weights are 0.3 and 0.7, respectively.

As a criterion for mode selection, the ISEI takes into account the periodicity and impact resistance of the signal and can more comprehensively reflect the fault characteristics of the bearing signal. The modal component with the largest ISEI index is selected for further analysis. Subsequently, the Hilbert transformation [29] is applied to calculate its envelope spectrum, which is then used to extract bearing fault characteristics. The fault diagnosis flowchart is shown in Figure 3.

After denoising the collected bearing signals, the modal components containing the most significant fault information are extracted. As early bearing fault signals are predominantly distributed in high-frequency bands, traditional Fourier transform methods often struggle to provide accurate analyses. The resonant frequency of the sensor modulates the bearing fault information into a high-frequency band, necessitating the use of the Hilbert transform to demodulate this high-frequency information into a lower frequency range. The specific procedure involves applying the Hilbert transform to the optimal modal component to obtain the analytical signal, from which the envelope curve is extracted. By taking the modulus of this curve, the analytical function is derived. Subsequently, the FFT Fourier transform is applied to the analytical function to generate the envelope spectrum, enabling the clear identification of fault characteristic frequencies and their multiples, thus facilitating the precise diagnosis of bearing faults. Utilizing the bearing parameters and the fault frequency calculation formula, the theoretical bearing fault characteristic frequency can be computed. A comparative analysis of the envelope spectrum against the corresponding fault frequency and its multiples is then conducted. If a peak in the envelope spectrum aligns with the theoretical fault frequency and its multiples, it confirms the presence of a corresponding fault in the bearing.

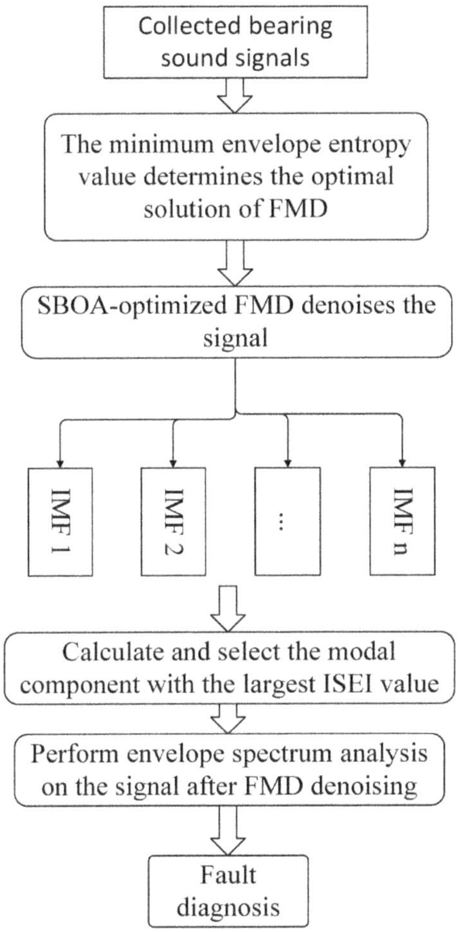

Figure 3. Flowchart of sound signal fault diagnosis.

4. Simulation Verification Analysis

By constructing a simulation signal of a fault in a rolling bearing [20], the feasibility of the SFMD method is verified. The simulation signal expression is:

$$\begin{cases} x(t) = s(t) + n(t) = \sum\limits_{i} A_i h(t - iT) + n(t) \\ A_i = 1 + A_0 \cos(2\pi f_r t) \\ h(t) = \exp(-Ct)\cos(2\pi f_n t) \end{cases} \tag{13}$$

where $s(t)$ is the periodic impact component;

The amplitude A_0 is 0.3;

The rotation frequency f_r is 30 Hz;

The attenuation coefficient C is 700;

The resonance frequency f_n is 4 khz;

The inner ring fault characteristic frequency is $f_i = 1/T = 120$ Hz;

$n(t)$ is Gaussian white noise; the signal-to-noise ratio of the noise-containing signal is -13 dB (calculation formula $\text{SNR} = 20\log_{10}(v_s/v_n)$; v_s and v_n are the effective values of the impact component and noise component, respectively); the sampling frequency f_s is 16 kHz; and the number of analysis points is 4096 points.

As shown in Figure 4, fault information cannot be identified in the time domain diagram, spectrogram, or envelope spectrum of the simulated signal after adding noise.

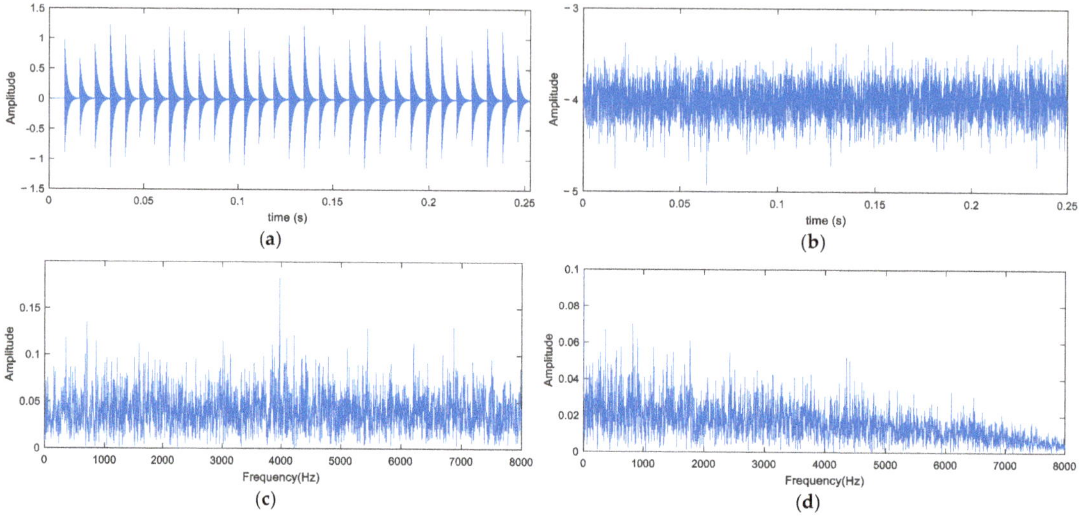

Figure 4. Simulated signals: (**a**) shock signal waveform; (**b**) time domain waveform; (**c**) spectrogram; (**d**) envelope spectrum.

Using the proposed method for signal denoising processing, the convergence curve of the SBOA obtained based on the minimum envelope entropy value is shown in Figure 5. The iterations converged after 13 times, and the optimal parameter combination was [250, 5, 5, 10].

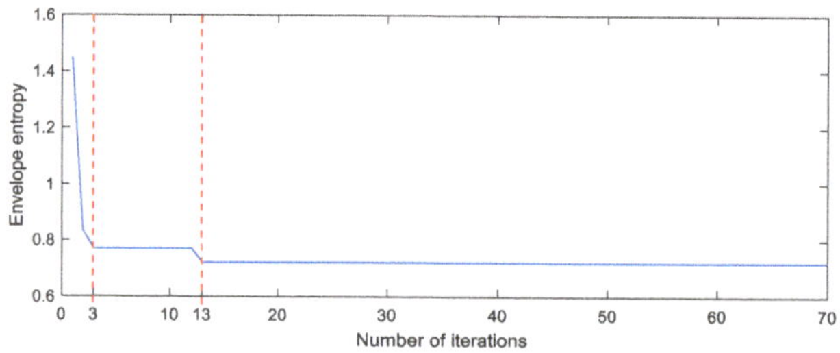

Figure 5. Convergence curve of parameter optimized by the SBOA (simulation with inner raceway fault).

FMD optimized by the SBOA is S-FMD, and its decomposed components are F-IMF. VMD optimized by CPO is C-VMD, and its decomposed components are V-IMF. Figure 6 shows the decomposed time domain diagram after optimizing FMD. Figure 7 shows the kurtosis, harmonic noise ratio, and ISEI of each F-IMF.

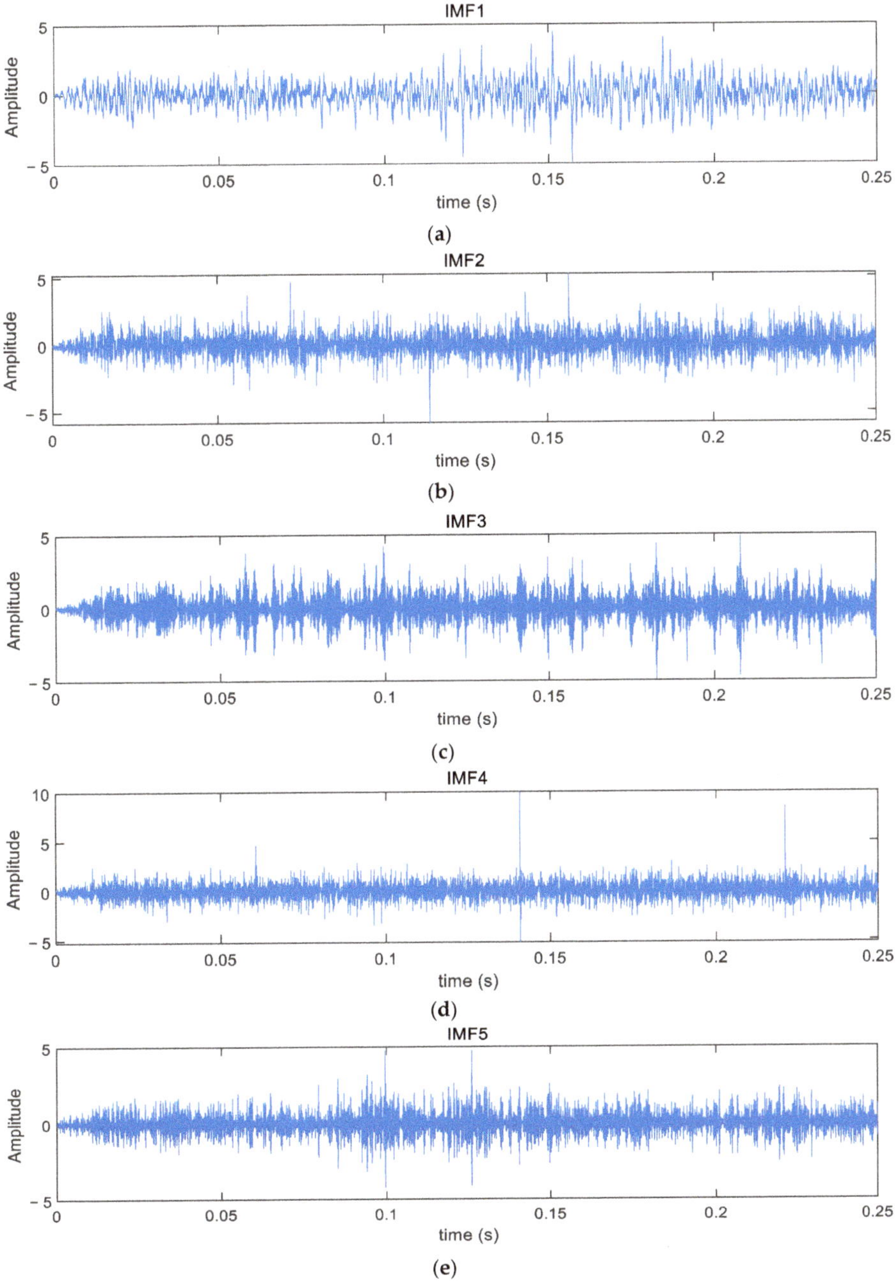

Figure 6. Parameter-optimized FMD results of the signal (simulation inner raceway fault). (**a**) Time domain diagram of F-IMF1; (**b**) time domain diagram of F-IMF2; (**c**) time domain diagram of F-IMF3; (**d**) time domain diagram of F-IMF4; (**e**) time domain diagram of F-IMF5.

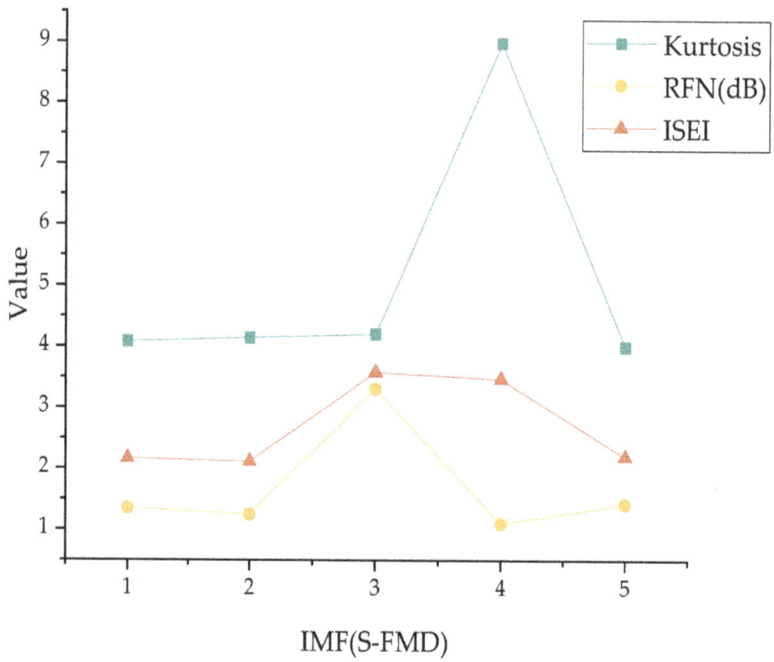

Figure 7. Feature values of the decomposed modes via S-FMD (simulation with inner raceway fault).

As shown in Figure 7, F-IMF4 has the largest kurtosis value. As shown in Figure 6, F-IMF4 (Figure 6d) has a large kurtosis because it contains three abnormal shocks, while RFN and ISEI indicate that F-IMF3 contains the most faults. An envelope spectrum analysis is performed on the F-IMF decomposed by each S-FMD. The envelope spectrum of F-IMF3 contains the most fault information.

The envelope spectrum of F-IMF3 in Figure 8 clearly shows the first eight harmonic components of the inner raceway fault frequency f_i 120 Hz and the rotation frequency f_r (30 Hz). This signifies a substantial reduction in the strong noise component added to the simulated signal at −13 dB, effectively mitigating noise interference and enhancing the extraction of fault information. This shows that using the kurtosis value as the selection criterion for the best mode is not applicable, but the ISEI comprehensive index is effective as the selection criterion. In addition, the RFN value of F-IMF3 is 3.31 dB, which is 16.31 dB higher than the −13 dB signal-to-noise ratio of the original simulation signal.

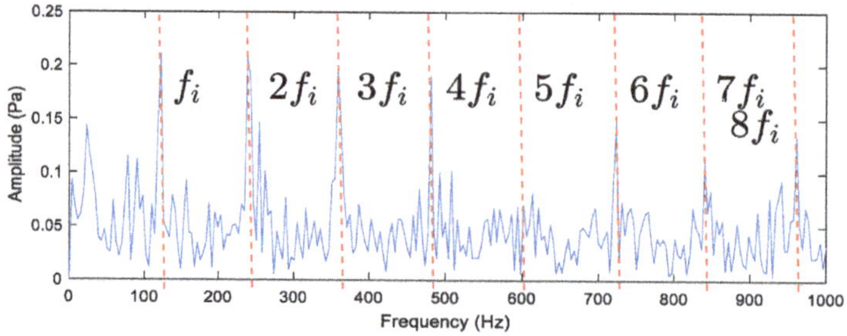

Figure 8. Envelope spectrum of F-IMF3 decomposed via parameter-optimized FMD (simulation with inner raceway fault).

The simulation signal is denoised separately using four methods including adaptive C-VMD. Based on the parameter settings in the literature [19], the fixed FMD parameters are set as follows: the filter length L is 40, the frequency band cutting number K is 5, the mode number n is 4, and the cycle period m is 15. The other methods include empirical mode decomposition (EMD) and complete ensemble empirical mode decomposition (CEEMD). Each method decomposes the simulated signal and calculates its RFN, kurtosis, and ISEI as illustrated in Figure 9.

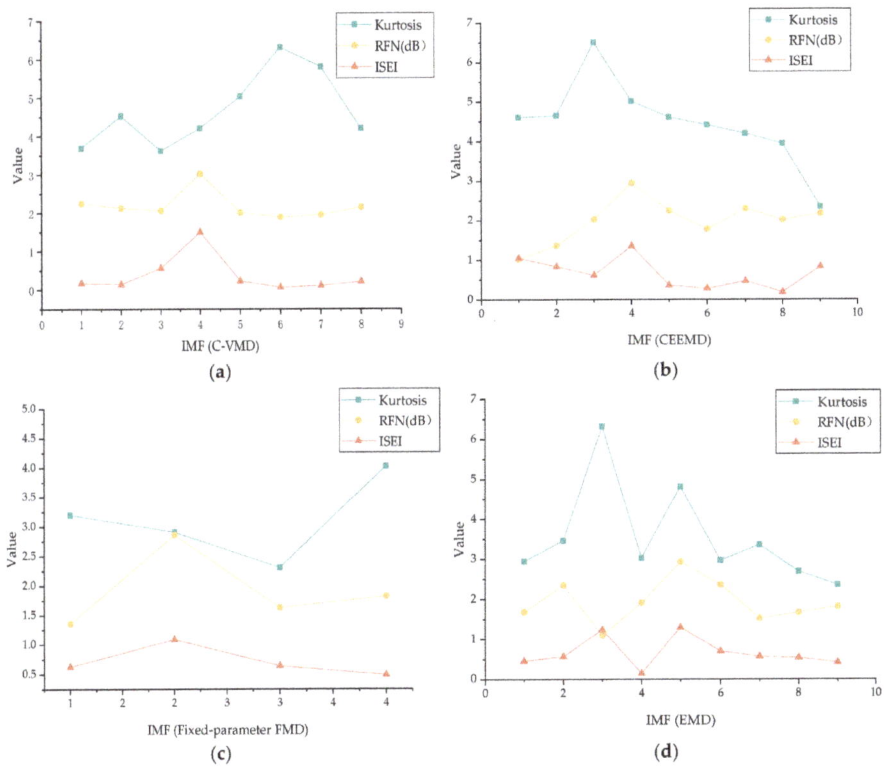

Figure 9. RFN, kurtosis, and ISEI of each component after decomposing the signal using each method (simulation with inner raceway fault). (**a**) The characteristic values for C-VMD's IMF components; (**b**) the characteristic values for CEEMD's IMF components; (**c**) the characteristic values for fixed-parameter FMD's IMF components; (**d**) the characteristic values for EMD's IMF components.

The noise tolerance τ and convergence error limit ε for VMD are set to the values reported in the literature [16], specifically $\tau = 0$ and $\varepsilon = 1 \times 10^{-7}$. The convergence curve of the CPO algorithm, derived from the minimum envelope entropy value, is shown in Figure 10. After 18 iterations, the envelope entropy value reaches its minimum, yielding the parameter combination K = 12 and α = 1217. The decomposed modal spectrum is shown in Figure 11, demonstrating that the VMD optimized by CPO effectively mitigates modal aliasing and endpoint effects. Each frequency band is concentrated around the center frequency, indicating a more thorough decomposition. The optimal component following C-VMD decomposition was selected based on the ISEI value, identifying V-IMF4 for envelope analysis. As shown in Figure 12, only f_i is visible, with its harmonic component being subtle. Under conditions of strong noise, the ability to extract fault information from the bearing remains limited.

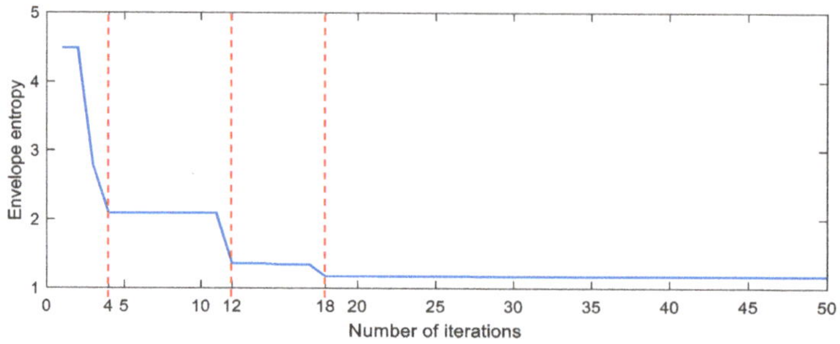

Figure 10. Convergence curve of parameters optimized by the CPO algorithm (simulation with inner raceway fault).

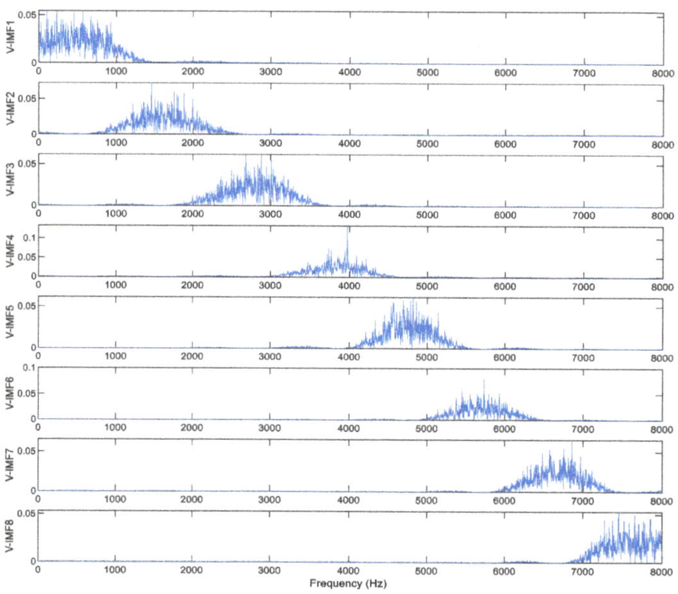

Figure 11. Spectrum of each IMF decomposed via parameter-optimized VMD (simulation with inner raceway fault).

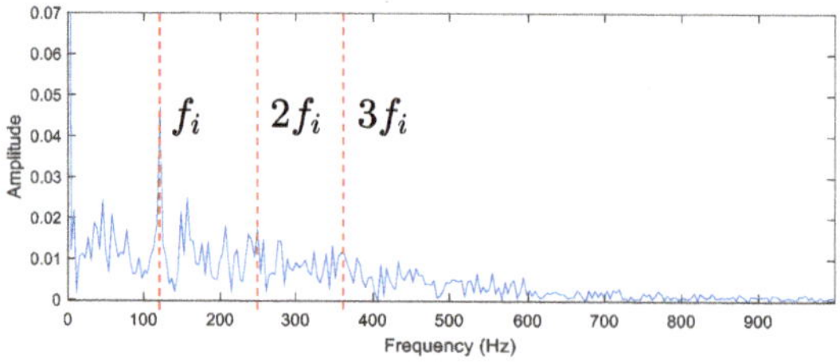

Figure 12. Envelope spectrum of V-IMF4 decomposed via parameter-optimized VMD (simulation with inner raceway fault).

The noise tolerance τ and convergence error limit ε for VMD are set to the values reported in the literature [16], specifically, $\tau = 0$ and $\varepsilon = 1 \times 10^{-7}$. The convergence curve of the CPO algorithm, derived from the minimum envelope entropy value, is shown in Figure 10. After 18 iterations, the envelope entropy value reaches its minimum, yielding the parameter combination $K = 12$ and $\alpha = 1217$.

The decomposed modal spectrum is shown in Figure 11, demonstrating that VMD optimized by CPO effectively mitigates modal aliasing and endpoint effects. Each frequency band is concentrated around the center frequency, indicating a more thorough decomposition. The optimal component following C-VMD was selected based on the ISEI value, identifying V-IMF4 for envelope analysis.

As shown in Figure 12, only f_i is visible, with its harmonic component being subtle. Under conditions of strong noise, the ability to extract fault information from the bearing remains limited. The fixed-parameter FMD, EMD, and CEEMD methods denoise the simulated signal to obtain the best components according to the ISEI value, subsequently performing envelope spectrum analysis. As shown in Figures 13–15, the envelope spectra of the best components were selected after fixed-parameter FMD, EMD, and CEEMD denoised the simulated signal, respectively. No obvious fault characteristics were detected; however, the effective information was obscured by strong noise. The fault information is buried within this noise, and fixed-parameter FMD methods demonstrate limited capabilities in fault extraction when compared to S-FMD and C-VMD, which yield more significant results. While the envelope spectrum of the optimal component is derived from the EMD and CEEMD methods, following the denoising of the simulated signal, can indicate the fault characteristic frequency of f_i, there is no discernible frequency doubling. This frequency doubling is entirely obscured by strong noise, which can easily lead to the misclassification of fault types. Furthermore, the fault extraction capabilities of the EMD and CEEMD methods for simulated signals are inherently constrained.

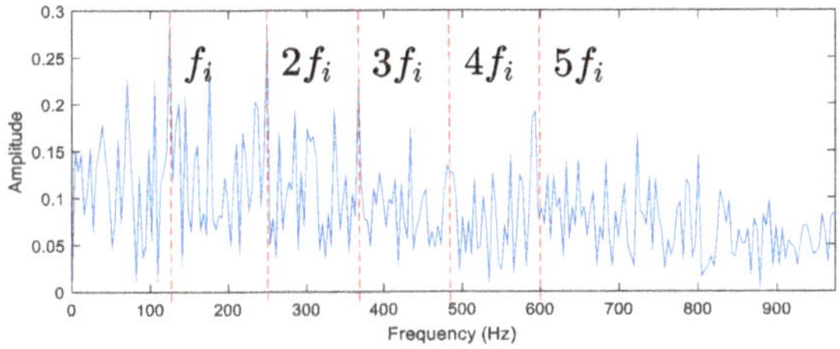

Figure 13. Envelope spectrum of F-IMF2 decomposed via fixed-parameter FMD (simulation with inner raceway fault).

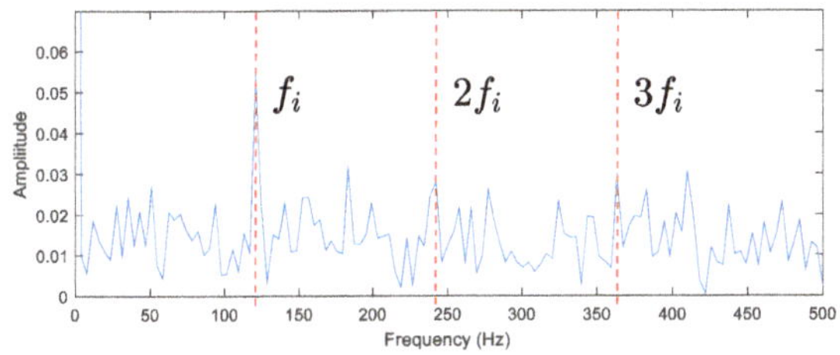

Figure 14. Envelope spectrum of the best IMF after EMD (simulation with inner raceway fault).

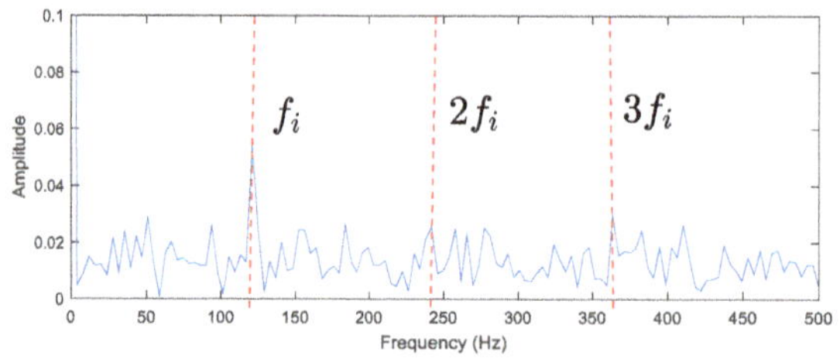

Figure 15. Envelope spectrum of the best IMF after CEEMD (simulation with inner raceway fault).

The proposed methods, i.e., S-FMD, C-VMD, EMD, CEEMD, and fixed-parameter FMD, presented in this article were employed to denoise the bearing inner raceway simulation signal after the introduction of noise. The RNF and ISEI obtained by the two methods in the figure have the same change trend, that is, the ISEI is as feasible as the RFN in selecting the best components. Each method yields multiple intrinsic mode function (IMF) components, from which the optimal components are selected for further analysis. The relative fault number (RFN) and Integrated Signal Energy Index (ISEI) are then calculated for these optimal components. Figure 16 illustrates the RFN and ISEI values of the optimal components obtained by each method. The figure indicates that the S-FMD method proposed in this article achieves the highest RFN and ISEI values, thereby retaining the most fault information. This finding aligns with the envelope spectrum derived from the proposed method, which distinctly reveals the fault characteristic frequency (f_i) along with multiple harmonics. Furthermore, it demonstrates that the fixed-parameter FMD method exhibits superior capabilities in extracting noise effects and fault features compared to C-VMD, EMD, and CEEMD. This suggests that the proposed method is effective for conducting bearing fault diagnosis in environments with strong noise, thereby affirming its feasibility for diagnosing faults based on rolling bearing sound signals. Additionally, it is noteworthy from Figure 16 that the adaptive C-VMD method shows higher RFN and ISEI values than EMD, CEEMD, and fixed-parameter FMD, indicating a better noise reduction effect compared to these three methods. Consequently, this article compares C-VMD, which demonstrates improved noise reduction effects, with the proposed S-FMD method for analyzing the measured sound signals.

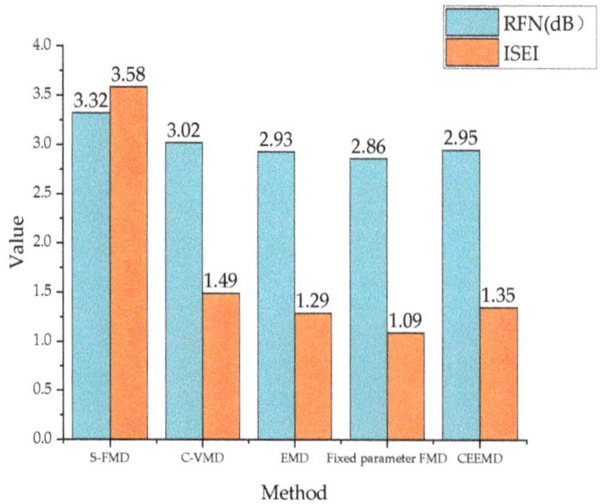

Figure 16. Each method decomposes the signal and selects the ISEI and RFN of the best components (simulation with inner raceway fault).

Relying solely on one set of parameters may impose certain limitations. To address this, we introduced two additional simulated experiments with Gaussian white noise levels of −11 dB and −9 dB, respectively, to mitigate deviations and provide a more comprehensive evaluation of the fault detection performance of the S-FMD method under varying noise conditions. Figures 17 and 18 illustrate the time domain diagrams corresponding to SNRs of −9 dB and −11 dB, while Figures 19 and 20 present the envelope spectrum following noise reduction processing using this method, clearly revealing the inner ring fault. The presence of the seventh harmonic frequency indicates that this method effectively identifies faults across different SNRs of the simulation signal.

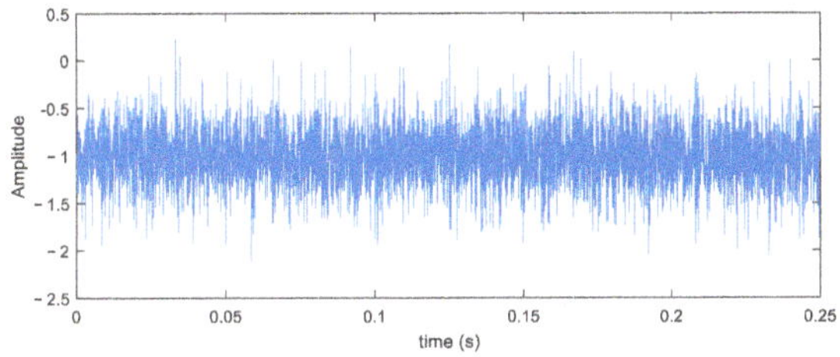

Figure 17. Time domain diagram with an SNR of −9 dB.

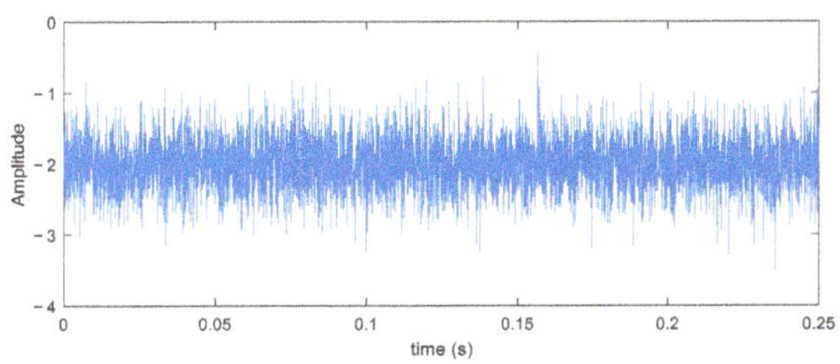

Figure 18. Time domain diagram with an SNR of −11 dB.

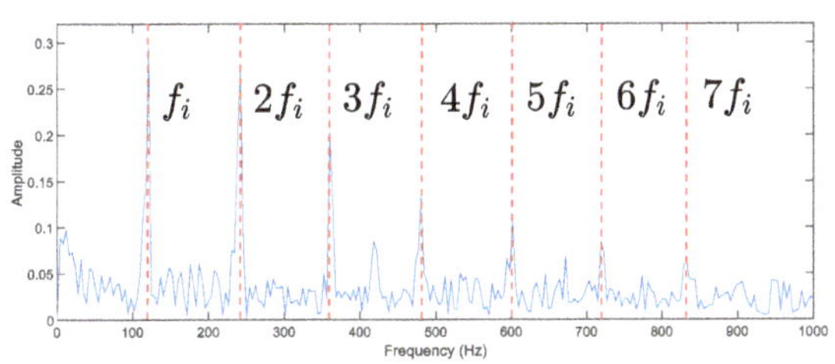

Figure 19. Envelope spectrum with an SNR of −9 dB.

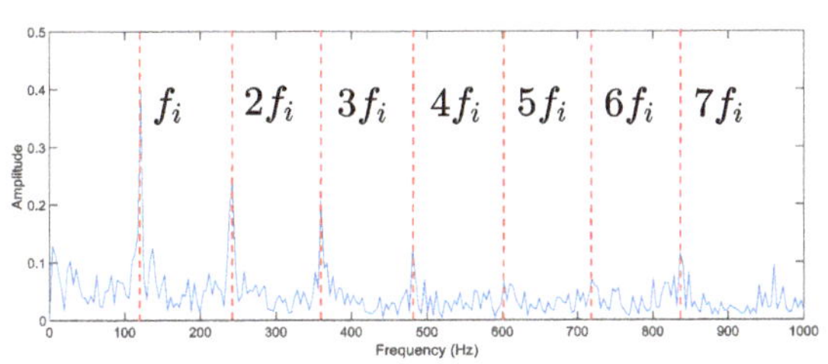

Figure 20. Envelope spectrum with an SNR of −11 dB.

5. Fault Diagnosis of Measured Bearing Sound Signals

5.1. Test Bearings and Test Equipment

The test object is the 7205 angular contact ball bearing, as shown in Figure 21. Its geometric dimensions and fault characteristic frequency are shown in Table 1.

Figure 21. Bearing 7205.

Table 1. The geometry and the fault characteristic frequencies of bearing 7205.

Geometric Dimensions	Size	Unit
Outer raceway diameter, *do*	52	mm
Inner raceway diameter, d_i	25	mm
Rolling element diameter, d	7.938	mm
Number of rolling elements, z	13	pieces
Pitch circle diameter, D	38.5	mm
Contact angle, θ	40°	(°)
Spindle rotation frequency, f_S/Hz	200	Hz
Fault characteristic frequency	Value	Unit
Inner raceway fault characteristic frequency, f_i	1505	Hz
Outer raceway fault characteristic frequency, f_o	1094	Hz
Rolling element failure characteristic frequency, f_b	945	Hz

The test device is a certain type of test bench from Schaeffler Trading (Shanghai, China) Co., Ltd.—Shanghai R&D Center, as shown in Figure 22. As shown in Figure 23, the red box is the position of the sound sensor, the radial position of the test bearing, and the radial data from the test bearing. The sound signal is picked up by a sound sensor composed of an AWA14423 microphone and an AWA14604 preamplifier. The data are sampled by NI acquisition card 9234. The sampling frequency and number of sampling points are both set to 51.2 kHz. The size and appearance of the microphone are shown in Figure 24. The sensor performance indicators are shown in Table 2.

Table 2. Microphone performance indicators.

Parameter	Value	Unit
Sensitivity, mV/Pa	50	mV/Pa
Dynamic range, dB	20~142	dB
Frequency range, Hz	10~20,000	Hz
Size, ϕ (mm)	12.7	mm
Frequency response characteristics	free field	

Figure 22. Rig of bearing 7205.

Figure 23. Sound sensor location.

Figure 24. Microphone sensor.

In this experiment, the relationship between the placement of the sensor and the bearing under test significantly impacts the fault detection results. As illustrated in the structural diagram of the test bench in Figure 25, the test bearings are symmetrically distributed on either side of the support bearings. The dimensions of the tested bearings are as follows: the outer raceway diameter (*do*) is equal to 52 mm, the inner raceway diameter (d_i) is equal to 25 mm, the rolling element diameter (d) is equal to 7.938 mm, the number of rolling elements (z) is equal to 13, the pitch circle diameter (D) is equal to 38.5 mm, and the contact angle (θ) is equal to 40°. In this paper, a disc spring is used for loading, with an axial load of 5 KN and a radial load of 2 KN. The drive mechanism is belt-driven. Bearings located near the belt side experience relatively large radial loads, and the fault characteristic signals are more pronounced in these locations. Consequently, the sound sensor is positioned at the bearing near the belt end to collect radial data, thereby enhancing the accuracy of fault characteristic signal capture. However, the experimental environment is characterized by considerable noise interference, which can lead to rapid attenuation of the sound signal during propagation. If the sound sensor is positioned too far from the test bench, the signal strength may be significantly diminished, and crucial fault information from the bearing may be obscured by noise, thereby compromising the accuracy of fault diagnosis. Conversely, if the sensor is placed too close, the vibrations and shaking of the bench during operation could potentially damage the front end of the contacting sound sensor, adversely affecting both the performance of the sensor and the stability of signal collection.

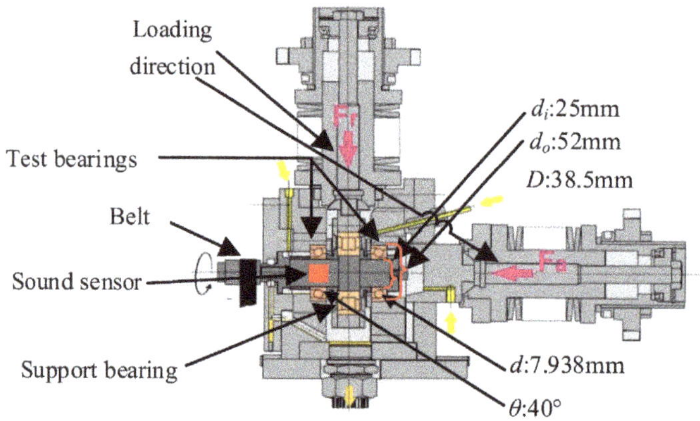

Figure 25. Structural diagram of the test bench.

To effectively address these challenges, extensive trials and experiments led to the conclusion that the sound sensor should be positioned as close as possible to the radial end of the bench's belt (as shown in Figure 26). This placement minimizes the signal propagation path and reduces noise interference. Additionally, 3D printing technology was employed to create a specialized shell (as shown in Figure 27) to secure the sound sensor, ensuring its stability throughout the experiment and preventing displacement or damage caused by bench vibrations. The shell shown in Figure 27 is made of polyamide (nylon). The principal dimensions of the shell are as follows: the total height is 36 mm, the inlet diameter is 16 mm, and the outlet diameter is 52 mm. Additionally, the cylindrical portion of the shell has a length of 20 mm and a wall thickness of 2 mm. The interior of the shell is filled with efficient noise-absorbing material for sealing, while the exterior is further reinforced with clay, providing dual protection. This design effectively diminishes external noise interference with sound signals, thereby ensuring the integrity and stability of sound signal transmission. Numerous experiments and tests [30,31] have demonstrated that these measures significantly enhance the accuracy and reliability of fault detection, establishing a robust data foundation for the early diagnosis of bearing faults. In this study, the primary noise sources identified were the vibrations of the bench and interference from surrounding equipment. To maintain the consistency of fault characteristic signals, the working conditions of the bearing (such as rotational speed and load) were kept constant during the experiments, ensuring that the sound sensor's performance remained stable and capable of accurately capturing sound signals.

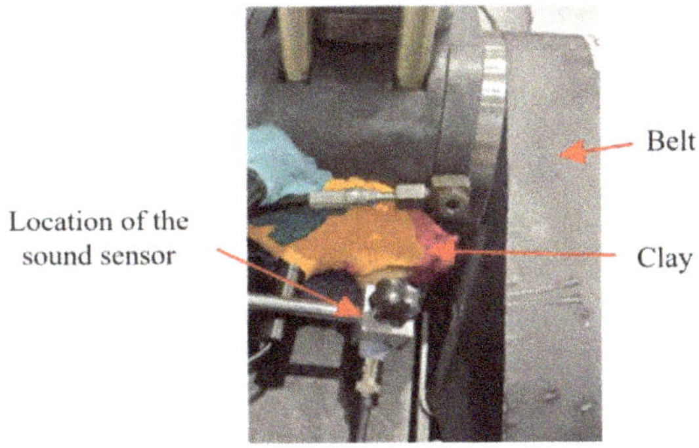

Figure 26. Location of the sound sensor.

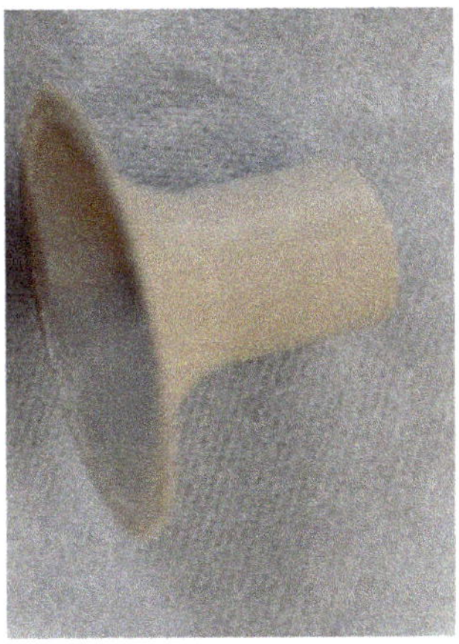

Figure 27. Shell.

5.2. Early Failure of Inner Raceway

Figure 28 shows the sound signal of an early failure of the bearing inner raceway. Since the sound signal is extremely susceptible to interference and the interference of strong noise from various bench operations in the test environment, it is difficult to find periodic pulse characteristics from the picture, and the bearing cannot confirm the health status.

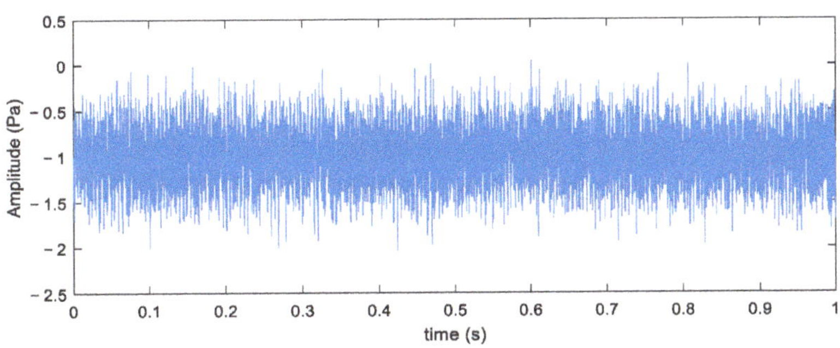

Figure 28. Time domain with incipient inner raceway fault.

The method proposed in this article and the adaptive VMD method were respectively used to perform noise reduction processing on the sound signal, as shown in Figure 28. The convergence curves of the SBOA and the CPO algorithm obtained based on the minimum envelope entropy value are shown in Figures 29 and 30. After 30 iterations of the SBOA, FMD reached the optimal combination. The filter length L is 34, the frequency band cutting number K is 14, the mode number n is 5, and the cycle period m is 34. CPO was iterated 21 times, and VMD reached the optimal combination. The number of decomposition modes K is 5, and the penalty factor α is 4985. The comprehensive index ISEI of each mode of the decomposition results is shown in Table 3.

Table 3. ISEI values of the decomposed modes via parameter-optimized FMD-VMD (inner raceway fault).

F-IMF	IMF1	IMF2	IMF3	IMF4
ISEI	1.28	1.60	2.49	1.19
V-IMF	IMF1	IMF2	IMF3	IMF4
ISEI	0.89	0.70	0.97	1.55

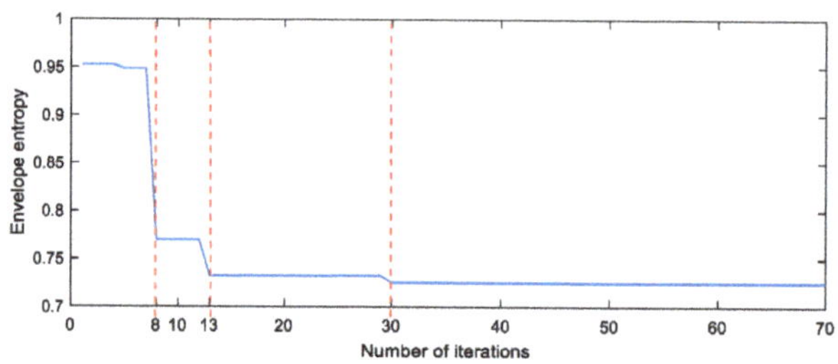

Figure 29. Convergence curve of the SBOA (inner raceway fault).

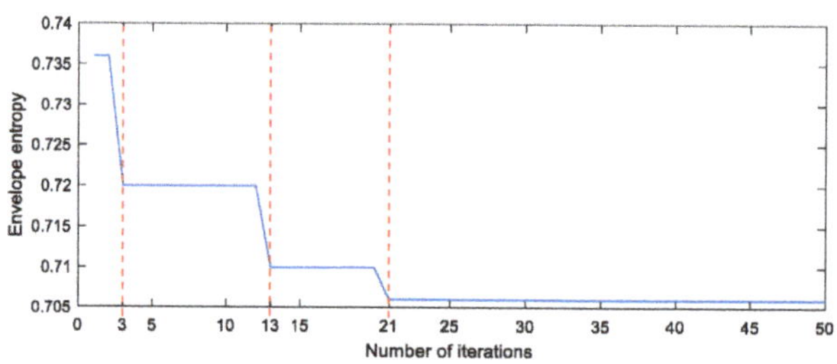

Figure 30. Convergence curve of the CPO algorithm (inner raceway fault).

As shown in Table 3, the F-IMF3 component in S-FMD has the largest ISEI value and contains the most fault information. The envelope spectrum analysis of F-IMF3 is shown in Figure 31. Two peaks with prominent amplitude energy are observed, corresponding to the inner raceway fault characteristic frequency f_i and its harmonic components, respectively. Although the signal noise floor is high and there are serious interference components near low frequencies, the proposed method can still effectively extract bearing inner raceway fault information, showing its strong anti-interference ability. The ISEI value of the V-IMF4 component in C-VMD is the largest. The envelope of this component is shown in Figure 32. The fault frequency is not intuitive, and f_i and its harmonic components cannot be observed.

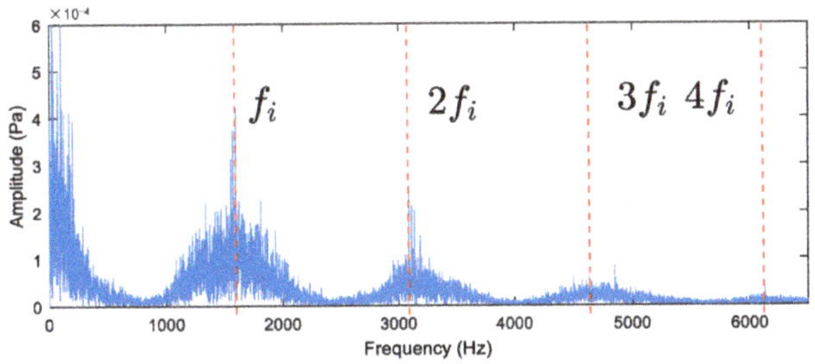

Figure 31. Envelope spectrum of F-IMF3 decomposed via parameter-optimized FMD (inner raceway fault).

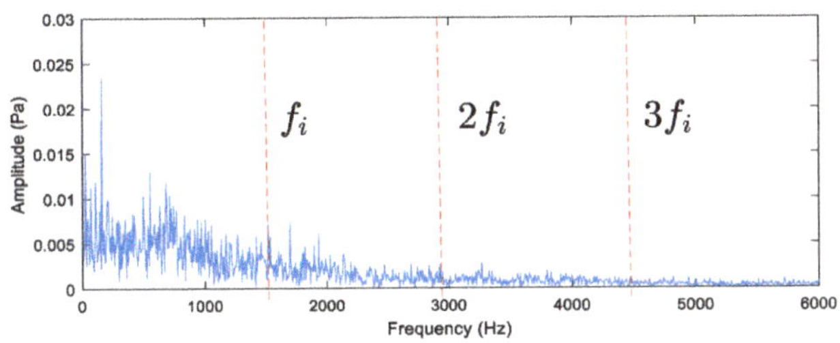

Figure 32. Envelope spectrum of V-IMF4 decomposed via parameter-optimized VMD (inner raceway fault).

5.3. Early Failure of Outer Raceway

Figure 33 shows the early fault sound signal of the bearing outer raceway. The analysis idea is the same as that of the bearing inner raceway. For noise reduction processing, the SBOA and CPO convergence curves obtained based on the minimum envelope entropy value are shown in Figures 34 and 35, respectively. After 47 iterations of the SBOA, FMD reached the optimal combination. The filter length L is 240, the frequency band cutting number K is 7, the mode number n is 7, and the cycle period m is 35. CPO was iterated 26 times, and VMD reached the optimal combination. The number of decomposition modes K is 5, and the penalty factor α is 5784. The ISEI values of each mode are shown in Table 4.

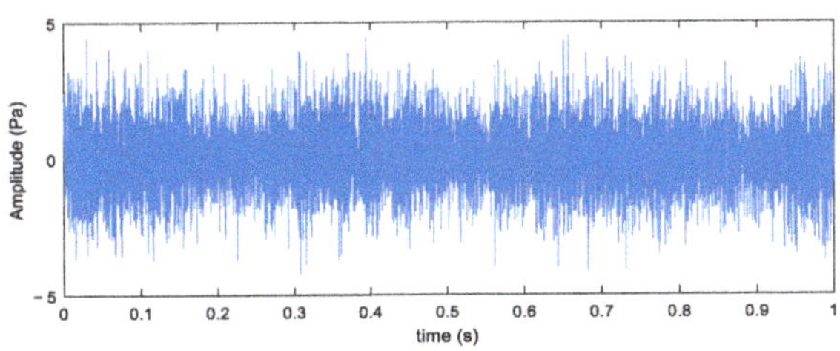

Figure 33. Time domain with incipient outer raceway fault.

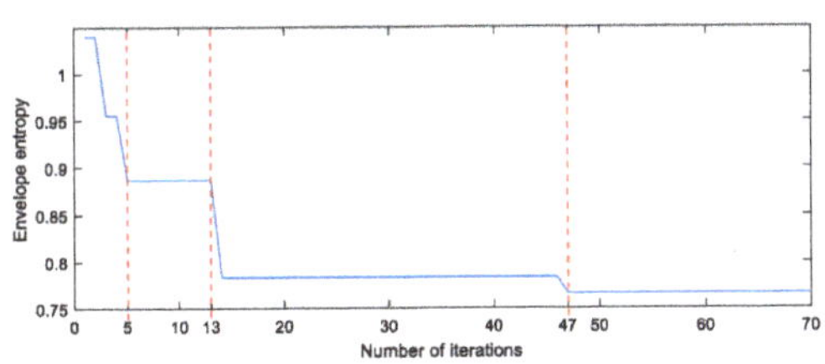

Figure 34. Convergence curve of the SBOA (outer raceway fault).

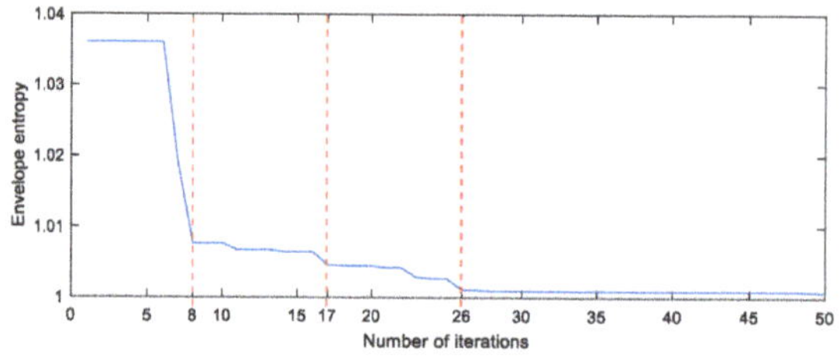

Figure 35. Convergence curve of the CPO algorithm (outer raceway fault).

Table 4. ISEI values of the decomposed modes via parameter-optimized FMD-VMD (outer raceway fault).

F-IMF	IMF1	IMF2	IMF3	IMF4	IMF5	IMF6	IMF7
ISEI	1.53	2.34	5.02	1.20	2.17	2.58	4.08
V-IMF	IMF1	IMF2	IMF3	IMF4	IMF5		
ISEI	1.48	1.46	1.40	2.12	1.98		

As shown in Table 4, F-IMF3 in S-FMD has the largest ISEI value and contains the most fault information. The envelope analysis of F-IMF3 is shown in Figure 36, and three peaks with prominent amplitude energy can be clearly seen, respectively corresponding to the outer raceway fault frequency f_o and harmonic components. The interference frequency of 25 Hz appears in the form of side frequency bands at f_o and $2f_o$, indicating that the noise interference is serious, but the impact of environmental noise is still significant. The ISEI of V-IMF4 in C-VMD is the largest. Its analysis is shown in Figure 37. The fault frequency is not intuitive, and the harmonic component of f_o is not obvious.

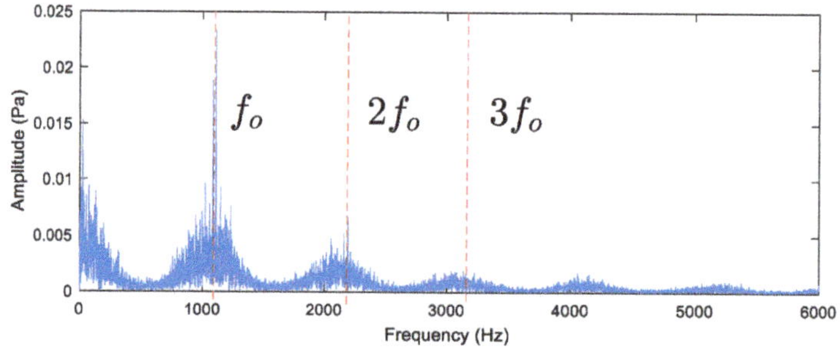

Figure 36. Envelope spectrum of F-IMF3 decomposed via parameter-optimized FMD (outer raceway fault).

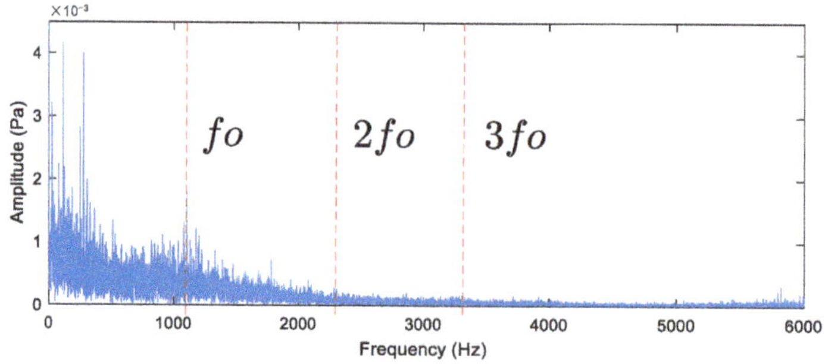

Figure 37. Envelope spectrum of V-IMF4 decomposed via parameter-optimized VMD (outer raceway fault).

5.4. Early Failure of Rolling Elements

The sound signal of the early weak fault of the bearing rolling element was selected for analysis, as shown in Figure 38, and the noise reduction process was performed. The SBOA and CPO convergence curves obtained based on the minimum envelope entropy value are shown in Figures 39 and 40. After 26 iterations of the SBOA, FMD reached the optimal combination. The filter length L is 120, the frequency band cutting number K is 7, the mode number n is 7, and the cycle period m is 15. CPO was iterated 15 times, and VMD reached the optimal combination. The number of decomposition modes K is 7, and the penalty factor α is 9697. The ISEI values of each mode are shown in Table 5.

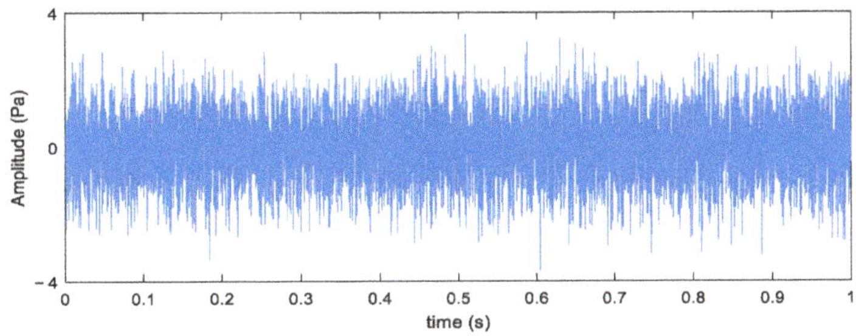

Figure 38. Time domain with incipient rolling element fault.

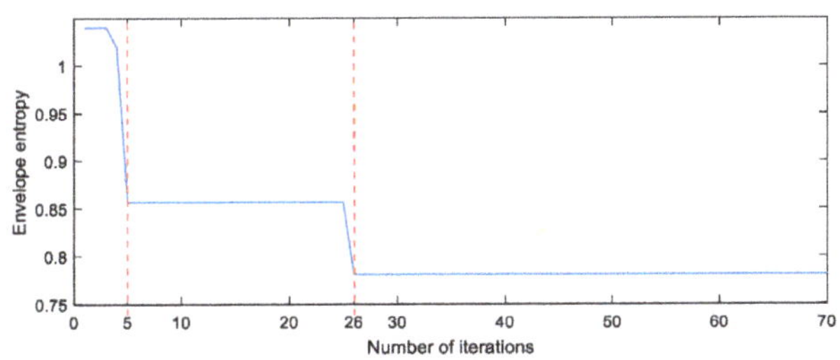

Figure 39. Convergence curve of the SBOA (rolling element fault).

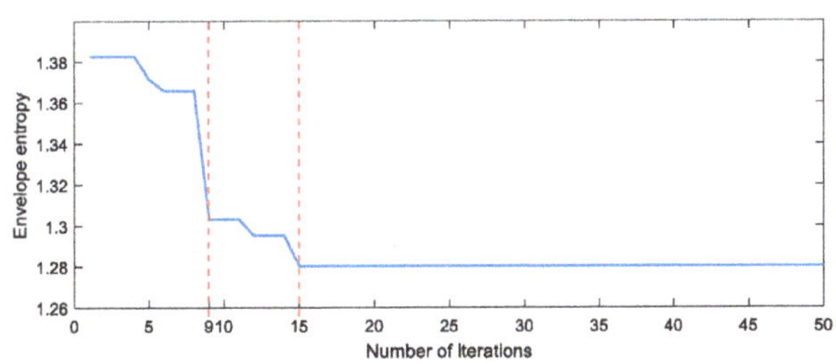

Figure 40. Convergence curve of the CPO algorithm (rolling element fault).

Table 5. ISEI values of the decomposed modes via parameter-optimized FMD-VMD (rolling element fault).

F-IMF	IMF1	IMF2	IMF3	IMF4	IMF	IMF6	IMF7
ISEI	3.93	5.24	2.65	2.34	2.13	1.78	2.12
V-IMF	IMF1	IMF2	IMF3	IMF4	IMF5	IMF6	IMF7
ISEI	1.96	2.74	1.24	1.10	1.15	2.41	1.30

As shown in Table 5, F-IMF2 in S-FMD has the largest ISEI value and contains the most fault information. F-IMF2 was selected for envelope analysis, as shown in Figure 41. From the envelope spectrum, four prominent amplitude energies can be clearly observed. The peak values respectively correspond to the bearing rolling element failure frequency f_b and the harmonic component, indicating that the bearing rolling element is faulty. It is observed from the envelope spectrum that there is a strong noise floor below 500 Hz, indicating that it is low-frequency interference noise. At the same time, there are sidebands of low-frequency interference noise near the f_b, $2f_b$, and $3f_b$ fault frequencies. The ISEI value of V-IMF2 in C-VMD is the largest. This analysis is shown in Figure 42. The fault frequency is not intuitive, and f_b and the harmonic components cannot be seen.

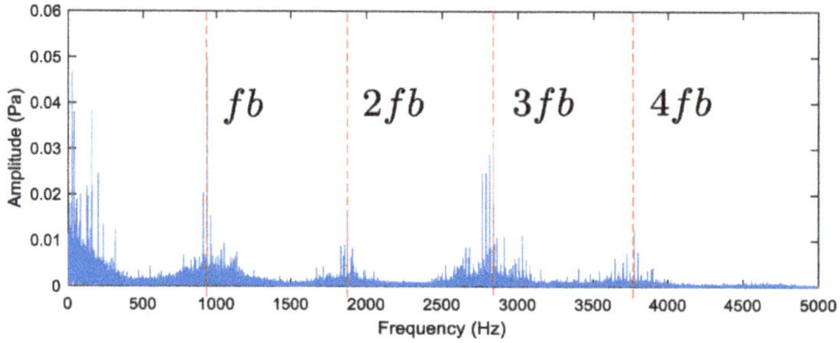

Figure 41. Envelope spectrum of F-IMF2 decomposed via parameter-optimized FMD (rolling element fault).

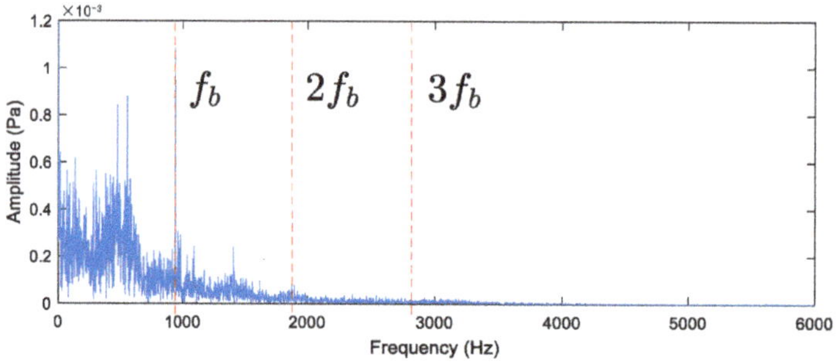

Figure 42. Envelope spectrum of V-IMF2 decomposed via parameter-optimized VMD (rolling element fault).

For further comparative analysis, the proposed methods S-FMD, C-VMD, EMD, CEEMD, and fixed-parameter FMD were used to perform noise reduction and decomposition on the simulated inner raceway fault signal and the measured sound signal of the inner raceway, outer raceway, and rolling element faults, and their ISEI values were selected respectively. The largest component was compared. The ISEI value is a comprehensive index obtained by merging kurtosis and the harmonic signal-to-noise ratio. Its size represents the degree of noise reduction effect of the bearing sound signal and the ratio of the bearing's fault characteristic frequency to the background noise energy. Different The ISEI values of the optimal components obtained by this method are shown in Figure 43.

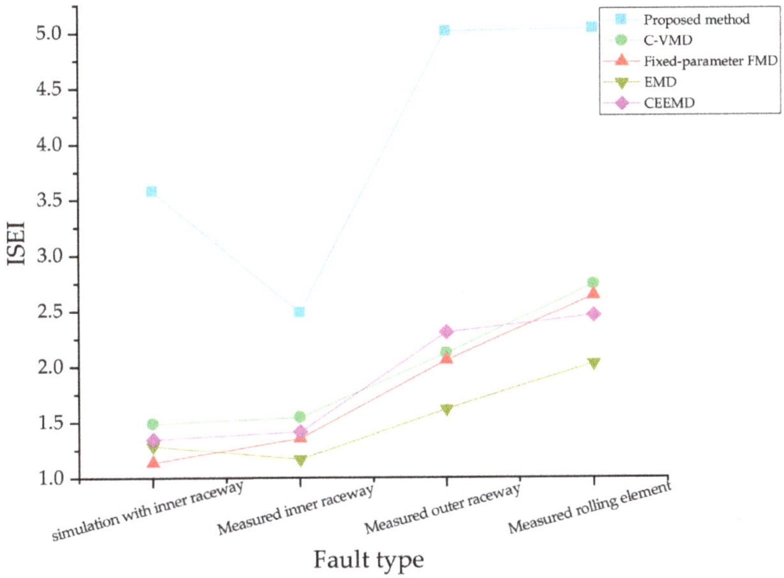

Figure 43. ISEI value of the optimal component obtained by different methods.

Different methods were applied for noise reduction to simulated inner raceway faults, measured inner raceway faults, measured outer raceway faults, and rolling element raceway faults. The component with the most fault-relevant information was then selected to calculate its ISEI. As illustrated in Figure 43, the ISEI value of the optimal component decomposed using the method proposed in this article is significantly higher than that of the best components obtained through the four methods: C-VMD, EMD, CEEMD, and fixed-parameter FMD. This finding demonstrates that the proposed method exhibits superior noise reduction and fault extraction capabilities in practical applications, as well as enhanced fault identification abilities, particularly in diagnosing weak faults.

6. Conclusions

Compared with vibration sensors, microphones have the advantages of non-contact detection, high sensitivity, low cost, and easy installation. In order to solve the problem of complex components and serious interference of rolling bearing sound signals, a rolling bearing acoustic signal fault diagnosis method based on the Secretary Bird Optimization Algorithm and SBOA optimization (Feature Mode Decomposition, FMD) was proposed. The method was verified through simulation and test signals. The following conclusions were obtained:

1. This paper proposes a method for bearing fault diagnosis based on sound signals. Compared with vibration signals, sound signals have the advantages of non-contact, economical price, and convenient remote monitoring. However, sound signals are extremely susceptible to noise interference, so noise reduction processing of sound signals is crucial.
2. The rolling bearing sound signal fault diagnosis method based on parameter-adaptive FMD studied in this article preprocesses the original sound signal through parameter-adaptive FMD, which can reduce the impact of noise and other interference components on diagnosis and then obtain the envelope spectrum for fault diagnosis.
3. Simulation and experimental analysis show that compared with the fixed parameter FMD and parameter adaptive VMD methods, the method studied in this paper can

effectively reduce noise interference, accurately extract fault information from the bearing sound signal, and realize bearing fault diagnosis.

Author Contributions: Conceptualization, H.W. and J.X.; methodology, H.W. and J.X.; software, H.W.; validation, H.W. and J.X.; formal analysis, H.W.; investigation, H.W.; resources, H.W.; data curation, H.W. and J.X.; writing—original draft preparation, J.X.; writing—review and editing, H.W.; visualization, J.X.; supervision, H.W.; project administration, H.W.; funding acquisition, H.W. All authors have read and agreed to the published version of the manuscript.

Funding: This work was supported by the Ningbo City Unveiled Project (2023T016).

Institutional Review Board Statement: Not applicable.

Informed Consent Statement: Not applicable.

Data Availability Statement: The data used to support the findings of this study are available from the corresponding author upon request.

Conflicts of Interest: The authors declare no conflicts of interest.

Abbreviations

Abbreviation	Full Form
SBOA	Secretary Bird Optimization Algorithm
FMD	Feature Mode Decomposition
ISEI	Integrated Signal Evaluation Index
SFMD	Secretary Bird Optimization Algorithm-optimized Feature Mode Decomposition
CNC	Computer Numerical Control
EMD	Empirical Mode Decomposition
EEMD	ensemble empirical mode decomposition
CEEMD	Complementary EEMD
VMD	variational mode decomposition
IMF	intrinsic mode function
CK	correlation kurtosis
EWT	Empirical Wavelet Transform
WOA	Whale Optimization Algorithm
SO	Snake Optimization Algorithm
SNR	signal-to-noise ratio
HNR	harmonic-to-noise ratio
AFMD	adaptive feature mode decomposition
FFR	feature frequency ratio
GMSV	Graph Modeling Singular Values
CC	correlation coefficient

References

1. Lei, Y.G.; Jia, F.; Kong, D.T.; Lin, J.; Xing, S.B. Opportunities and challenges of machinery intelligent fault diagnosis in big data era. *J. Mech. Eng.* **2018**, *54*, 94–104. [CrossRef]
2. Patel, S.; Patel, S. Research progress on bearing fault diagnosis with signal processing methods for rolling element bearings. *Noise Vib. Worldw.* **2024**, *55*, 96–112. [CrossRef]
3. Chen, S.; Peng, Z.; Zhou, P. Review of signal decomposition theory and its applications in machine fault diagnosis. *J. Mech. Eng.* **2020**, *56*, 91–107.
4. Pandiyan, M.; Babu, T.N. Systematic Review on Fault Diagnosis on Rolling-Element Bearing. *J. Vib. Eng. Technol.* **2024**, *12*, 8249–8283. [CrossRef]
5. Chen, J.; Xu, T.L.; Huang, Z.; Sun, T.H.; Li, X.Y.; Ji, L.; Yang, H.J. Fault diagnosis of rolling bearings based on acoustic signals. *J. Vib. Shock.* **2023**, *42*, 237–244.

6. Daraz, A.; Alabied, S.; Gu, F.; Ball, A.D. A comparative study of acoustic and vibration signals for bearing fault detection and diagnosis based on MSB analysis. In Proceedings of the International Conference on Maintenance Engineering; Springer: Cham, Switzerland, 2020; pp. 854–868.
7. Wu, S.; Jiang, W.; Zhao, L. Research on KPCA Fault Diagnosis Method Based on Sound Signal. *Mach. Tool Hydraul.* **2016**, *44*, 184–187.
8. Liu, H.; Li, L.; Ma, J. Rolling Bearing Fault Diagnosis Based on STFT-Deep Learning and Sound Signals. *Shock. Vib.* **2016**, *2016*, 6127479. [CrossRef]
9. Aswin, M.; Wardana, I.; Irawan, Y.S.; Suyono, H. Bearing damage detection based on sound signal. *Appl. Mech. Mater.* **2014**, *548*, 698–702. [CrossRef]
10. Wu, Z.; Huang, N.E. Ensemble empirical mode decomposition: A noise-assisted data analysis method. *Adv. Adapt. Data Anal.* **2009**, *1*, 1–41. [CrossRef]
11. Yeh, J.-R.; Shieh, J.-S.; Huang, N.E. Complementary ensemble empirical mode decomposition: A novel noise enhanced data analysis method. *Adv. Adapt. Data Anal.* **2010**, *2*, 135–156. [CrossRef]
12. Gu, J.; Peng, Y. An improved complementary ensemble empirical mode decomposition method and its application in rolling bearing fault diagnosis. *Digit. Signal Process.* **2021**, *113*, 103050. [CrossRef]
13. Sahu, P.K.; Rai, R.N. Fault diagnosis of rolling bearing based on an improved denoising technique using complete ensemble empirical mode decomposition and adaptive thresholding method. *J. Vib. Eng. Technol.* **2023**, *11*, 513–535. [CrossRef]
14. Bendjama, H. Feature extraction based on vibration signal decomposition for fault diagnosis of rolling bearings. *Int. J. Adv. Manuf. Technol.* **2024**, *130*, 821–836. [CrossRef]
15. Dragomiretskiy, K.; Zosso, D. Variational mode decomposition. *IEEE Trans. Signal Process.* **2013**, *62*, 531–544. [CrossRef]
16. Wang, H.-D.; Deng, S.-E.; Yang, J.-X.; Liao, H.; Li, W.-B. Parameter-Adaptive VMD Method Based on BAS Optimization Algorithm for Incipient Bearing Fault Diagnosis. *Math. Probl. Eng.* **2020**, *2020*, 5659618. [CrossRef]
17. Wang, X.; Wen, J. Fault diagnosis of rolling bearings based on noise signal and improved VMD. *Noise Vib. Control* **2021**, *41*, 118–124.
18. Li, H.; Liu, T.; Wu, X.; Chen, Q. An optimized VMD method and its applications in bearing fault diagnosis. *Measurement* **2020**, *166*, 108185. [CrossRef]
19. Miao, Y.; Zhang, B.; Li, C.; Lin, J.; Zhang, D. Feature mode decomposition: New decomposition theory for rotating machinery fault diagnosis. *IEEE Trans. Ind. Electron.* **2022**, *70*, 1949–1960. [CrossRef]
20. Jia, G.; Meng, Y. Study on Fault Feature Extraction of Rolling Bearing Based on Improved WOA-FMD Algorithm. *Shock. Vib.* **2023**, *2023*, 5097144. [CrossRef]
21. Chauhan, S.; Vashishtha, G.; Kumar, R.; Zimroz, R.; Gupta, M.K.; Kundu, P. An adaptive feature mode decomposition based on a novel health indicator for bearing fault diagnosis. *Measurement* **2024**, *226*, 114191. [CrossRef]
22. Yan, X.; Yan, W.-J.; Xu, Y.; Yuen, K.-V. Machinery multi-sensor fault diagnosis based on adaptive multivariate feature mode decomposition and multi-attention fusion residual convolutional neural network. *Mech. Syst. Signal Process.* **2023**, *202*, 110664. [CrossRef]
23. Yan, X.; Jia, M. Bearing fault diagnosis via a parameter-optimized feature mode decomposition. *Measurement* **2022**, *203*, 112016. [CrossRef]
24. Li, C.; Zhou, J.; Wu, X.; Liu, T. Adaptive feature mode decomposition method for bearing fault diagnosis under strong noise. *Proc. Inst. Mech. Eng. Part C J. Mech. Eng. Sci.* **2024**, *239*, 508–519. [CrossRef]
25. Wen, X.; Lu, G.; Liu, J.; Yan, P. Graph modeling of singular values for early fault detection and diagnosis of rolling element bearings. *Mech. Syst. Signal Process.* **2020**, *145*, 106956. [CrossRef]
26. Fu, Y.; Liu, D.; Chen, J.; He, L. Secretary bird optimization algorithm: A new metaheuristic for solving global optimization problems. *Artif. Intell. Rev.* **2024**, *57*, 1–102. [CrossRef]
27. Hashim, F.A.; Hussien, A.G. Snake Optimizer: A novel meta-heuristic optimization algorithm. *Knowl.-Based Syst.* **2022**, *242*, 108320. [CrossRef]
28. Mirjalili, S.; Lewis, A. The whale optimization algorithm. *Adv. Eng. Softw.* **2016**, *95*, 51–67. [CrossRef]
29. Kschischang, F.R. The hilbert transform. *Univ. Tor.* **2006**, *83*, 277.
30. Yan, J.; Kim, M.-S.; Kang, K.-M.; Joo, K.-H.; Kang, Y.-J.; Ahn, S.-H. Evaluation of PP/clay composites as soundproofing material. *Polym. Polym. Compos.* **2014**, *22*, 65–72. [CrossRef]
31. Shen, W.; Sun, J.Q. Study of shell interior noise control. In Proceedings of the Smart Structures and Materials 1997: Smart Structures and Integrated Systems; SPIE: San Francisco, CA, USA, 1997; pp. 812–818.

applied sciences

MDPI

Article

Multi-Patch Time Series Transformer for Robust Bearing Fault Detection with Varying Noise

Sangkeun Ko and Suan Lee *

Department of Computer Science, Semyung University, Jecheon 27136, Republic of Korea; kskrnfjrl@semyung.ac.kr
* Correspondence: suanlee@semyung.ac.kr; Tel.: +82-43-649-1273

Abstract: In time-series studies involving bearing sensor data, Gaussian noise and white noise techniques are commonly employed to evaluate model robustness. However, these conventional noise techniques are limited in their applicability to real-world industrial environments. This paper proposes three novel noise techniques—electrical interference noise, harmonic noise, and random shock noise—that more accurately reflect the complex noise encountered in industrial settings. Additionally, a new deep learning model, MultiPatchTST, is introduced, demonstrating robust performance under various noise conditions. Experimental results reveal that Gaussian noise has minimal impact on model performance, whereas the proposed noise techniques significantly affect performance, providing a more realistic evaluation of noise robustness. The proposed MultiPatchTST model achieves superior performance across all metrics in the presence of all four noise types, confirming its robustness and reliability.

Keywords: bearing fault detection; multi patch; time series; transformer; noise

1. Introduction

Bearings are a critical component in mechanical systems, significantly influencing the performance and reliability of machines. Consequently, anomaly detection and condition classification based on sensor data have become key areas of research. However, industrial environments present various forms of noise that adversely affect data analysis and model performance, posing a significant challenge to the effectiveness of current methodologies.

Previous studies have primarily employed Gaussian noise [1] and white noise techniques to evaluate model robustness to noise. While useful in controlled scenarios, these techniques fail to capture the complexity and variability of real-world industrial noise. Industrial environments introduce a diverse range of noise sources, including electrical interference, machine wear, mechanical vibration, harmonic disturbances, and environmental factors. These noise types are difficult to represent accurately using simple noise models, highlighting the need for new approaches that better simulate realistic noise conditions.

To address this gap, this study proposes novel noise techniques that overcome the limitations of conventional Gaussian and white noise methods. These techniques aim to reflect the complex noise patterns encountered in industrial environments, enabling the generation of more realistic data. Furthermore, we introduce a new deep learning model specifically designed to classify bearing-sensor anomalies effectively, even when subjected to these noise conditions.

The significance of this study lies in addressing the challenges posed by complex industrial noise, which has been insufficiently explored in previous research. The proposed noise techniques and deep learning model contribute to the development of robust systems capable of maintaining a stable performance across various noisy environments. In addition, we analyze the impact of different noise types on sensor data and evaluate the robustness of the proposed model under these conditions.

The main objectives of this study are twofold:

- To develop novel noise techniques that accurately reflect the diverse and realistic noise characteristics of industrial environments, generating synthetic data that closely mirrors real-world conditions.
- To propose and evaluate a new deep learning model capable of effectively classifying bearing-sensor anomalies in noisy environments, demonstrating the utility and robustness of the proposed techniques and model.

2. Related Work

2.1. Bearing Fault Detection Research in Noisy Environments

Moysidis et al. conducted extensive research on bearing fault detection and diagnosis systems in industrial electrical machinery, emphasizing the challenges of achieving accurate fault diagnosis in noisy environments [2]. Their study compared the performance of various machine learning and deep learning models, including Support Vector Machines (SVM), LeNet-5, 1D-CNN, and 2D-CNN, under noisy conditions. Signal preprocessing techniques such as time-series analysis, wavelet transform (CWT), waveform decomposition (WPD), and signal-image conversion (SIC) were employed to evaluate the efficiency of these models. The findings indicated that CWT-based features were particularly effective for fault detection, with the 2D-CNN model outperforming others in noise-affected scenarios. By introducing Gaussian noise, the study examined its impact on fault detection performance and observed that higher Signal-to-Noise Ratios (SNR) resulted in better model accuracy. This work underscores the importance of designing diagnostic systems capable of maintaining robustness in noisy conditions and highlights the utility of Gaussian noise simulation for developing reliable fault detection systems.

Fan Li et al. proposed an adaptive multi-scale fully convolutional network (AMFCN) to enhance bearing fault diagnosis in varying noise environments [3]. The AMFCN model introduced two key strategies to improve noise adaptability. First, random sampling was utilized to reduce noise while intentionally introducing data incompleteness, which improved the model's robustness during training. Second, large-kernel convolutional layers were employed to extract wide temporal features, effectively suppressing high-frequency noise. The AMFCN architecture dynamically adjusted feature weights across multiple scales, emphasizing critical features while mitigating noise interference. Performance evaluation under different SNR conditions, ranging from −8 dB to 8 dB, demonstrated that AMFCN consistently achieved high fault diagnosis accuracy, even under severe noise conditions. Notably, the model achieved an accuracy exceeding 97% at an SNR of −8 dB. These results highlight the AMFCN's robustness against noise and its potential as a reliable solution for fault diagnosis in real-world industrial applications.

2.2. Time-Frequency-Based Anomaly Classification Research

Kulevome et al. developed a novel data augmentation technique leveraging analytical wavelets to enhance time-frequency-based data for bearing fault diagnosis, addressing challenges such as limited availability of fault data and the specialization of existing augmentation methods for image-based studies [4]. Their approach employed analytical wavelets to generate scalograms while preserving the original data's characteristics. This was implemented in the preprocessing stage of a simple CNN model, comprising three convolutional layers and one fully connected layer. By tuning parameters that control the decay rate in the time domain and the bandwidth in the frequency domain (set to three and twenty, respectively), the augmented data provided significant improvements in model performance. When applied to the CWRU dataset, the proposed method achieved an impressive accuracy of 99.49% and demonstrated robust performance (90.14% accuracy) even in low SNR conditions. While this study highlights the effectiveness of wavelet-based data augmentation for time-frequency-specific tasks, it also notes limitations related to the sensitivity of parameter selection in the wavelet design process.

Chen et al. introduced a Fourier Markov Transition Field (FMTF)-CNN-based method for short-time-series fault diagnosis to overcome the limitations of traditional Markov Transition Field (MTF) techniques in feature extraction and their slow neural network training speeds [5]. The FMTF method transforms time-domain signals into the frequency domain using the Fast Fourier Transform (FFT), followed by the generation of two-dimensional feature images via the MTF technique. This approach enhances the accuracy and efficiency of feature extraction compared to conventional MTF methods. Using the Jiangnan University public bearing dataset, each signal was divided into 256 sample points, and 128 data points containing frequency information within the range of 0–25 kHz were extracted via FFT. The MTF then generated 128 $\times$ 128 feature images, which were input into a classification model with 16 convolutional filters of size 5 $\times$ 5. Experimental results showed that FMTF-CNN improved accuracy by 23.61% and reduced iteration steps by 47.77% compared to baseline methods. Within the same training time (100 s), FMTF-CNN achieved an accuracy of 85.65%. However, this study's reliance on a single dataset limits its ability to generalize across diverse industrial environments and noise conditions.

2.3. *Raw Data-Based Anomaly Classification Research*

Knap et al. highlighted the critical role of bearing fault diagnosis in wind farms while addressing the high computational costs associated with signal processing-based methods [6]. To overcome these challenges, they proposed a Boosted CNN algorithm that directly utilizes raw data for efficient classification of bearing faults without the need for extensive signal preprocessing. The dataset was collected using a PLC controller and categorized into healthy bearings, inner ring defects, and outer ring defects. For comparison, the study evaluated a spectrogram-based model leveraging the MobileNet-v2 network structure alongside the Boosted CNN and MLP models, both utilizing raw data as the input. Experiments conducted on a Raspberry Pi 4B minicomputer demonstrated that the Boosted CNN model achieved a 5.7% higher accuracy and a 42% reduction in execution time compared to the spectrogram-based model. These results underline the efficiency and practicality of the Boosted CNN model for real-time processing in low-power, low-memory environments, showcasing its potential applicability in industrial fault diagnosis systems.

Ahsan et al. proposed a hybrid DCNN-LSTM model that combines the strengths of a Deep Convolutional Neural Network (DCNN) and a Long Short-Term Memory (LSTM) network to diagnose bearing faults directly from raw vibration signals [7]. The study employed the IMS variable speed dataset and the CWRU bearing dataset, segmenting the signals into 1000-unit windows using a sliding-window technique. Experimental results indicated that the DCNN-LSTM model achieved a high accuracy exceeding 99% on both datasets, demonstrating its robustness and effectiveness for raw signal-based fault diagnosis. These findings highlight the potential of the DCNN-LSTM model for application in real-world industrial fault diagnosis systems.

Tian YU et al. proposed the Reparameterised Residual Denoising Diffusion Probability Model (ReF-DDPM), a novel data augmentation method to handle unbalanced rotating bearing fault data and improve diagnostic accuracy [8]. The ReF-DDPM improves the DDPM structure to enhance the quality and diversity of data generation and addresses the problems of mode collapse and learning instability. The proposed model combines multi-scale residual network (RepRFN) and multi-head recurrent layer attention mechanism (MRLA) to enhance local and global features and effectively solve the problem of imbalanced data. The ReF-DDPM faithfully reproduces the characteristics of the original data in the time-frequency domain and outperforms the existing WGAN-GP and DDPM in evaluation metrics such as JSD, MMD, and COSINE. In particular, experiments on various datasets such as CWRU, MFPT, and MFS demonstrated high data generation quality and excellent diagnostic performance even in imbalanced data and achieved a high accuracy of 99.5% in balanced data. However, due to the slow sampling process of DDPM-based models, there are limitations in large-scale data processing and real-time application, and the complexity of the model structure increases, which may limit its application to real industrial datasets. In addition, the evaluation metrics limited to JSD, MMD, COSINE, etc., do not elaborately describe the impact of the generated data on the actual diagnostic performance, and the initial diagnostic performance is lower than that of the GAN-based model in situations with extreme data imbalance ratios. To overcome these limitations, future research should optimize sampling time, improve real-time processing capability, validate application in industrial environments, and develop new evaluation metrics. Table 1 summarizes the Related Work.

Table 1. Summary of related research.

Dataset	Methods	Results	Limitations	Reference
CWRU	TDA, WPD, CWT, SIC, SVM, CNN	• CWT most effective; SIC (image-based) performs poorly. • Validates reliability in noisy environments using Gaussian noise.	• Relies on a single dataset and metric, limiting generalizability.	[2]
CWRU, PU	AMFCN design, random sampling	• Superior performance under varying SNR conditions. • Applicable in real-world environments.	• Single metric prevents precise performance comparisons.	[3]
CWRU	Analytical wavelets, data augmentation, CNN	• Effective for time-frequency fault diagnosis. • 99.49% accuracy in low SNR.	• Sensitive to wavelet parameter selection. • Relies on a single dataset.	[4]
Jiangnan	FMTF image conversion, CNN	• 23.61% accuracy improvement and 47.77% faster training compared to traditional methods.	• Limited generalizability due to single dataset and metric reliance.	[5]
Custom datasets	MobileNet-V2, Boosted CNN, MLP	• 5.7% higher accuracy and 42% faster execution time.Real-time operation in low-power environments.	• Limited generalizability due to use of a single dataset.	[6]
IMS, CWRU	DCNN-LSTM	• High accuracy (>99%) across datasets.	• Lacks precise performance comparison due to reliance on a single metric.	[7]
CWRU, MFPT, MFS	Ref-DDPM	• Outperforms traditional WGAN-GP and DDPM on evaluation metrics such as JSD, MMD, COSINE, etc. • Achieves accuracy as high as 99.5% on balanced data	• Slow sampling process limits large-scale data processing or real-time application • Increasing complexity of model structures and limitations in applying to real-world industry datasets • Initial diagnostic performance is lower than GAN-based models in situations with extreme data imbalance ratios	[8]

2.4. The Distinctiveness of This Study

Previous studies on bearing fault detection have utilized various techniques for analyzing time-series data, often incorporating Gaussian noise to evaluate the robustness of their models. However, these approaches face limitations when translating to real-world industrial environments, where noise characteristics differ significantly from Gaussian assumptions.

This study addresses this gap by introducing three novel-noise-modeling techniques that better represent the noise encountered in industrial scenarios. Furthermore, it proposes a new deep learning model designed specifically for robust performance under these realistic noise conditions. By bridging the gap between theoretical noise assumptions and practical industrial challenges, this study aims to advance the reliability and applicability of bearing fault detection systems.

3. Noise Techniques

3.1. Novel Noise Techniques Reflecting Industrial Noise

There are various causes and types of noise generated in actual industrial sites. Mechanical factors include wear and tear, defects, imbalance, misalignment, and misalignment of mechanical components, and electrical factors include electromagnetic interference caused by poor power supply quality, motor transformers, etc. Finally, environmental factors are external impacts, such as temperature and humidity changes in the environment where the equipment is installed. While general noise usually means that numerical data fluctuates irregularly, such as noise caused by external shocks, noise caused by mechanical and electrical factors is often more periodic because the source of noise is generated by repetitive and periodic motions or patterns. To address the limitations of existing noise methods, this study introduces three new noise techniques that better represent the characteristics of industrial noise environments by reflecting one source of each of the three factors described above.

3.1.1. Electrical Interference Noise

Electrical interference noise is a periodic noise commonly observed near electronic devices or power facilities. It manifests as a sine wave with specific frequency and amplitude, appearing as a periodic signal concentrated at certain frequencies, distinguishable from other signals. This type of noise can be modeled as:

$$E(t) = A \cdot \sin(2\pi f t + \varnothing) \quad (1)$$

- A: Amplitude of the interference.
- f: Frequency of the interfering signal.
- t: Time, corresponding to each data point.
- $\varnothing$: Phase offset, indicating the starting point of the sine wave.

3.1.2. Harmonic Noise

Harmonic noise comprises signal components that are integer multiples of a fundamental frequency, creating a complex periodic signal. These components increase in frequency but decrease in amplitude as the harmonic order rises, contributing to a wider frequency spectrum. Harmonic noise can blur or complicate the original signal by overlapping multiple frequency components. It can be represented as:

$$H(t) = \sum_{n}^{N=1} \frac{A}{n} \cdot \sin(2\pi f \cdot n \cdot t + \varnothing_n) \tag{2}$$

- A: Base amplitude.
- n: Order of harmonics.
- f: Fundamental frequency.
- $\varnothing_n$: Phase offset for each harmonic component.

3.1.3. Random Shock Noise

Random shock noise simulates sudden and localized anomalies that disrupt the signal briefly before returning to the original pattern. It introduces shocks at random points, affecting a few data values rather than altering the overall signal pattern. This noise is modeled by:

$$S(t) = A \cdot U(t) \cdot I(t) \tag{3}$$

$$I(t) = \begin{cases} 1, & \textit{with probability } p \\ 0, & \textit{with probability } p \end{cases} \tag{4}$$

- A: Maximum magnitude of the shock.
- $U(t)$: Random variable defining the size and direction of the shock.
- $I(t)$: Indicator function determining the occurrence of the shock based on probability p.

These techniques capture the diverse and complex noise patterns present in industrial environments, providing a more realistic basis for evaluating fault detection models.

3.2. Causes of Noise and the Necessity of Its Reflection

Industrial noise arises from various factors, each impacting system behavior in unique ways. This section discusses the causes of three types of noise—electrical interference noise, harmonic noise, and random shock noise—and highlights the necessity of reflecting them in fault detection systems.

- Electrical Interference Noise

  Electrical interference noise is caused by uneven fluctuations in current flow due to bearing failures such as pitting or wear. These failures alter the electrical contact area and resistance, leading to inconsistent contact between the bearing race and rolling elements. This inconsistency increases the load required for operation, impacting the entire system and altering its vibration pattern. Simulating these conditions using electrical interference noise techniques aids in developing early fault detection and classification systems, offering insights into system anomalies.

- Harmonic Noise

  Harmonic noise arises from factors such as component wear, cracks, or misalignment, often triggered by excessive loads, thermal expansion in high-temperature environments, or prolonged usage. These conditions produce not only fundamental frequencies but also harmonic frequencies (e.g., doubles, triples) of the fundamental frequency. By utilizing harmonic noise techniques, early abnormal conditions can be detected, and harmonic component analysis can help predict failures, enabling proactive maintenance in industrial settings.

- Random Shock Noise

  Random shock noise results from sudden shocks or collisions between machine parts due to abnormal forces. These impacts often precede machine failures, creating strong disturbances in sensor data. Simulating random shock noise facilitates the design of systems capable of preventing accidents or mitigating the impact of such events. This noise type is particularly useful for triggering emergency responses, such as system shutdowns, to prevent large-scale damage.

Although these noise types may originate from similar sources, they exhibit distinct signal characteristics, such as varying frequency bands and patterns. Electrical interference noise enables early warning systems by identifying initial anomalies. Harmonic noise supports fault progression monitoring and maintenance scheduling to optimize system responses. Random shock noise allows for immediate emergency measures, such as halting system operations to prevent extensive damage. By analyzing these noise types in a multi-dimensional manner, a fine-grained response system can be built, enhancing the ability to address industrial site failures effectively and efficiently.

4. Deep Learning Models for Anomaly Classification

This study utilizes six RNN-based and five CNN-based deep learning models, along with a newly proposed model based on the best-performing model, for bearing sensor data anomaly classification. A total of 12 models are evaluated in the experiments.

4.1. RNN-Based Deep Learning Models

The RNN-based models include LSTM [9], GRU [10], LSTMAttention [11], GRUAttention [12], TSSequencer [13], and PatchTST [14]. Each model offers distinct features and advantages for handling time-series data:

- LSTM and GRU Models

  LSTM (Long Short-Term Memory) models address the long-term dependency problem by utilizing a cell state and gate structure (Forget, Input, and Output gates). These gates control how much information is retained, added, or discarded, allowing LSTMs to efficiently process sequential data with complex dependencies. GRUs (Gated Recurrent Units) simplify this structure by using only two gates (Reset and Update), making them more computationally efficient while still capable of learning long-term dependencies. However, GRUs may struggle with highly complex dependencies compared to LSTMs.

- LSTMAttention and GRUAttention Models

  These models enhance LSTM and GRU architectures by incorporating an attention mechanism, which assigns weights to specific time steps, enabling the model to focus on the most important hidden states for extracting relevant information.

- TSSequencer Model

  The TSSequencer model processes input data in non-overlapping patches using a BiLSTM (Bidirectional LSTM) layer and MLP (Multilayer Perceptron) for channel mixing. Unlike Transformers, which explicitly compute dependencies between all tokens, the BiLSTM layer captures temporal dependencies sequentially, making it a lightweight alternative for time-series data.

- PatchTST Model

  PatchTST is a Transformer-based architecture designed for multivariate time-series prediction and representation learning. It separates each time series into independent channels and uses a Transformer encoder to process each channel individually. The key innovation lies in its patching mechanism, which divides each channel into patch sequences of length PPP and stride SSS, reducing computational complexity while preserving long-term temporal correlations. The Multi-Head Attention mechanism learns dependencies between patches, and the final output is refined using BatchNorm and residual concatenation.

Additionally, PatchTST employs a self-supervised learning approach using a Masked Autoencoder, which masks patches instead of individual time steps, enabling deeper learning of temporal patterns. This method enhances the model's ability to generalize by learning robust feature representations from unlabeled data.

4.2. CNN-Based Deep Learning Models

For CNN-based deep learning models, we utilized the Fully Convolutional Network (FCN) [15], ResNet [16], InceptionTime [17], XceptionTime [18], and an Explainable Convolutional Neural Network (XCM) [19]. Each model offers unique approaches for handling time-series data:

- Fully Convolutional Network (FCN)

  FCN is designed for time-series classification using convolutional layers exclusively. Unlike traditional CNNs, FCN removes fully connected layers, passing outputs through global average pooling, significantly reducing the number of model parameters while retaining performance.

- ResNet

  ResNet addresses the vanishing gradient problem in deep neural networks by introducing residual connections. These connections allow the output of one layer to bypass intermediate layers and feed directly into subsequent layers, enabling stable learning even in very deep networks.

- InceptionTime

  Inspired by the Inception module, InceptionTime utilizes bottleneck layers to reduce the dimensionality of time-series data before applying multiple convolutional filters of varying lengths in parallel. This architecture captures both short- and long-term dependencies, with MaxPooling applied in parallel to enhance feature extraction. InceptionTime is an ensemble of five Inception networks trained with different initial weights, improving robustness and accuracy.

- XceptionTime

  Based on InceptionTime and the Xception architecture, XceptionTime incorporates depthwise separable convolutions to improve efficiency. It applies multiple 1D filters to extract features of varying lengths and utilizes three depthwise separable convolution filters to process short- and long-term features in parallel. By separating input channels and recombining them, XceptionTime reduces computational overhead. Additionally, a bottleneck layer is used to decrease the dimensionality of input data, enhancing parametric efficiency.

- Explainable Convolutional Neural Network (XCM)

  XCM is designed for multivariate time-series classification, prioritizing both high prediction performance and interpretability. It employs 2D and 1D convolutional filters in parallel to extract observation-specific and time-dependent features, ensuring comprehensive learning without information loss. Grad-CAM is integrated into XCM to provide clear and precise explanations for prediction results, enhancing its applicability in scenarios where interpretability is critical.

4.3. MultiPatchTST Model

This paper proposes the MultiPatchTST model, a model based on the previously introduced PatchTST architecture, for bearing sensor data anomaly classification. The MultiPatchTST model addresses key limitations of the original PatchTST model, where patch length is arbitrarily set by the user, potentially overlooking optimal configurations. Additionally, the best-performing patch lengths may vary across different datasets, making it labor intensive to experiment with multiple patch lengths manually.

The MultipatchTST model solves this problem by generating multiple PatchTST models with patch lengths from one to the user-set sequence length (seq_len). For example, if seq_len=120, the model generates 120 PatchTST models, each corresponding to a patch size from one to one-hundred-and-twenty. Each model processes the data independently, and the results from each model are passed to the MLP layer to produce the final output value, which is called the result. This is the output of the MultiPatchTST model.

This approach eliminates the need for users to manually specify patch lengths, automatically identifying and applying the optimal patch size for each dataset. Consequently, the MultiPatchTST model enhances performance by adapting to the characteristics of the data without additional experimentation. Figure 1 shows the architecture of the MultiPatchTST model. The bottom left image is an example image of the input time-series data, the top box is the backbone of the PatchTST model, the purple box in the bottom center represents the PatchTST models with sizes up to n, and the dotted box on the right represents the MLP structure that receives the output of each PatchTST model as an input. and Algorithm 1 is the pseudocode for the MultiPatchTST model.

Algorithm 1: Pseudocode for MultiPatchTST model.

Input: DataLoader (dls), Input channels (c_in), Output channels (c_out), Sequence length (seq_len), Hidden dimensions (hidden_dim)
Output: Final model predictions
Steps:
1. Initialize the parent class.
2. Generate valid patch lengths based on the sequence length:
    Set self.patch_lengths to valid patch lengths derived from seq_len.
3. Create multiple PatchTST models with different patch lengths:
    For each patch length in self.patch_lengths:
        Create a PatchTST model with the given parameters.
4. Define an MLP to combine outputs from all PatchTST models:
    Set self.hidden_dim to hidden_dim.
    Initialize self.mlp as:
        A linear layer with input size (len(patch_lengths) * c_out) and output size (hidden_dim).
        A ReLU activation.
        A linear layer with input size (hidden_dim) and output size (c_out).
5. Define the forward function:
    Initialize outputs as an empty list.
    For each model in self.patch_models:
        Append model's prediction on x to outputs.
    Concatenate all outputs into a single tensor.
    Apply self.mlp on concatenated outputs to generate final output.
    Return the final output.
6. Define the get_valid_patch_lengths function:
    Generate valid patch lengths based on the sequence length.
    Return a list of integers from 1 to seq_len (inclusive).

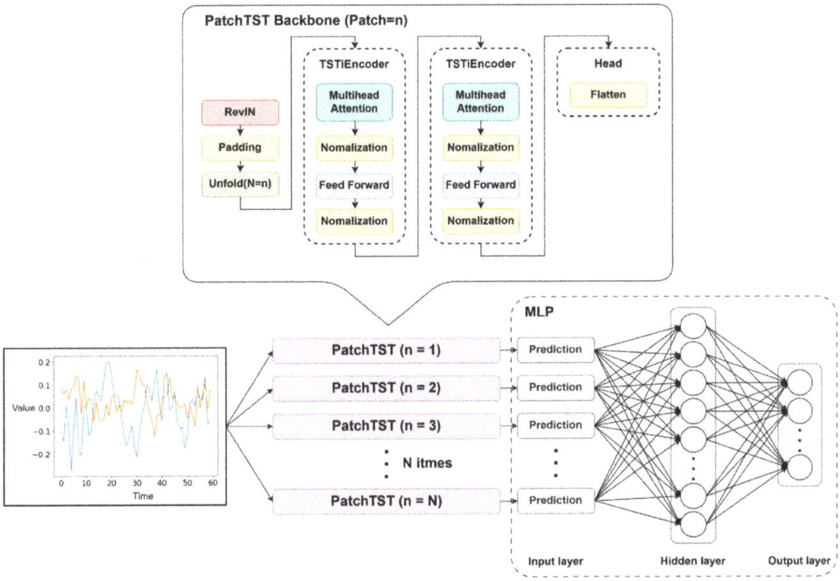

Figure 1. MultiPatchTST model architecture. Top Box: components of the PatchTST backbone. Bottom left image: visualization of the input time-series data. Purple boxes: represent the PatchTST modules configured for various patch sizes. Dashed box: shows the process of combining the predicted outputs from each PatchTST module through the MLP structure.

5. Experiments

5.1. Dataset

The Case Western Reserve University Bearing Data Centre (CWRU) dataset [20] is a widely used standard for bearing sensor data research. It records changes in vibration signals based on the presence, location, and severity of bearing damage. The dataset includes three failure types: inner race failure, ball failure, and outer race failure. Each failure type is artificially induced and measured at defect sizes of 7, 14, 21, and 28 mils. The data, sampled at 12,000 vibrations per second, includes measurements from sensors at the drive end and fan end of the test apparatus. For outer race defects, the data considers defect positions at 3 o'clock, 6 o'clock, and 12 o'clock to account for location-specific impacts. Some data exclusions apply, such as the absence of ball and inner race defect data at 28 mils and limited outer race defect data at 14 mils for specific positions. All measurements are labeled for anomaly classification tasks. Figure 2 is an image of an example bearing element and bearing system from the CWRU data set.

Figure 2. *Cont.*

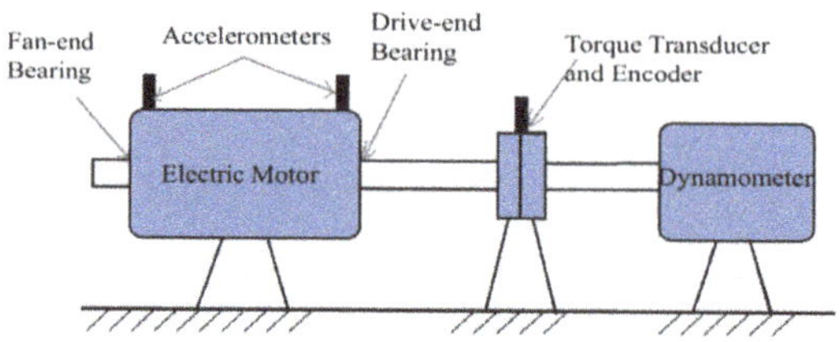

Figure 2. Bearing elements and an experimental setup of the CWRU bearing test rig for a ball bearing system. Reprinted with permission from Ref. [21]. Copyright 2021, IEEE.

The MFPT dataset [22], provided by the Society for Machinery Failure Prevention Technology, was collected using a test rig with NICE bearings (roller diameter: 0.235 inches; pitch diameter: 1.245 inches; number of elements: 8; contact angle: 0). The dataset includes data for tops, inner race defects, and outer race defects. Tops were sampled at a rate of 97,656 samples per second (SPS) for 6 s, while inner and outer race defects were sampled at 48,828 SPS for 3 s. Figure 3 is an image to illustrate the defects in the MFPT dataset.

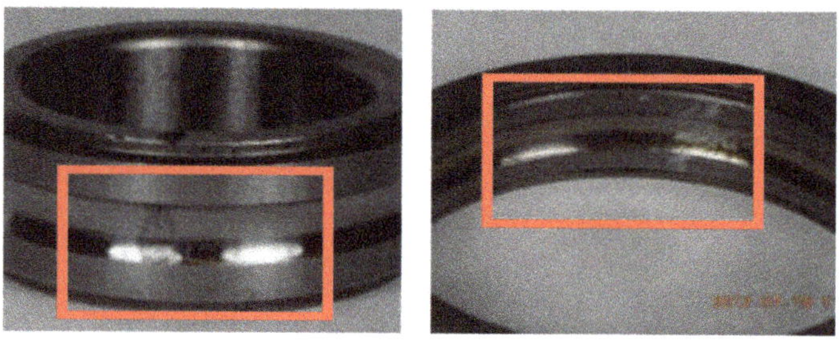

Figure 3. MFPT dataset types of bearing defects. Left image: inner race defects, right image: outer race defects [22,23].

The Paderborn University (PU) dataset [24] contains high-resolution vibration data from experiments on six healthy bearings and 26 damaged bearings. Of the damaged bearings, 12 were artificially damaged, and 14 were subjected to accelerated life tests to simulate real-world damage conditions. Figure 4 is an image that describes each element of the equipment that collected the PU data set.

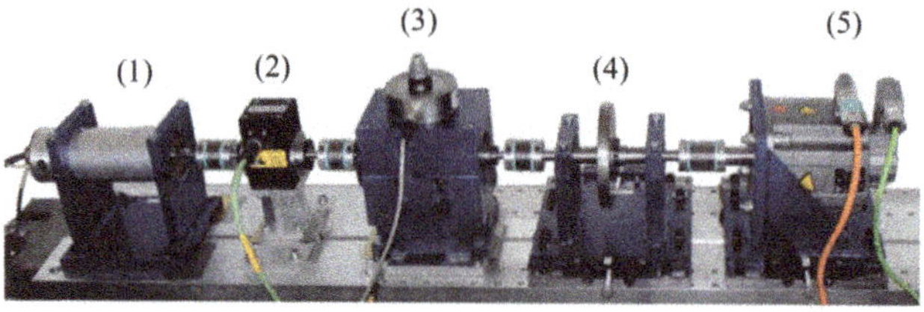

Figure 4. Experimental setup of the PU dataset test rig. (1) Electric motor, (2) torque measuring shaft, (3) rolling bearing test module, (4) flywheel, and (5) load motor Reprinted with permission from Ref. [25]. Copyright 2020, IEEE.

5.2. Experimental Setup

The experiments were conducted using the CWRU, MFPT, and PU datasets introduced earlier. Raw data served as the input for all models, with only MinMax normalization and a sliding window applied as preprocessing techniques. A sliding window of 0.01 s was used for all datasets to enable precise analysis.

For the CWRU dataset, which has a sampling rate of 12,000 measurements per second, the window size was set to 120, and the step size was set to 60. For the MFPT dataset, healthy data was originally sampled at 97,656 samples per second (SPS), while defect data was sampled at 48,828 SPS. To ensure consistency, the healthy data was downsampled to match the defect data's sampling rate. The window size was set to 488, and the step size to 244. For the PU dataset, sampled at 64,000 SPS (64 kHz), the window size was set to 640, and the step size to 320.

The training parameters were set as follows: 10 epochs, a batch size of 2048, and a learning rate of 1×10^{-3}. To evaluate the model's robustness against noise, experiments were conducted using electrical interference noise, harmonic noise, and random shock noise, each applied at four intensity levels. Gaussian noise was also added for comparison purposes. And we ran these tests on Ubuntu 20.04, Intel (R) Core (TM) i9-10900X CPU @ 3.70 GHz, 192 GB of RAM, and one RTX 3090 (Intel, Santa Clara, CA, USA).

5.3. Experimental Results

Experiments were conducted using the CWRU, MFPT, and PU datasets, and the detailed performance metrics for each model under various noise conditions—including noiseless original data, Gaussian noise, electrical interference noise, harmonic noise, and random shock noise—are summarized in Tables 2–4.

Table 2. Experimental results on CWRU dataset with mixed noise.

Noise Type	Model	Accuracy	Precision	Recall	F1
No noise	LSTM	0.993	0.9966	0.9939	0.9952
	GRU	0.992	0.9962	0.9934	0.9948
	FCN	0.993	0.9966	0.994	0.9953
	ResNet	0.9919	0.9958	0.9932	0.9945
	InceptionTime	0.9923	0.9971	0.9932	0.9952
	XceptionTime	0.9939	0.9977	0.9951	0.9964
	XCM	0.9928	0.9967	0.9939	0.9953
	LSTMAttention	0.9943	0.9975	0.9948	0.9962
	GRUAttention	0.9943	0.9971	0.9948	0.996
	TSSequencer	**0.9951**	**0.9975**	**0.9961**	**0.9968**
	PatchTST	0.9932	0.9971	0.994	0.9956
	MultiPatchTST	0.9942	0.9973	0.9954	0.9963
Gaussian noise	LSTM	0.9809	0.9857	0.9842	0.9846
	GRU	0.9801	0.9856	0.9846	0.9848
	FCN	0.9786	0.9848	0.9828	0.9832
	ResNet	0.9786	0.9826	0.9833	0.9826
	InceptionTime	0.9767	0.9819	0.9828	0.9821
	XceptionTime	0.9782	0.9826	0.9842	0.9828
	XCM	0.9797	0.9832	0.9854	0.9841
	LSTMAttention	0.9787	0.9827	0.9833	0.9826
	GRUAttention	0.981	0.9854	0.9867	0.9858
	TSSequencer	0.9768	0.9839	0.9819	0.9826
	PatchTST	0.9795	0.9847	0.9834	0.984
	MultiPatchTST	**0.9832**	**0.9863**	**0.9889**	**0.9875**

Table 2. *Cont.*

Noise Type	Model	Accuracy	Precision	Recall	F1
Electrical interference noise	LSTM	0.9884	0.9914	0.9899	0.9907
	GRU	0.988	0.992	0.9891	0.9905
	FCN	0.9878	0.9928	0.989	0.9909
	ResNet	0.9883	0.9925	0.9894	0.9909
	InceptionTime	0.989	0.993	0.9905	0.9918
	XceptionTime	0.9878	0.9926	0.9896	0.9911
	XCM	0.9882	0.993	0.9897	0.9913
	LSTMAttention	0.9864	0.991	0.9879	0.9894
	GRUAttention	0.9894	0.993	0.9901	0.9916
	TSSequencer	0.9866	0.9918	0.988	0.9899
	PatchTST	0.9874	0.9923	0.9885	0.9904
	MultiPatchTST	**0.9907**	**0.993**	**0.9918**	**0.9927**
Harmonic noise	LSTM	0.984	0.9903	0.9858	0.988
	GRU	0.9838	0.9912	0.9849	0.988
	FCN	0.9831	0.9906	0.9847	0.9877
	ResNet	0.9845	0.9909	0.9864	0.9886
	InceptionTime	0.9818	0.989	0.9839	0.9864
	XceptionTime	0.9829	0.9894	0.9848	0.987
	XCM	0.9865	0.9914	0.9878	0.9898
	LSTMAttention	0.9842	0.9917	0.9857	0.9887
	GRUAttention	0.9817	0.9912	0.9838	0.9875
	TSSequencer	0.9818	0.9894	0.9835	0.9864
	PatchTST	0.9842	0.9918	0.9859	0.9888
	MultiPatchTST	**0.9868**	**0.992**	**0.9883**	**0.99**
Random shock noise	LSTM	0.9545	0.968	0.9598	0.9638
	GRU	0.9534	0.9668	0.9584	0.9625
	FCN	0.9521	0.9667	0.9581	0.9623
	ResNet	0.9534	0.9667	0.9597	0.9631
	InceptionTime	0.9526	0.9661	0.9586	0.9623
	XceptionTime	0.9536	0.9676	0.9592	0.9634
	XCM	0.9541	0.968	0.9583	0.9631
	LSTMAttention	0.9479	0.962	0.9553	0.9586
	GRUAttention	0.9546	0.9685	0.9601	0.9643
	TSSequencer	0.9522	0.9653	0.9582	0.9617
	PatchTST	0.9538	0.9677	0.9602	0.9639
	MultiPatchTST	**0.9549**	**0.9685**	**0.9602**	**0.9644**

Table 3. Experimental results on MFPT dataset with mixed noise.

Noise Type	Model	Accuracy	Precision	Recall	F1
No noise	LSTM	0.9961	0.9965	0.9961	0.9963
	GRU	0.9961	0.9961	0.9965	0.9963
	FCN	0.9954	0.9961	0.9954	0.9958
	ResNet	0.9954	0.9958	0.9961	0.996
	InceptionTime	0.9961	0.9968	0.9961	0.9965
	XceptionTime	0.9958	0.9958	0.9965	0.9961
	XCM	0.9961	0.9961	0.9965	0.9963
	LSTMAttention	0.9965	0.9968	0.9968	0.9968
	GRUAttention	0.9961	0.9965	0.9972	0.9968
	TSSequencer	0.9961	0.9961	0.9965	0.9963
	PatchTST	0.9965	0.9972	0.9965	0.9968
	MultiPatchTST	**0.9972**	**0.9972**	**0.9975**	**0.9973**
Gaussian noise	LSTM	0.9789	0.9816	0.9821	0.9815
	GRU	0.9797	0.9821	0.9826	0.9821
	FCN	0.9834	0.985	0.9855	0.985
	ResNet	0.9829	0.9848	0.9845	0.9845
	InceptionTime	0.9802	0.9823	0.9823	0.9821
	XceptionTime	0.9818	0.9832	0.9842	0.9834
	XCM	0.9767	0.979	0.9789	0.9785
	LSTMAttention	0.9783	0.9811	0.981	0.9806
	GRUAttention	0.9834	0.9847	0.9847	0.9845
	TSSequencer	0.9799	0.9807	0.9834	0.9819
	PatchTST	0.9831	0.9844	**0.9869**	0.9854
	MultiPatchTST	**0.9837**	**0.986**	0.985	**0.9854**
Electrical interference noise	LSTM	0.9649	0.9658	0.9656	0.9657
	GRU	0.9691	0.9706	0.9698	0.9702
	FCN	0.967	0.9682	0.9674	0.9678
	ResNet	0.964	0.9651	0.9649	0.965
	InceptionTime	0.9627	0.9629	0.9631	0.963
	XceptionTime	0.9667	0.9678	0.9671	0.9674
	XCM	0.9618	0.963	0.9621	0.9625
	LSTMAttention	0.9627	0.9634	0.9635	0.9634
	GRUAttention	0.9615	0.9614	0.9619	0.9616
	TSSequencer	0.9672	0.9681	0.9676	0.9679
	PatchTST	0.9663	0.967	0.9675	0.9672
	MultiPatchTST	**0.97**	**0.97**	**0.9704**	**0.9702**

Table 3. *Cont.*

Noise Type	Model	Accuracy	Precision	Recall	F1
Harmonic noise	LSTM	0.9857	0.9867	0.986	0.9863
	GRU	0.9877	0.988	0.9881	0.9881
	FCN	0.9857	0.9861	0.9863	0.9862
	ResNet	0.985	0.9854	0.9857	0.9855
	InceptionTime	0.9858	0.9864	0.9866	0.9865
	XceptionTime	0.9853	0.986	0.9853	0.9857
	XCM	0.9872	0.9876	0.9877	0.9876
	LSTMAttention	0.9859	0.9864	0.9861	0.9862
	GRUAttention	0.9846	0.9852	0.9849	0.9851
	TSSequencer	0.986	0.987	0.9867	0.9868
	PatchTST	0.9862	0.9872	0.9869	0.987
	MultiPatchTST	**0.9881**	**0.9898**	**0.9883**	**0.989**
Random shock noise	LSTM	0.9717	0.9738	0.9723	0.9723
	GRU	0.9681	0.9695	0.969	0.9687
	FCN	0.9691	0.9711	0.9694	0.9696
	ResNet	0.9758	0.9768	0.9764	0.9763
	InceptionTime	0.9727	0.9739	0.9733	0.9731
	XceptionTime	0.9782	0.9791	0.9784	0.9784
	XCM	0.978	0.9789	0.9782	0.9783
	LSTMAttention	0.9801	0.98	0.9802	0.9802
	GRUAttention	0.9767	0.9778	0.9769	0.9769
	TSSequencer	0.9784	0.9794	0.9788	0.9788
	PatchTST	0.9814	0.982	0.9819	0.9817
	MultiPatchTST	**0.9865**	**0.987**	**0.9869**	**0.9869**

Table 4. Experimental results on PU dataset with mixed noise.

Noise Type	Model	Accuracy	Precision	Recall	F1
No noise	LSTM	0.9844	0.9854	0.9851	0.9853
	GRU	0.9865	0.987	0.9869	0.986
	FCN	0.9854	0.9868	0.9862	0.9865
	ResNet	0.9861	0.9871	0.9868	0.9869
	InceptionTime	0.9842	0.9853	0.9847	0.9849
	XceptionTime	0.9849	0.9855	0.9859	0.9857
	XCM	0.9855	0.9864	0.9864	0.9864
	LSTMAttention	0.9855	0.9863	0.9864	0.9864
	GRUAttention	0.9852	0.9867	0.986	0.9863
	TSSequencer	0.9859	0.9867	0.987	0.9868
	PatchTST	0.9868	0.9873	0.9874	0.9874
	MultiPatchTST	**0.9881**	**0.9887**	**0.9887**	**0.9887**

Table 4. *Cont.*

Noise Type	Model	Accuracy	Precision	Recall	F1
Gaussian noise	LSTM	0.9826	0.9842	0.9839	0.984
	GRU	0.9841	0.9852	0.9853	0.9854
	FCN	0.9836	0.9853	0.9845	0.9848
	ResNet	0.9826	0.9841	0.9836	0.9838
	InceptionTime	0.984	0.9849	0.9851	0.9849
	XceptionTime	0.9819	0.9835	0.9834	0.9834
	XCM	0.9836	0.9854	0.9845	0.9849
	LSTMAttention	0.984	0.9854	0.9852	0.9852
	GRUAttention	0.9827	0.9841	0.9835	0.9837
	TSSequencer	0.9825	0.9841	0.9835	0.9837
	PatchTST	0.9838	0.9853	0.9847	0.9849
	MultiPatchTST	**0.9845**	**0.9863**	**0.9856**	**0.9857**
Electrical interference noise	LSTM	0.7596	0.8292	0.7672	0.7695
	GRU	0.7607	0.8292	0.7678	0.7704
	FCN	0.7587	0.83	0.7659	0.7687
	ResNet	0.7628	0.8317	0.7688	0.7718
	InceptionTime	0.7611	0.831	0.7685	0.7711
	XceptionTime	0.7619	0.8316	**0.7703**	0.7724
	XCM	0.7595	0.8273	0.7674	0.7694
	LSTMAttention	0.7529	0.8268	0.7609	0.7622
	GRUAttention	0.7611	0.8316	0.7673	0.7701
	TSSequencer	0.7581	0.8289	0.7652	0.7686
	PatchTST	0.7613	0.8303	0.7691	0.7715
	MultiPatchTST	**0.7631**	**0.8327**	0.7697	**0.7734**
Harmonic noise	LSTM	0.8084	0.8576	0.8141	0.8192
	GRU	0.8056	0.8559	0.8102	0.815
	FCN	0.806	0.8556	0.8118	0.8153
	ResNet	0.804	0.8544	0.8106	0.8142
	InceptionTime	0.8001	0.8539	0.8061	0.8112
	XceptionTime	0.8058	0.8587	0.8119	0.8176
	XCM	0.8044	0.8544	0.8111	0.8156
	LSTMAttention	0.81	0.857	0.8153	0.8216
	GRUAttention	0.8057	0.8562	0.8109	0.8155
	TSSequencer	0.8082	0.8587	0.8147	0.8197
	PatchTST	0.8086	0.8612	0.8138	0.8199
	MultiPatchTST	**0.8112**	**0.8613**	**0.8169**	**0.8216**

Table 4. *Cont.*

Noise Type	Model	Accuracy	Precision	Recall	F1
Random shock noise	LSTM	0.9114	0.9203	0.9158	0.918
	GRU	0.9163	0.9245	0.9221	0.9232
	FCN	0.9164	0.9242	0.9203	0.9222
	ResNet	0.9108	0.9197	0.9152	0.9174
	InceptionTime	0.9131	0.9229	0.9188	0.9208
	XceptionTime	0.9145	0.9227	0.9189	0.9207
	XCM	0.915	0.9231	0.9198	0.9214
	LSTMAttention	0.9145	0.9231	0.9191	0.9211
	GRUAttention	0.9149	0.9216	0.9186	0.9201
	TSSequencer	0.9128	0.9209	0.9166	0.9188
	PatchTST	0.9176	0.9263	0.9215	0.9239
	MultiPatchTST	**0.9204**	**0.9283**	**0.9249**	**0.9261**

The experiments on the original noiseless datasets demonstrated that all models achieved high performance on the CWRU and MFPT datasets, with accuracy, precision, recall, and F1 metrics exceeding 0.98. These results indicate no significant performance differences between the models under noiseless conditions. Notably, the TSSequencer model showed the best performance on the CWRU dataset, whereas the MultiPatchTST model achieved the highest scores on both the MFPT and PU datasets.

When a Gaussian noise of amplitude 0.5 was applied, the performance degradation across all datasets was minimal. The top-performing models achieved metrics above 97%, indicating that the impact of Gaussian noise on classification performance was negligible. However, the minimal effect of Gaussian noise highlights its limitations as a method for evaluating the noise robustness of models. Among the models, PatchTST and MultiPatchTST, which utilize patch-based learning, consistently exhibited superior performance, reinforcing their robustness.

To evaluate the validity of the models under more realistic industrial noise conditions, the original datasets were combined with noise reflecting electrical interference, harmonic, and random shock noise at four intensity levels. The CWRU dataset consistently maintained strong performance across all noise conditions. Similarly, the MFPT dataset demonstrated high performance, sustaining metrics above 95% in most scenarios. However, the PU dataset exhibited higher sensitivity to noise. Under electrical interference noise, the PU dataset's performance dropped below 80%, and although it improved slightly under harmonic noise, it still remained below 85%. In contrast, random shock noise had a different impact, with the PU dataset achieving over 92% performance when processed with the MultiPatchTST model, demonstrating its robustness to aperiodic noise.

Across all datasets and experimental conditions, the MultiPatchTST model consistently achieved the highest scores on accuracy, precision, recall, and F1 metrics. These findings emphasize that the inclusion of raw and noisy data enhances robustness against random shock noise compared to periodic noise. Moreover, the high sensitivity of the PU dataset to noise underscores its vulnerability in mixed-noise environments.

The results suggest that aperiodic noise, such as random shock noise, can be effectively handled when mixed with original data, while periodic noise poses greater challenges. This experiment reaffirms the efficacy of patch-based learning, as implemented in the MultiPatchTST model, establishing it as a reliable methodology for bearing fault classification in noisy environments. The detailed experiments and results are shown in Appendix A.

6. Conclusions

This paper addresses the limitations of Gaussian noise and white noise, which have been predominantly used in bearing sensor-based fault classification research, highlighting their inadequacy for representing real-world industrial noise. To overcome these limitations, three novel noise techniques—electrical interference noise, harmonic noise, and random shock noise—were proposed, reflecting noise characteristics observed in industrial environments. Additionally, the study compared the performance of RNN-based and CNN-based deep learning models, and introduced a new deep learning model, MultiPatchTST, for bearing sensor data anomaly classification.

The experimental results demonstrate that all three proposed noise types significantly impact model performance. Among these, random shock noise causes the greatest performance degradation when applied independently. However, when mixed with original data, models perform best under random shock noise, debunking the notion that periodic noise inherently leads to overfitting due to the model learning its periodicity.

The proposed MultiPatchTST model outperforms all other models across various noise environments, confirming its robustness to noise. The models used in this study can be categorized into three groups: sequential models (LSTM, GRU, LSTMAttention, GRUAttention), CNN-based models (e.g., FCN, ResNet), and patch-based models (PatchTST, MultiPatchTST). The MultiPatchTST model consistently delivered superior performance, while LSTMAttention and GRUAttention performed the worst in some experiments, indicating that attention mechanisms alone are insufficient for optimal performance. Instead, patch-based learning emerged as a key factor for effective bearing fault classification in vibration sensor data.

The main strength of the MultiPatchTST model is its ability to capture the temporal and frequency patterns of data in detail through a learning strategy that utilizes multiple patch sizes, which is a fundamental difference from traditional models that rely on a single patch size, enabling it to reliably learn the key characteristics of data, especially in complex and atypical industrial noise environments. By using a combination of different patch sizes to learn the multi-layered nature of the data, it effectively overcomes the noise sensitivity issues of traditional RNN and CNN-based models.

In addition, the MultiPatchTST model demonstrated high information extraction capability even in noisy data through the combination of a self-attention mechanism and patch-based learning, which contributed to its excellent performance even in complex noise environments such as electrical interference noise, harmonic noise, and random impact noise. The experimental results showed that the MultiPatchTST model outperformed all other comparison models, with the largest performance advantage in the random impulse noise environment. The model's multi-patch ensemble learning approach has a practical advantage in that it can automatically learn the optimal patch size based on the characteristics of the data without requiring the user to manually set the patch size.

The findings of this study provide valuable insights for developing robust fault classification models and are expected to contribute significantly to industrial field systems, improving their efficiency and reliability. In future research, more diverse data sets will be used to verify and improve the performance of the model in different environments. In addition, the noise technique proposed in this paper will be tested using datasets of other mechanical elements as noise that can affect not only bearings but also other mechanical equipment elements, so that it can be developed into a general-purpose model that shows good performance in various mechanical equipment rather than a bearing-specific model.

Author Contributions: Conceptualization, S.L.; data curation, S.L.; formal analysis, S.K.; funding acquisition, S.L.; investigation, S.L.; methodology, S.L.; project administration, S.L.; resources, S.K. and S.L.; software, S.K.; supervision, S.L.; validation, S.K. and S.L.; visualization, S.K.; writing—original draft, S.K. and S.L.; writing—review and editing, S.L. All authors have read and agreed to the published version of the manuscript.

Funding: This paper was supported by Semyung University's University Innovation Support Project in 2023.

Institutional Review Board Statement: Not applicable.

Informed Consent Statement: Not applicable.

Data Availability Statement: The datasets used in this study are openly available as follows: The CWRU Dataset can be accessed at the Case Western Reserve University Bearing Data Center website [19] (http://csegroups.case.edu/bearingdatacenter, accessed on 25 November 2024). The MFPT Dataset is available through the Condition-Based Maintenance Fault Database for Testing of Diagnostic and Prognostics Algorithms [20] (https://www.mfpt.org/fault-data-sets/, accessed on 25 November 2024). The PU Dataset is provided by the Paderborn University Bearing Data Center website [21] (https://mb.uni-paderborn.de/kat/forschung/kat-datacenter/bearing-datacenter/data-sets-and-download, accessed on 25 November 2024).

Conflicts of Interest: The authors declare no conflicts of interest.

Appendix A

Appendix A.1 Experimental Results After Applying Electrical Interference Noise

Electrical interference noise was configured at four levels (Levels 1, 2, 3, and 4) with a fixed frequency of 50 and amplitudes of 0.03, 0.1, 0.2, and 0.5, respectively. The experimental results indicate that the CWRU and MFPT datasets maintained high performance (over 90%), while the PU dataset achieved over 80% accuracy. Across all three datasets, performance degradation increased with higher noise intensity, but the differences between RNN-based and CNN-based approaches were not significant.

The GRUAttention model performed the worst on the CWRU dataset, the LSTMAttention model on the MFPT dataset, and the XceptionTime model on the PU dataset. While the TSSequencer model narrowly achieved the highest Level 4 precision metric on the MFPT dataset, the MultiPatchTST model consistently outperformed other models on all remaining metrics for the MFPT dataset and on all metrics for the CWRU and PU datasets.

These results suggest that patch-based learning and self-attention structures, as implemented in the MultiPatchTST model, are robust to electrical interference noise. These structures enable the retention of critical information even in the presence of significant noise, whereas other models without these features show more pronounced performance degradation. Detailed performance metrics for each model are presented in Tables A1–A3, with noise levels converted to Signal-to-Noise Ratio (SNR).

Table A1. Experimental results on CWRU dataset with electrical interference noise.

Model	SNR	Accuracy	Precision	Recall	F1
LSTM	27 dB	0.98	0.9872	0.9819	0.9845
	17 dB	0.9666	0.9828	0.9702	0.9764
	11 dB	0.9313	0.9672	0.938	0.9518
	3 dB	0.8757	0.9503	0.8883	0.9171
GRU	27 dB	0.981	0.9877	0.9835	0.9855
	17 dB	0.9715	0.9866	0.9739	0.9806
	11 dB	0.9337	0.9668	0.9405	0.9532
	3 dB	0.8788	0.9498	0.8919	0.9186

Table A1. *Cont.*

Model	SNR	Accuracy	Precision	Recall	F1
FCN	27 dB	0.9816	0.9876	0.984	0.9858
	17 dB	0.9735	0.9854	0.9775	0.9813
	11 dB	0.9469	0.974	0.9509	0.962
	3 dB	0.8782	0.9474	0.8861	0.9145
ResNet	27 dB	0.9816	0.9876	0.9849	0.9862
	17 dB	0.9704	0.9851	0.9739	0.9793
	11 dB	0.9312	0.97	0.9373	0.953
	3 dB	0.8712	0.9475	0.8837	0.9136
InceptionTime	27 dB	0.9786	0.9879	0.981	0.9843
	17 dB	0.9691	0.9847	0.9724	0.9784
	11 dB	0.934	0.9673	0.9399	0.953
	3 dB	0.9012	0.9574	0.9113	0.9334
XceptionTime	27 dB	0.9777	0.9842	0.9819	0.9829
	17 dB	0.9674	0.9841	0.9709	0.9773
	11 dB	0.931	0.9664	0.9361	0.9508
	3 dB	0.8859	0.9527	0.8965	0.923
XCM	27 dB	0.9811	0.9864	0.9846	0.9855
	17 dB	0.9675	0.983	0.971	0.9769
	11 dB	0.9367	0.9716	0.9432	0.9568
	3 dB	0.8799	0.9491	0.8929	0.9192
LSTMAttention	27 dB	0.9802	0.987	0.9822	0.9845
	17 dB	0.9733	0.9833	0.97	0.9811
	11 dB	0.9394	0.9721	0.9463	0.9588
	3 dB	0.8889	0.9508	0.903	0.9256
GRUAttention	27 dB	0.9816	0.9868	0.9837	0.9852
	17 dB	0.9699	0,9842	0.9737	0.9788
	11 dB	0.9416	0.9728	0.9454	0.9586
	3 dB	0.8689	0.9511	0.8779	0.9115
TSSequencer	27 dB	0.98	0.9868	0.9843	0.9855
	17 dB	0.9664	0.9818	0.971	0.9763
	11 dB	0.9424	0.9701	0.9487	0.959
	3 dB	0.8918	0.9531	0.8994	0.9247
PatchTST	27 dB	0.9805	0.9888	0.982	0.9852
	17 dB	0.9724	0.9862	0.9758	0.9809
	11 dB	0.943	0.9743	0.9495	0.9615
	3 dB	0.8959	0.9553	0.9043	0.9283
MultiPatchTST	27 dB	0.9852	0.9906	0.9867	0.9886
	17 dB	0.9735	0.9866	0.9772	0.9819
	11 dB	0.9478	0.9762	0.9527	0.9615
	3 dB	0.9021	0.9582	0.9114	0.9337

Table A2. Experimental results on MFPT dataset with electrical interference noise.

Model	SNR	Accuracy	Precision	Recall	F1
LSTM	27 dB	0.993	0.9932	0.9943	0.9937
	17 dB	0.977	0.9794	0.9774	0.9782
	11 dB	0.9496	0.953	0.951	0.9519
	3 dB	0.9315	0.9375	0.9354	0.9362
GRU	27 dB	0.9903	0.9912	0.9914	0.9911
	17 dB	0.9753	0.9765	0.9774	0.9769
	11 dB	0.9541	0.9563	0.9559	0.9561
	3 dB	0.925	0.9315	0.9319	0.9317
FCN	27 dB	0.9887	0.9897	0.9893	0.9893
	17 dB	0.9767	0.9777	0.977	0.9774
	11 dB	0.9527	0.9546	0.9552	0.9549
	3 dB	0.9288	0.9354	0.934	0.9346
ResNet	27 dB	0.9922	0.9926	0.9933	0.9929
	17 dB	0.9802	0.9804	0.9809	0.9805
	11 dB	0.9444	0.9472	0.9472	0.9472
	3 dB	0.9302	0.9339	0.9357	0.9348
InceptionTime	27 dB	0.9903	0.9914	0.9909	0.991
	17 dB	0.9746	0.9759	0.9746	0.9753
	11 dB	0.9468	0.9495	0.9482	0.9487
	3 dB	0.9288	0.9346	0.9354	0.9349
XceptionTime	27 dB	0.9922	0.9924	0.9946	0.9934
	17 dB	0.9725	0.9737	0.9743	0.9737
	11 dB	0.9447	0.9497	0.9472	0.9484
	3 dB	0.9309	0.9398	0.9347	0.9372
XCM	27 dB	0.9909	0.9919	0.9917	0.9917
	17 dB	0.9791	0.9798	0.9805	0.9802
	11 dB	0.9486	0.9517	0.9506	0.9512
	3 dB	0.9343	0.9403	0.9378	0.9389
LSTMAttention	27 dB	0.9911	0.9926	0.9914	0.9919
	17 dB	0.9802	0.9805	0.9812	0.9809
	11 dB	0.9493	0.9518	0.9524	0.9521
	3 dB	0.9211	0.9282	0.9295	0.9288
GRUAttention	27 dB	0.9938	0.9939	0.9943	0.9941
	17 dB	0.9829	0.983	0.984	0.9835
	11 dB	0.9489	0.9515	0.951	0.9512
	3 dB	0.9319	0.9376	0.9364	0.937
TSSequencer	27 dB	0.9914	0.9917	0.9922	0.9918
	17 dB	0.9822	0.983	0.9829	0.9829
	11 dB	0.9468	0.9501	0.9489	0.9494
	3 dB	0.9309	0.9414	0.9333	0.9373

Table A2. *Cont.*

Model	SNR	Accuracy	Precision	Recall	F1
PatchTST	27 dB	0.9927	0.9942	0.9927	0.9934
	17 dB	0.9847	0.9852	0.985	0.985
	11 dB	0.9489	0.9539	0.9503	0.9521
	3 dB	0.9326	0.9382	0.9378	0.9379
MultiPatchTST	27 dB	0.9949	0.9952	0.9958	0.9955
	17 dB	0.9864	0.9864	0.9864	0.9864
	11 dB	0.9524	0.9549	0.9541	0.9545
	3 dB	0.9368	0.941	0.9416	0.9413

Table A3. Experimental results on PU dataset with electrical interference noise.

Model	SNR	Accuracy	Precision	Recall	F1
LSTM	27 dB	0.9253	0.9324	0.9281	0.9302
	17 dB	0.8096	0.8357	0.8182	0.8269
	11 dB	0.7915	0.8231	0.7992	0.8106
	3 dB	0.7857	0.8198	0.796	0.8076
GRU	27 dB	0.9244	0.9303	0.9279	0.9291
	17 dB	0.8146	0.837	0.8244	0.8306
	11 dB	0.7898	0.8251	0.7994	0.8114
	3 dB	0.7905	0.8209	0.7992	0.8098
FCN	27 dB	0.9267	0.9329	0.9292	0.9311
	17 dB	0.808	0.8371	0.817	0.8269
	11 dB	0.7967	0.8267	0.8061	0.8159
	3 dB	0.7909	0.8197	0.8006	0.8099
ResNet	27 dB	0.9263	0.9318	0.9289	0.9303
	17 dB	0.8125	0.8371	0.821	0.8289
	11 dB	0.791	0.8241	0.8002	0.8117
	3 dB	0.7948	0.8239	0.8037	0.8136
InceptionTime	27 dB	0.9241	0.9307	0.927	0.9288
	17 dB	0.8154	0.8364	0.8238	0.8299
	11 dB	0.7923	0.8227	0.8016	0.8116
	3 dB	0.7954	0.8234	0.8056	0.8144
XceptionTime	27 dB	0.9258	0.9338	0.9294	0.9315
	17 dB	0.8131	0.8373	0.8217	0.8293
	11 dB	0.7901	0.8227	0.7978	0.8098
	3 dB	0.7861	0.8148	0.7956	0.8049
XCM	27 dB	0.9255	0.9325	0.9271	0.9298
	17 dB	0.8157	0.838	0.8242	0.831
	11 dB	0.7928	0.8265	0.8024	0.8138
	3 dB	0.7893	0.8192	0.7992	0.809

Table A3. *Cont.*

Model	SNR	Accuracy	Precision	Recall	F1
LSTMAttention	27 dB	0.9258	0.9319	0.9285	0.9302
	17 dB	0.8135	0.8371	0.8216	0.8292
	11 dB	0.7929	0.8251	0.8016	0.8128
	3 dB	0.7923	0.8208	0.8003	0.8102
GRUAttention	27 dB	0.9269	0.9338	0.93	0.9319
	17 dB	0.8062	0.8334	0.8161	0.8246
	11 dB	0.7955	0.828	0.8052	0.8162
	3 dB	0.7974	0.8241	0.8059	0.8148
TSSequencer	27 dB	0.9267	0.9327	0.9292	0.9309
	17 dB	0.8112	0.8354	0.8211	0.8282
	11 dB	0.7931	0.8259	0.8049	0.8149
	3 dB	0.7881	0.8192	0.7968	0.8078
PatchTST	27 dB	0.9286	0.9339	0.9319	0.9329
	17 dB	0.8164	0.8376	0.8243	0.8309
	11 dB	0.7981	0.8327	0.8069	0.8192
	3 dB	0.7985	0.827	0.8066	0.8166
MultiPatchTST	27 dB	0.9328	0.9385	0.9351	0.9368
	17 dB	0.8198	0.8457	0.8275	0.8365
	11 dB	0.8066	0.8376	0.8148	0.8256
	3 dB	0.8	0.8275	0.81	0.8182

Appendix A.2 Experimental Results After Applying Harmonic Noise

Experiments with harmonic noise were conducted across four levels (Levels 1–4) with varying parameters: Level 1 (amplitude 0.03, fundamental frequency 20, harmonic set [2,3]), Level 2 (amplitude 0.1, fundamental frequency 40, harmonic set [2–4]), Level 3 (amplitude 0.2, fundamental frequency 40, harmonic set [2–4]), and Level 4 (amplitude 0.5, fundamental frequency 60, harmonic set [2–5]). The experimental results, as illustrated in Figures A1 and A2, and Table A4, show a significant decrease in model performance as noise intensity increases.

In the MFPT and PU datasets, there was no substantial difference in performance among the models. However, in the CWRU dataset, the TSSequencer model performed the worst, with performance degradation becoming more pronounced as noise intensity increased. Additionally, the FCN model showed the poorest performance for the MFPT and PU datasets. At Level 3 on the CWRU dataset, the XCM model achieved the best performance, while the MultiPatchTST model consistently outperformed others at Levels 1, 2, and 4 across all metrics for all three datasets.

Sequential models such as LSTM and GRU were significantly affected by harmonic noise, particularly at Levels 3 and 4, showing considerable performance degradation. While GRUAttention and LSTMAttention, enhanced with self-attention mechanisms, performed better than standard sequential models, they remained more sensitive to noise compared to Transformer-based models. The PatchTST and MultiPatchTST models, leveraging patch-based learning and Transformer architectures, demonstrated superior accuracy and stability, particularly at higher noise levels (Levels 3 and 4).

In Table A4, the noise level is expressed as a signal-to-noise ratio (SNR).

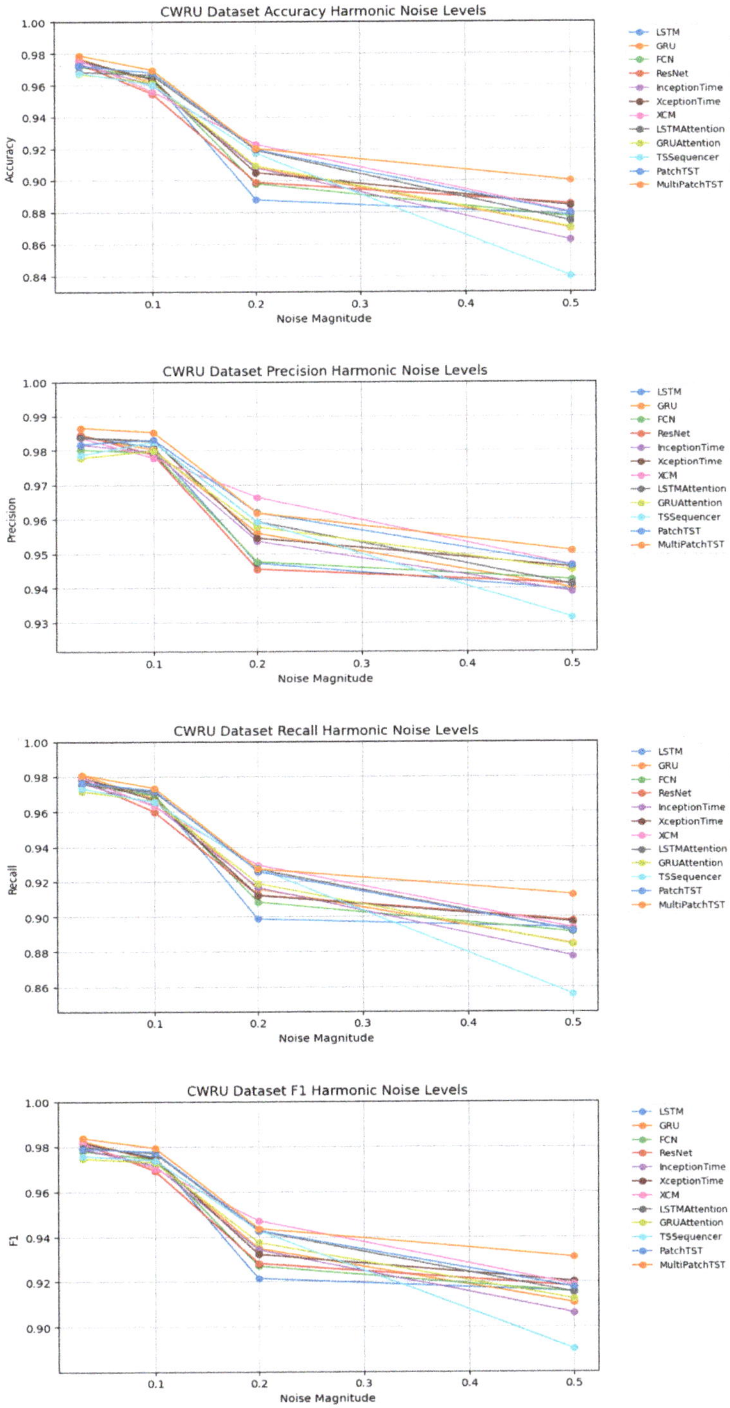

Figure A1. Experimental results by harmonic noise level for the CWRU dataset.

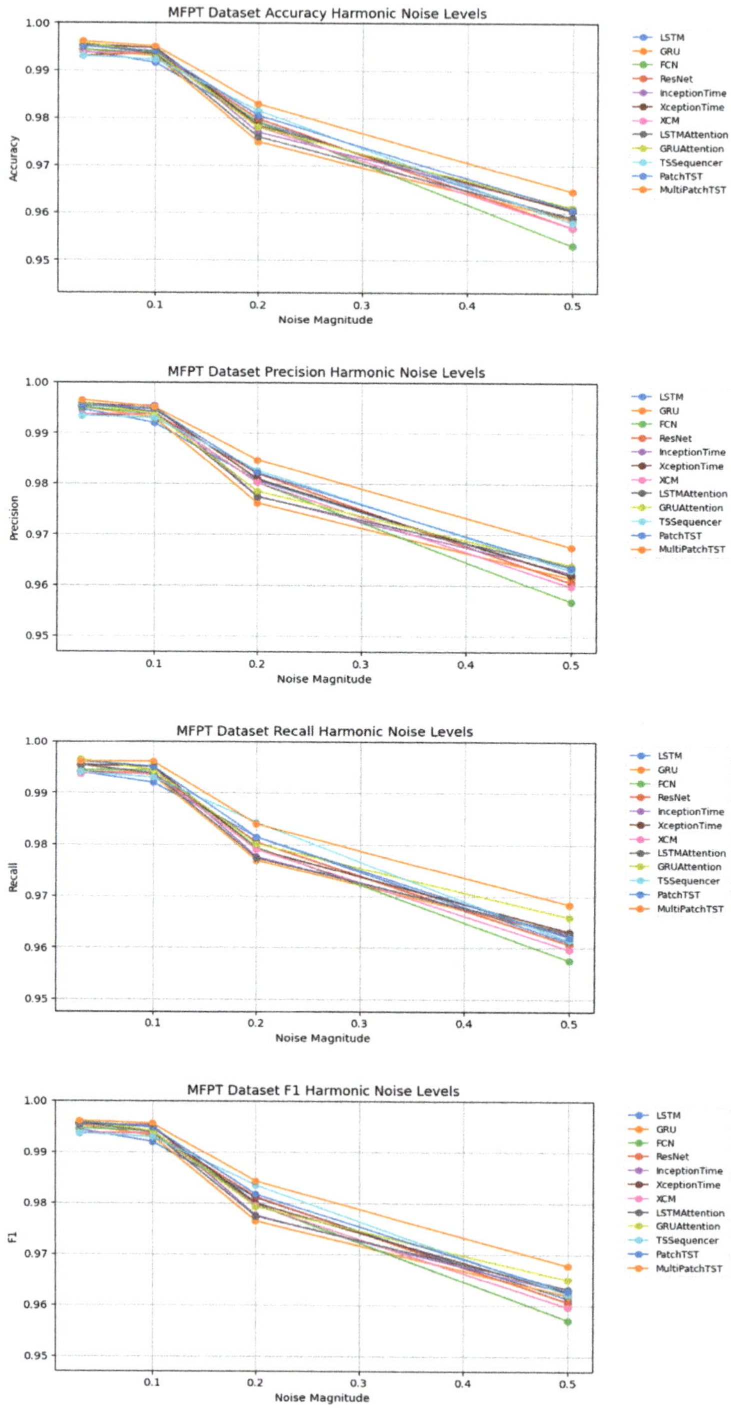

Figure A2. Experimental results by harmonic noise level for the MFPT dataset.

Table A4. Experimental results on PU dataset with harmonic noise.

Model	SNR	Accuracy	Precision	Recall	F1
LSTM	27 dB	0.959	0.9609	0.9604	0.9607
	17 dB	0.8629	0.8768	0.8691	0.8727
	11 dB	0.8424	0.8571	0.849	0.853
	3 dB	0.786	0.8191	0.7972	0.8079
GRU	27 dB	0.9579	0.9608	0.9597	0.9602
	17 dB	0.869	0.8819	0.8743	0.878
	11 dB	0.8284	0.843	0.836	0.8394
	3 dB	0.787	0.82	0.796	0.8078
FCN	27 dB	0.9581	0.961	0.9596	0.9603
	17 dB	0.871	0.8809	0.875	0.8778
	11 dB	0.8374	0.8526	0.8452	0.8486
	3 dB	0.7835	0.8161	0.7928	0.8042
ResNet	27 dB	0.9548	0.9586	0.9562	0.9574
	17 dB	0.8693	0.8817	0.8744	0.878
	11 dB	0.8342	0.8519	0.8405	0.8458
	3 dB	0.7922	0.8238	0.8011	0.8123
InceptionTime	27 dB	0.9547	0.9583	0.9571	0.9577
	17 dB	0.8762	0.8855	0.8801	0.8827
	11 dB	0.8317	0.8512	0.8394	0.8452
	3 dB	0.7855	0.8191	0.7961	0.8073
XceptionTime	27 dB	0.9566	0.9595	0.958	0.9587
	17 dB	0.8682	0.8819	0.8725	0.877
	11 dB	0.8336	0.8487	0.8407	0.8447
	3 dB	0.7853	0.8185	0.795	0.8067
XCM	27 dB	0.9575	0.9601	0.959	0.9595
	17 dB	0.8704	0.8817	0.8768	0.8791
	11 dB	0.8344	0.8492	0.8417	0.8454
	3 dB	0.7973	0.8261	0.8055	0.8156
LSTMAttention	27 dB	0.9585	0.961	0.9595	0.9602
	17 dB	0.8713	0.8809	0.8759	0.8783
	11 dB	0.8368	0.8542	0.8449	0.8494
	3 dB	0.7865	0,8182	0.7946	0.8062
GRUAttention	27 dB	0.9593	0.9615	0.951	0.9612
	17 dB	0.8662	0.878	0.8716	0.8747
	11 dB	0.8388	0.8537	0.845	0.8493
	3 dB	0.7921	0.8257	0.8009	0.8131
TSSequencer	27 dB	0.9555	0.9584	0.9575	0.9579
	17 dB	0.8718	0.8834	0.87	0.8801
	11 dB	0.8317	0.8485	0.8394	0.8437
	3 dB	0.7953	0.8236	0.8038	0.8135
PatchTST	27 dB	0.9591	0.9621	0.9602	0.9611
	17 dB	0.873	0.8857	0.8786	0.8821
	11 dB	0.8418	0.8575	0.8478	0.8525
	3 dB	0.7955	0.8264	0.8072	0.8167

Table A4. *Cont.*

Model	SNR	Accuracy	Precision	Recall	F1
MultiPatchTST	27 dB	0.9609	0.9632	0.9624	0.9628
	17 dB	0.8753	0.8908	0.8785	0.8851
	11 dB	0.8439	0.8579	0.8517	0.8547
	3 dB	0.7968	0.8305	0.8067	0.8184

Appendix A.3 Experimental Results After Applying Random Shock Noise

Random shock noise experiments were conducted with a fixed shock occurrence probability of 0.1 and shock amplitudes set to 1.0, 1.5, 2.0, and 3.0 for Levels 1, 2, 3, and 4, respectively. The results indicate that, unlike periodic noise methods, random shock noise, as a non-periodic noise type, has a significantly greater impact on model performance. Performance decreases as noise intensity increases, with the degradation being more pronounced than that observed with the previous two noise methods.

For the MFPT and PU datasets, the performance differences between the models are minimal. However, for the CWRU dataset, the MultipatchTST model shows superior performance as the noise intensity increases. The PU dataset shows a significant performance degradation for all models in the Level 4 condition, indicating increased sensitivity to random impulse noise. Nevertheless, the patch-based learning models, PatchTST and MultipatchTST, show greater resilience to random impulse noise compared to the other models, highlighting the effectiveness of patch-based learning in mitigating the adverse effects of aperiodic noise.

Figures A3 and A4 show graphs of the experimental results for the CWRU and PU datasets, and Table A5 shows a table of the experimental results performance metrics for the mfpt dataset.

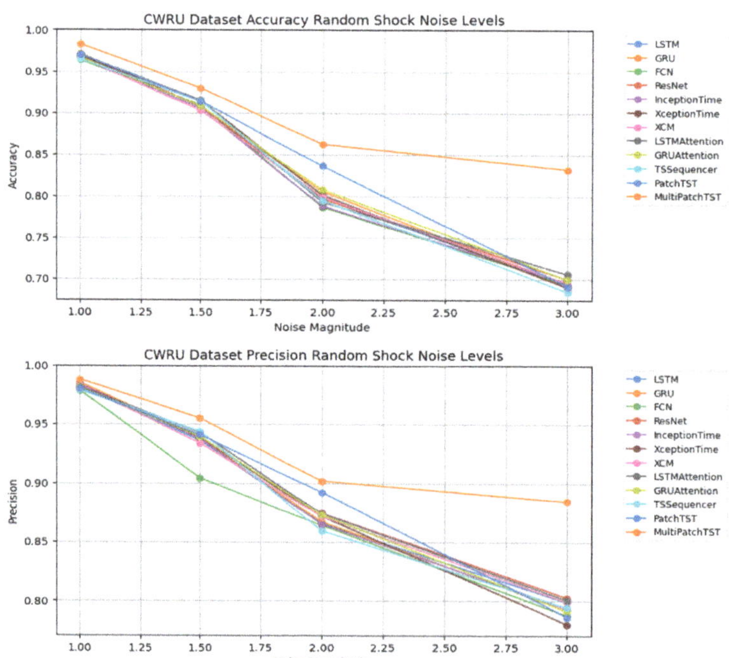

Figure A3. *Cont.*

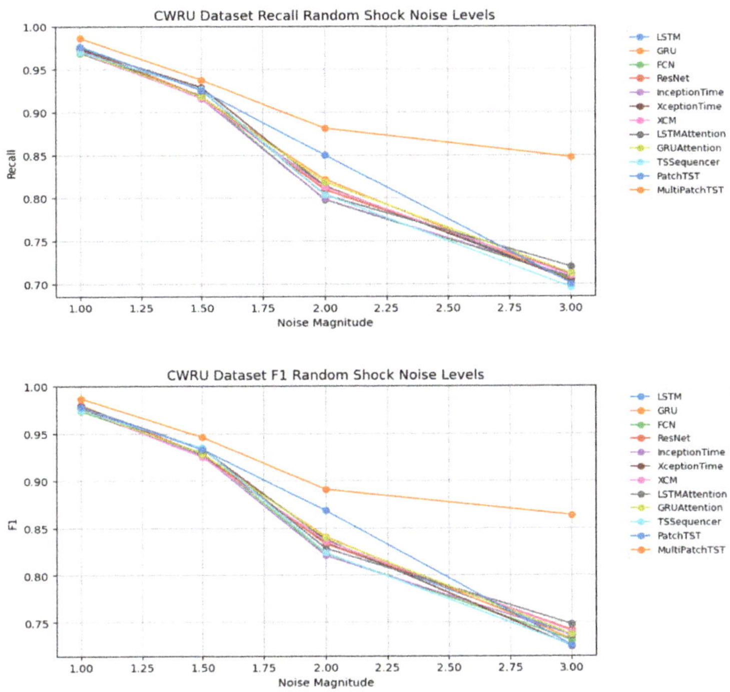

Figure A3. Experimental results by random shock noise level for the CWRU dataset.

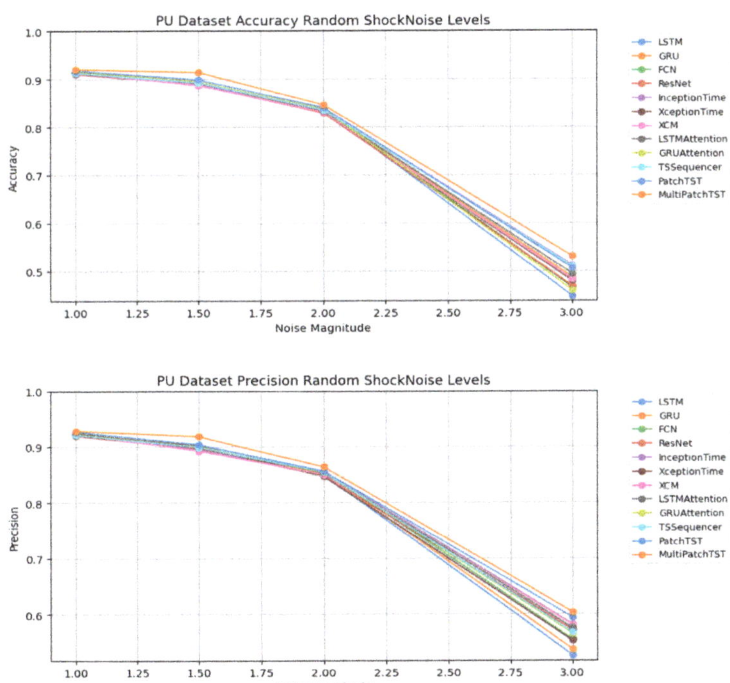

Figure A4. *Cont.*

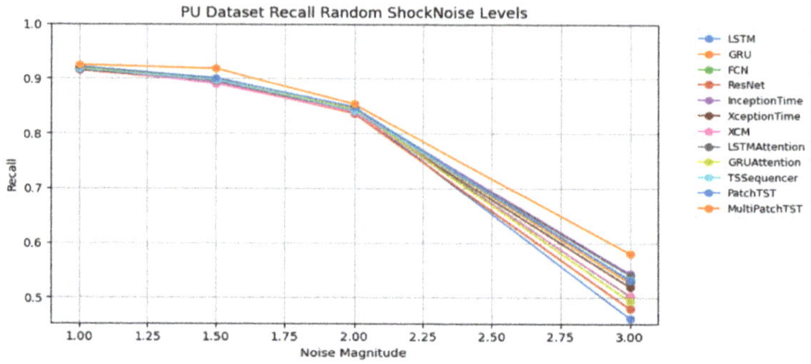

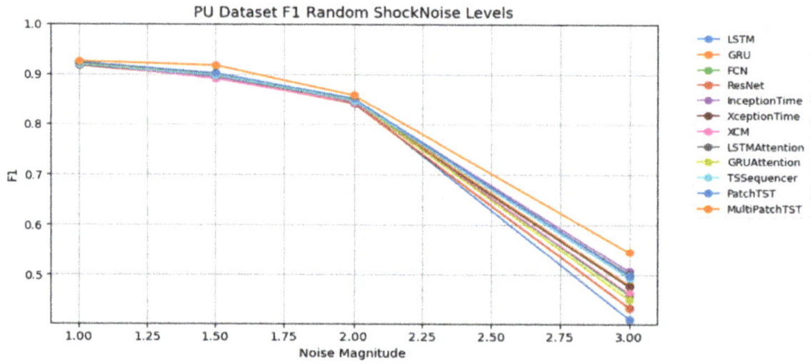

Figure A4. Experimental results by random shock noise level for the PU dataset.

Table A5. Experimental results on MFPT dataset with random shock noise.

Model	SNR	Accuracy	Precision	Recall	F1
LSTM	27 dB	0.9579	0.9609	0.9638	0.9622
	17 dB	0.9239	0.9278	0.9298	0.9287
	11 dB	0.7586	0.8731	0.7594	0.6995
	3 dB	0.65	0.825	0.6506	0.5347
GRU	27 dB	0.95	0.956	0.9534	0.9546
	17 dB	0.9173	0.9204	0.9222	0.9212
	11 dB	0.7692	0.8796	0.7695	0.721
	3 dB	0.6506	0.8251	0.6506	0.5348
FCN	27 dB	0.9552	0.9597	0.9579	0.9586
	17 dB	0.9173	0.9199	0.9222	0.9207
	11 dB	0.8164	0.8905	0.8169	0.8073
	3 dB	0.6506	0.8252	0.6506	0.5349
ResNet	27 dB	0.959	0.9635	0.9628	0.9631
	17 dB	0.9371	0.9397	0.9413	0.9405
	11 dB	0.7706	0.8763	0.7735	0.7243
	3 dB	0.6506	0.8251	0.6513	0.5364
InceptionTime	27 dB	0.9531	0.9607	0.9559	0.958
	17 dB	0.9315	0.9354	0.9385	0.9367
	11 dB	0.7644	0.8746	0.7674	0.7135
	3 dB	0.6513	0.8253	0.6513	0.5364

Table A5. *Cont.*

Model	SNR	Accuracy	Precision	Recall	F1
XceptionTime	27 dB	0.9506	0.9546	0.9555	0.955
	17 dB	0.926	0.9315	0.9263	0.9288
	11 dB	0.7711	0.8788	0.7711	0.7228
	3 dB	0.6503	0.825	0.6503	0.5341
XCM	27 dB	0.9531	0.9567	0.9562	0.9564
	17 dB	0.919	0.921	0.9225	0.9216
	11 dB	0.7623	0.8739	0.7644	0.7083
	3 dB	0.6506	0.8251	0.6506	0.5348
LSTMAttention	27 dB	0.9517	0.9578	0.9534	0.9556
	17 dB	0.9218	0.9271	0.9246	0.9257
	11 dB	0.7799	0.8791	0.7842	0.742
	3 dB	0.6506	0.8251	0.6506	0.5348
GRUAttention	27 dB	0.9541	0.9587	0.9579	0.9583
	17 dB	0.9236	0.9264	0.9295	0.9279
	11 dB	0.7778	0.8783	0.7858	0.7436
	3 dB	0.6506	0.8251	0.651	0.5356
TSSequencer	27 dB	0.9555	0.9607	0.9583	0.9595
	17 dB	0.9375	0.9421	0.9388	0.9404
	11 dB	0.7759	0.8779	0.7789	0.7334
	3 dB	0.651	0.8252	0.6513	0.5364
PatchTST	27 dB	0.9572	0.9605	0.96	0.96
	17 dB	0.9385	0.9418	0.9423	0.9412
	11 dB	0.7898	0.8821	0.8	0.7654
	3 dB	0.6513	0.8253	0.6513	0.5364
MultiPatchTST	27 dB	0.9861	0.9868	0.9878	0.9873
	17 dB	0.9486	0.9492	0.9524	0.9508
	11 dB	0.8257	0.8908	0.8245	0.8114
	3 dB	0.652	0.8256	0.652	0.538

References

1. Wen, Q.; Sun, L.; Yang, F.; Song, X.; Gao, J.; Wang, X.; Xu, H. Time series data augmentation for deep learning: A survey. *arXiv* **2020**, arXiv:2002.12478. [CrossRef]
2. Moysidis, D.A.; Karatzinis, G.D.; Boutalis, Y.S.; Karnavas, Y.L. A Study of Noise Effect in Electrical Machines Bearing Fault Detection and Diagnosis Considering Different Representative Feature Models. *Machines* **2023**, *11*, 1029. [CrossRef]
3. Li, F.; Wang, L.; Wang, D.; Wu, J.; Zhao, H. An adaptive multiscale fully convolutional network for bearing fault diagnosis under noisy environments. *Measurement* **2023**, *216*, 112993. [CrossRef]
4. Kulevome, D.K.B.; Wang, H.; Cobbinah, B.M.; Mawuli, E.S.; Kumar, R. Effective time-series Data Augmentation with Analytic Wavelets for bearing fault diagnosis. *Expert Syst. Appl.* **2024**, *249*, 123536. [CrossRef]
5. Chen, X.; Zhang, Y.; Su, Y.; Zhou, Y.; Gong, W. A short time series rolling bearing fault diagnosis method based on FMTF-CNN. *Eng. Res. Express* **2024**, *6*, 025346. [CrossRef]
6. Knap, P.; Lalik, K.; Bałazy, P. Boosted Convolutional Neural Network Algorithm for the Classification of the Bearing Fault Form 1-D Raw Sensor Data. *Sensors* **2023**, *23*, 4295. [CrossRef] [PubMed]

7. Ahsan, M.; Salah, M.M. Efficient DCNN-LSTM model for fault diagnosis of raw vibration signals: Applications to variable speed rotating machines and diverse fault depths datasets. *Symmetry* **2023**, *15*, 1413. [CrossRef]
8. Yu, T.; Li, C.; Huang, J.; Xiao, X.; Zhang, X.; Li, Y.; Fu, B. ReF-DDPM: A novel DDPM-based data augmentation method for imbalanced rolling bearing fault diagnosis. *Reliab. Eng. Syst. Saf.* **2024**, *251*, 110343. [CrossRef]
9. Graves, A. Long short-term memory. In *Supervised Sequence Labelling with Recurrent Neural Networks*; Springer: Berlin/Heidelberg, Germany, 2012; pp. 37–45. [CrossRef]
10. Chung, J.; Gulcehre, C.; Cho, K.; Bengio, Y. Empirical evaluation of gated recurrent neural networks on sequence modeling. *arXiv* **2014**, arXiv:1412.3555. [CrossRef]
11. Zhao, B.; Xing, H.; Wang, X.; Song, F.; Xiao, Z. Rethinking attention mechanism in time series classification. *Inf. Sci.* **2023**, *627*, 97–114. [CrossRef]
12. Jia, Y.; Fan, S.; Li, Z.; Xia, K. A Fast Detection Method of Turbulent Gases Based on Gated Recurrent Unit and Attention Mechanism. *IEEE Sens. J.* **2023**, *23*, 5974–5987. [CrossRef]
13. Tatsunami, Y.; Taki, M. Sequencer: Deep lstm for image classification. *Adv. Neural Inf. Process. Syst.* **2022**, *35*, 38204–38217. [CrossRef]
14. Nie, Y.; Nguyen, N.H.; Sinthong, P.; Kalagnanam, J. A time series is worth 64 words: Long-term forecasting with transformers. *arXiv* **2022**, arXiv:2211.14730. [CrossRef]
15. Wang, Z.; Yan, W.; Oates, T. Time series classification from scratch with deep neural networks: A strong baseline. In Proceedings of the 2017 International Joint Conference on Neural Networks (IJCNN), Anchorage, AK, USA, 14–19 May 2017; pp. 1578–1585.
16. He, K.; Zhang, X.; Ren, S.; Sun, J. Deep residual learning for image recognition. In Proceedings of the IEEE Conference on Computer Vision and Pattern Recognition, Las Vegas, NV, USA, 27–30 June 2016. [CrossRef]
17. Fawaz, H.I.; Lucas, B.; Forestier, G.; Pelletier, C.; Schmidt, D.F.; Weber, J.; Webb, G.I.; Idoumghar, L.; Muller, P.-A.; Petitjean, F. Inceptiontime: Finding alexnet for time series classification. *Data Min. Knowl. Discov.* **2020**, *34*, 1936–1962. [CrossRef]
18. Rahimian, E.; Zabihi, S.; Atashzar, S.F.; Asif, A.; Mohammadi, A. Xceptiontime: A novel deep architecture based on depthwise separable convolutions for hand gesture classification. *arXiv* **2019**, arXiv:1911.03803. [CrossRef]
19. Fauvel, K.; Lin, T.; Masson, V.; Fromont, É.; Termier, A. XCM: An explainable convolutional neural network for multivariate time series classification. *Mathematics* **2021**, *9*, 3137. [CrossRef]
20. Case Western Reserve University Bearing Data Center Website. Available online: http://csegroups.case.edu/bearingdatacenter (accessed on 25 November 2024).
21. Neupane, D.; Kim, Y.; Seok, J. Bearing fault detection using scalogram and switchable normalization-based CNN (SN-CNN). *IEEE Access* **2021**, *9*, 88151–88166. [CrossRef]
22. Bechhoefer, E. Condition Based Maintenance Fault Database for Testing of Diagnostic and Prognostics Algorithms. Available online: https://www.mfpt.org/fault-data-sets/ (accessed on 25 November 2024).
23. Zhang, Q.; Deng, L. An intelligent fault diagnosis method of rolling bearings based on short-time Fourier transform and convolutional neural network. *J. Fail. Anal. Prev.* **2023**, *23*, 795–811. [CrossRef]
24. Lessmeier, C.; Kimotho, J.K.; Zimmer, D.; Sextro, W. Condition monitoring of bearing damage in electromechanical drive systems by using motor current signals of electric motors: A benchmark data set for data-driven classification. In Proceedings of the European Conference of the Prognostics and Health Management Society, Bilbao, Spain, 5–8 July 2016; Citeseer: Princeton, NJ, USA, 2016; pp. 5–8.
25. Hou, L.; Jiang, R.; Tan, Y.; Zhang, J. Input feature mappings-based deep residual networks for fault diagnosis of rolling element bearing with complicated dataset. *IEEE Access* **2020**, *8*, 180967–180976. [CrossRef]

applied sciences

MDPI

Article

Drainage Pipeline Multi-Defect Segmentation Assisted by Multiple Attention for Sonar Images

Qilin Jin [1], Qingbang Han [1,*], Jianhua Qian [2], Liujia Sun [1], Kao Ge [1] and Jiayu Xia [1]

1 College of Information Science and Engineering, Hohai University, Changzhou 213200, China; jinqilin@126.com (Q.J.); liujia_sun@163.com (L.S.); 230222010002@hhu.edu.cn (K.G.); 231323010012@hhu.edu.cn (J.X.)

2 Auto Subsea Vehicles Inc., Shanghai 201306, China; jhqian@autosubsea.com

* Correspondence: 20111841@hhu.edu.cn; Tel.: +86-13815057632

Abstract: Drainage pipeline construction projects are vulnerable to a range of defects, such as branch concealed joints, variable diameter, two pipe mouth significances, foreign object insertion, pipeline rupture, and pipeline end disconnection, generated during long-term service in a complex environment. This paper proposes two enhancements to multiple attention learning to detect and segment multiple defects. Firstly, we collected numerous samples of drainage pipeline sonar defect videos. Then, our multiple attention segmentation network was used for target segmentation. The test precision and accuracy of MAP@50 reach 96.0% and 90.9%, respectively, in the segmentation prediction. Compared to the coordinate attention and convolutional block attention module attention models, it had a significant precision advantage, and the weight file size is merely 7.0 MB, which is far smaller than the Yolov9 model segmentation weight size. The multiple attention method proposed in this paper was adopted for detection, instance segmentation, and pose detection in different public datasets, especially in the object detection of the coco128-seg dataset under the same condition. Map@50:95 has increased by 13.0% assisted by our multiple attention mechanism. The results indicated the memory efficiency and high precision of the integration of the multiple attention model on several public datasets.

Keywords: multiple attention learning; enhance attention learning; sonar multi-defect detection; instance segmentation; YoloV11

1. Introduction

The drainage pipeline is one of the important infrastructures of the city, which is responsible for the collection and transportation of urban sewage and wastewater. Sewage from the branch pipe goes into the trunk pipe, and then into the main pipe. The pipes range from small to large similar to human body blood vessels, and their operational status directly affects people's daily lives and the operation of the city. However, as drainage pipes age and the surrounding environment erodes, the integrity of the pipes is highly susceptible to damage and the formation of various defects. According to the relevant national standards, defects can be specifically categorized into structural defects and functional defects, accompanied by some special structures. Structural defects include concealed joints (AJ), misalignment (CK), rupture (PL), disconnection (TJ), etc., which directly affect the structural integrity and bearing capacity of the pipeline; functional defects, such as penetration of foreign objects (CR), crippled walls and dams (CQ), and obstacles (ZW), etc., which mainly affect the drainage function of the pipeline; and special structures, such as reducer (BJ), and concealed wells (MJ). Various drainage pipe defects seriously impede the pipeline's conveying capacity, and in serious cases, these defects can even cause major ground subsidence, jeopardizing people's lives and property. Therefore, it is particularly important to carry out regular defect detection of sewage pipes.

Traditional inspection methods include the observation method, the reflector method, and the diving inspection method. The observation method determines whether there are structural defects such as clogging and rupture of the pipeline by manually observing the water level and water quality composition in the manholes between the same pipelines. The reflector method utilizes reflectors to reflect daylight into the interior of the pipeline and observes the clogging and misalignment of the pipeline and other conditions through the mirror. The diving inspection method requires professional diving inspectors to dive into the pipeline to observe and determine whether the pipeline is clogged or misaligned. Although these methods are easy to operate, they are mostly applicable to sewage pipes with large diameters and low water levels, and the scope of the inspection is limited, even requiring inspectors to enter the pipeline to directly observe and record pipeline defects, which is not only inefficient but also poses certain safety risks.

At present, the optically based closed-circuit television (CCTV) method is commonly used, where a camera carried by the robot is used. The optical image has a large field of view and high resolution, and the defects can be identified by the image [1,2]. Some scholars have combined deep learning methods with CCTV inspection methods to greatly improve the inspection efficiency of drainage pipes. Ji et al. provided an efficient self-attention mechanism model for sewer optical image defect classification via a transformer, which enables more effective detection and maintenance of underground drainage systems and reduces the risk of accidents due to sewer pipe defects. Wang et al. proposed a lightweight AutoSewerNet network structure for accurate classification that can prevent system failures in time and prevent environmental pollution caused by sewage leakage, thus safeguarding public health and ensuring the normal operation of urban infrastructures. Dang et al. proposed an automated sewage pipe defect segmentation framework based on DeepLabV3+, which is capable of recognizing the type, location, geometrical information, and severity of the defects. However, optical methods also face many difficulties. For example, the scattering of turbid water seriously affects the imaging quality. In addition, water absorbs a lot of light, especially in deep water areas, and it is difficult to obtain good images.

With the maturity of sonar imaging technology and the development of underwater unmanned detection technology, sonar plays an increasingly important role in underwater target detection. Due to the good characteristics of acoustic waves in underwater media, sonar is also very effective in the field of pipeline defect detection, and it is more effective than cameras or radar [3]. It can provide more accurate and detailed information, even in dark and turbid situations. Furthermore, a sonar image has the advantages of high frequency, long distance, high resolution, multiple beams, and strong real-time performance, and it has been widely used in the field of underwater target detection [4,5] and segmentation [6,7].

In previous deep learning target detection networks, the YOLO (You Only Look Once) network framework was widely used for object detection and recognition in drainage pipeline images [8–14]. Chen [12] adopted lightweight yolo5s and swin transformer networks to detect underwater pipeline sonar target images and achieve high precision. Louise and Gurrala Chandrashekar [15,16] introduced CycleGAN [17] network or data augmentation to overcome the data-sparsity problem in the sonar domain, and the results improved performance compared to training on solely simulated images. Meanwhile, some researchers applied attention mechanism learning methods to enhance counting or recognition accuracy in sonar images. To boost sonar image-based fish counting accuracy, Duan [18] adopted a novel network with global attention and point supervision, and the results demonstrated that GPNet achieves state-of-the-art performance both in counting accuracy and noise removal. Cheng [19] also proposed an instance segmentation model through dynamic attention graph optimization, and the proposed method achieved a performance gain of 4.2% in terms of detection accuracy on ARACATI 2017 datasets. Zhao presented a feature pyramid u-net network for forward-looking sonar image semantic segmentation. The method had advantages in small-object and few sample classes [20]. Huang [21] simultaneously proposed a double split attention which fuses spatial attention and channel attention effectively, and the experimental results achieved 78.4% mAP@0.5 on the SCTD dataset.

However, little research has been conducted on the detection and segmentation of multi-defects in drainage pipeline sonar based on multiple attention mechanism learning; furthermore, the current segmentation accuracy and efficiency need further improvement to accommodate the deployment constraints of compact autonomous underwater vehicles. A delicate equilibrium must be struck between the size of the weight coefficient and the accuracy obtained during training. This involves a trade-off between the efficiency of the model, which is often inversely proportional to its size, and the precision of predictions it can make, which is influenced by the complexity of the training process. The key objective is to optimize the model's compactness while maintaining acceptable levels of accuracy for practical deployment scenarios.

The main problem to be solved in this paper is how to locate and segment target defects in side-scan sonar images more quickly and accurately. Furthermore, there is currently a gap in the literature on measuring the quality of multi-defects in sonar pipe images, and by segmenting and quantifying sonar pipe defects, it is more conductive to assessing and repairing drainage pipeline defect areas. This article is prepared to complete the following tasks.

(1) To gather sufficient drainage pipeline multi-defect sonar datasets taken from drainage pipeline ROV inspection equipment for several years. Combining RFA [22], CA [23], significance attention, and improved traditional image enhancement methods, we proposed an enhanced ENAM model, which has an excellent trade-off between noise immunity and accuracy. We compared various attention network models for the detection and recognition of multiple defects in drainage pipelines and further verified the superiority of the ENAM module.
(2) We generated a multiple attention target recognition and instance segmentation algorithm, which was applied to an ENAM-SEG network model based on the YOLOv8n or YOLOv11n structure. The ENCM and ENAM modules were combined with the backbone and head network. This model improved the accuracy and speed of drainage pipeline multi-defect detection, and all class's average precision and MAP@50 values were raised to 96.0% and 90.9% in collected datasets in the segmentation prediction results.

2. Methods and Materials

To identify and segment underwater sonar multi-defect images in this article, a multi-network architecture was proposed. The original images were captured by the same device, so their pixel sizes are consistent. The detailed processing method diagram is presented in Figure 1.

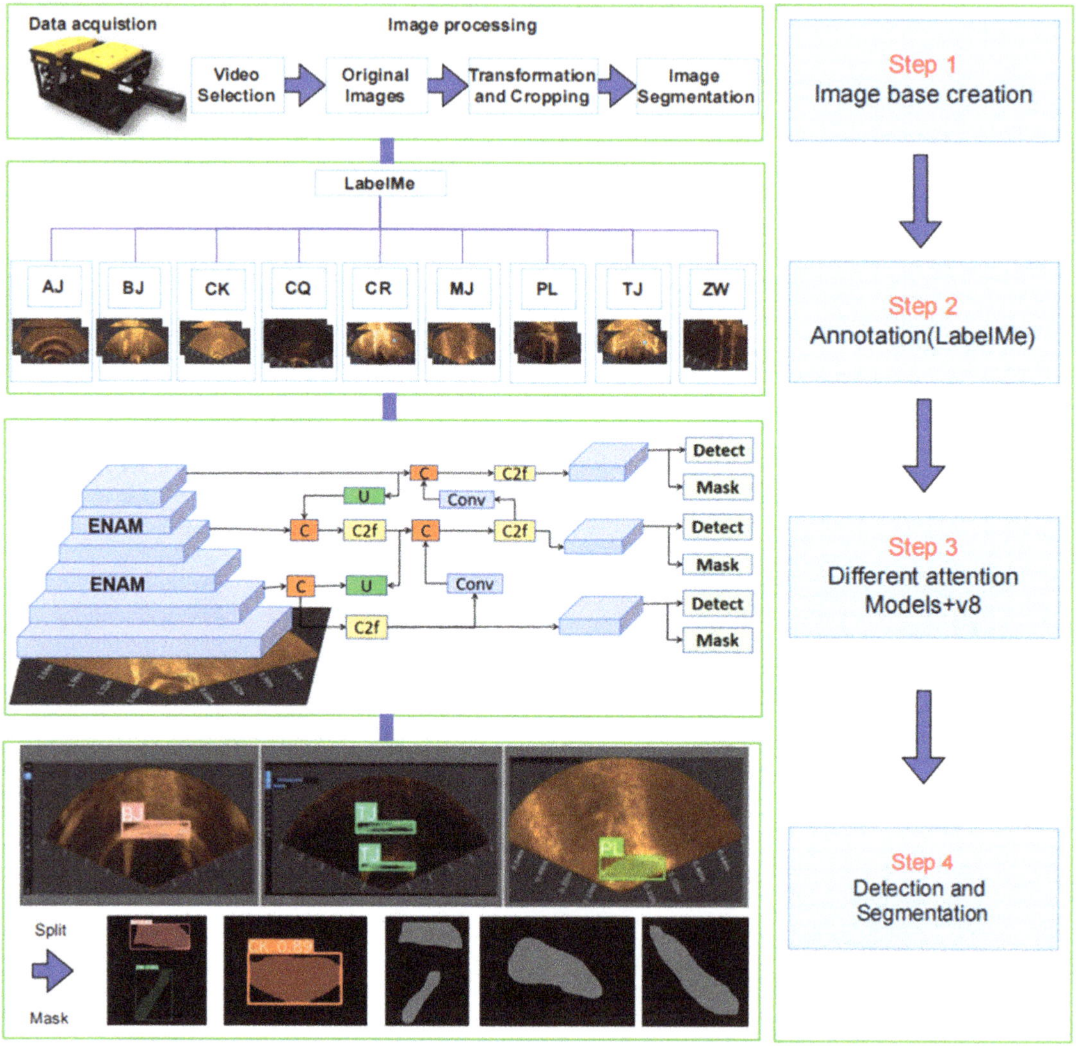

Figure 1. The processing of sonar multi-defect detection and segmentation. Defects include branch concealed joints (AJ), variable diameter (BJ), two pipe mouth significances (CK), residual walls and dam roots (CQ), foreign object insertion (CR), hidden wells (MJ), pipeline rupture (PL), pipeline end disconnection (TJ), and obstacles (ZW).

Firstly, we extracted the original images from the collected sonar video sequences following specific protocols, selecting a subset of images that exhibit defects. Subsequently, we expanded the original dataset through the application of transformations and data augmentation techniques to generate the requisite sonar defect dataset. In the subsequent phase, the dataset was annotated using LabelMe V4.5.13 software, undergoing format conversion processes, and partitioning to ensure it aligns with the training requirements for our models. The ENAM-SEG model was then employed for training and validation purposes in the third segment. This model includes three parts: backbone, neck, and head. The backbone downsampling network is applied for feature extraction, and the proposed ENAM multi-attention mechanism is added at different scales to ensure that channel, spatial, significance, and positional information can be obtained, while preliminary feature fusion is performed. The neck network continues to integrate different scales of backbone information to obtain richer semantic target features while enhancing the fused features. The head is responsible for feature extraction and prediction, including detection and masking (segmentation), predicting the specific location and category of the target object. Finally, we derived prediction outcomes and employed a synergistic approach combining optical character recognition (OCR) and morphological analysis methods to quantify the defects.

2.1. General Multiple Attention Feature

The sonar defect in this article is manifested by the features of the image, and the attention mechanism is good at extracting image features from different dimensions. Therefore, this article adopts a hybrid attention mechanism to highlight it. From Figure 2, we summarize attention into the following parts: channel attention, significance attention, position attention, and receptive field attention (multi-convolutional kernel extraction of spatial feature information). Any combination of two or more is called mixed attention. We make this distinction because most attention mechanisms are confined to extracting features of a single or merely two types, which can lead to the overlooking of other types of contextual feature information, particularly salient features.

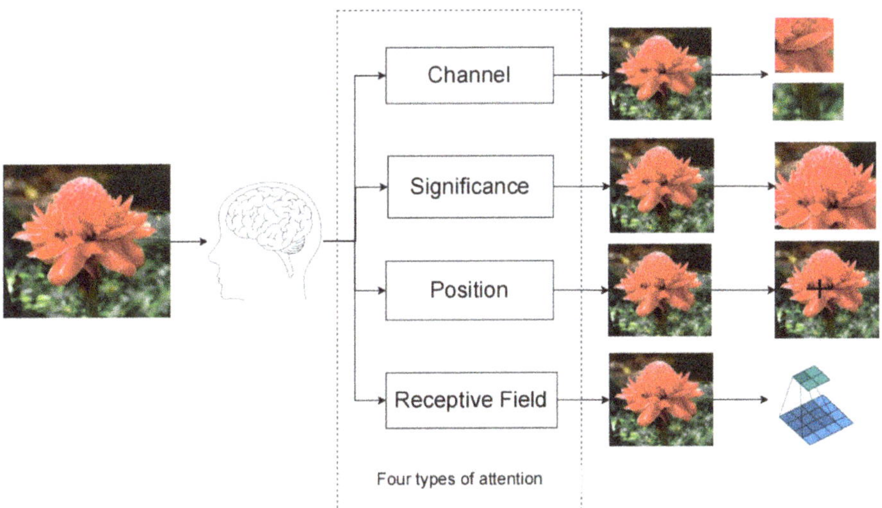

Figure 2. Multiple attention framework.

Firstly, let $X \in R^{B\times C\times H\times W}$ denote the input feature maps, where B, C, H, and W represent the number of batch sizes and channels as well as the height and width of the feature map, respectively.

To simplify the expression X, we assume that $B = 1$, the intermediate attention feature $X_{C\_B}(H,W)$ is defined as

$$X_{C\_B}(H,W) = \prod_{i=1}^{M} R(F_i(X_C(H,W))) \tag{1}$$

where F is a sequential attention model, including linear, a combination of Conv2d, MaxPool2d, Linear, and other operations. R is the activation function, and M is the number of the initial sequential model. Take the intermediate feature map as input and feed it into another set of attention feature extractions. Finally, the output of our general attention block Y can be written as

$$y_c(H,W) = X_{C\_B}(H,W)\prod_{j=1}^{N}\sigma(F_j(X_{C\_B}(H,W))) \tag{2}$$

Or

$$y_c(H,W) = X_{C\_B}(H,W)\sum_{j=1}^{N} w_j\sigma(F_j(X_{C\_B}(H,W))) \tag{3}$$

where w_j is the weight coefficients of the attention features in the j sequential model, N is the number of attention feature models, and σ is the sigmoid function. w_j can be denoted as

$$w_j = \frac{\left(\sum_{H,W}\left|f_j(X_{C\_B}(H,W)) - X_D\right| + eps\right)^{\frac{1}{p}}}{\sum_{j=1}^{J}\left(\sum_{H,W}\left|f_j(X_{C\_B}(H,W)) - X_D\right| + eps\right)^{\frac{1}{p}}} \tag{4}$$

where eps is a constant 1×10^{-6}, X_D is a tensor of dimension $X_{C\_B}(H,W)$ with a constant of 0.5, and p is the power exponent of the distance metric, defaulting to 2.

2.2. Multiple Attention

Attention modules with spatial and channel-wise attention, such as CBAM [24] (channel-wise attention and spatial attention) and CA [23] (coordinate attention), have the potential to capture spatial feature information in the receptive field. To enhance this, researchers have proposed RFCBAM [22] (residual fusion of channel-wise and spatial attention for CBAM) and RFCA [22] (residual fusion of channel attention for CA), which integrate the two mechanisms by separately applying convolution for channel and spatial domains. The motivation behind RFCBAM lies in simultaneously weighting channel and spatial attention to ensure diverse spatial focus per channel. However, RFCA, though it combines channel attention, lacks the emphasis on distinct feature information. Thus, we proposed two fully fusion significance models, namely ENAM and ENCM, as illustrated in Figures 3 and 4.

a. The overall structure of our proposed enhance attention module (ENAM) is shown in Figure 3. ENAM consists of four interrelated components of attention: CA (coordinate attention), DA (significance attention), ECA (efficient channel attention), and RFA (receptive-filed attention) [22]. In this section, we discuss the distinctive mechanism of ENAM, which learns effective channel representations without resorting to dimensionality reduction in convolutional operations, and consequently, enhances pixel-level attention for high-level feature maps. First, an averaging operation is performed on the entire input features to obtain the significance features. Then, a convolution operation is performed on the anomaly features, which involves a 3 × 3 convolution kernel followed by a nonlinear sigmoid activation. Specifically, we select the final weight feature from the CA module and multiply the CA weight result with the significance weight feature to obtain the final ENAM weight feature. The ENAM model mainly consists of the following parts: Part A is receptive-filed attention, and the overall special F_a expression is as follows:

$$F_a = Softmax(g^{i \times i}(AvgPool(X(C,H,W)))) \times \mathrm{ReLU}(BatchNorm(g^{k \times k}(X(C,H,W)))) \quad (5)$$

where X is the input feature map, F_a is the output feature map, $g^{i \times i}$ representing grouped convolutions of size $i \times i$, k represents the size of the convolution kernel, *BathNorm* represents data normalization, and *AvgPool* represents average pooling operation.

b. Part B is efficient channel attention, and local cross-channel interaction was implemented using 1D convolution to extract inter-channel dependencies. The feature F_b expression is as follows:

$$F_b = \sigma(C_1 D_k(X(KH, KW))) \quad (6)$$

where D_k is average pooling, C_1 is calculated by 1×1 convolution, and σ represents the sigmoid activation function.

c. Part C is coordinate position attention, which is the high average pooling and sigmoid activation function, wide average pooling, convolution, sigmoid activation function, and high width feature fusion, ultimately resulting in a feature map of $C \times KH \times KW$ size. The calculation expression is as follows:

$$F_c = X_C(KH, KW) \times \sigma(F_{KH}(f^{KH})) \times \sigma(F_{KW}(f^{KW})) \quad (7)$$

$$f = \delta(F_1([z^h, z^w])) \quad (8)$$

In the formula, X is the input feature vector, and f^{KH} and f^{KW} are transformed into tensors with the same channel C as X using 1×1 convolution, representing the sigmoid activation function. F is the intermediate feature map containing both horizontal and vertical spatial dimensions, representing the nonlinear activation function. [.,.] represents the linking operation along the spatial dimension.

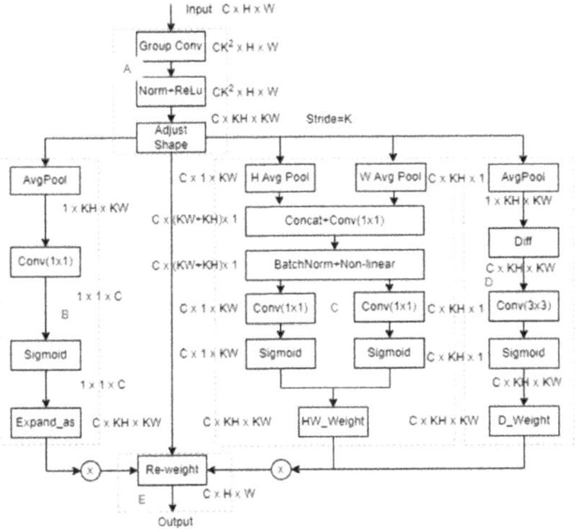

Figure 3. Enhance attention module (ENAM) (A. Receptive field attention. B. Efficient channel attention. C. Coordinate attention. D. Significance attention. E. Fusion of multiple attention).

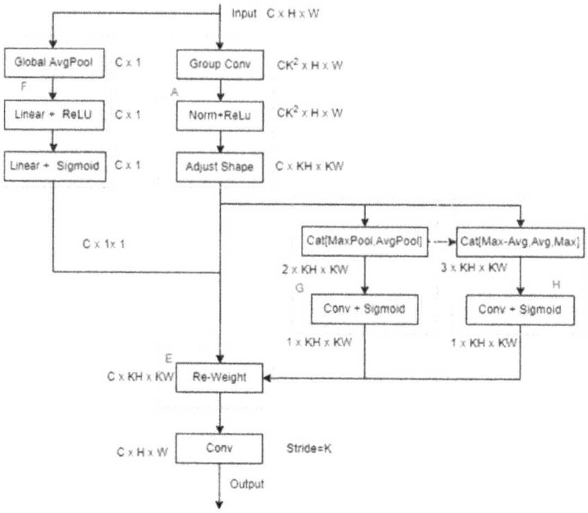

Figure 4. Enhance CBAM module (ENCM) (A. Receptive field attention E. Fusion of multiple attention F. Squeeze and excitation attention G. CBAM attention H. Significance attention).

d. Part D is deviation attention or significance attention, significance D represents as follows:

$$D = U - X = \frac{1}{C}\sum_{i=0}^{C} X(C, KH, KW) - X(C, KH, KW) \tag{9}$$

In the formula, U represents the mean of all channels C under the X feature vector, X is the input feature vector, and the formula for calculating the significance attention feature map is as follows:

$$F_d = \sigma(Conv_{3\times3}(D(C, KH, KW))) \tag{10}$$

where σ is the sigmoid activation function, and $Conv_{3\times3}$ represents a convolution operation with a kernel size of 3×3.

e. Part E is the fusion of multiple attention weights, and the calculation expression is the same as Formula (2).

The overall structure of our proposed enhanced CBAM module (ENCM) is shown in Figure 4. ENCM consists of three interrelated components of attention, referred to as CBAM (convolutional block attention module) [24], SA (significance attention), and RFA (receptive-filed attention) [22]. Part F in Figure 4 obtains channel feature information and part G obtains spatial feature information, forming the CBAM module. The detailed expressions of several parts are as follows.

f. Part F is the fusion of squeeze and excitation (SE) [25] weights, and it's also channel attention. The feature map is as follows.

$$F_f = \sigma(W_2\delta(W_1\frac{1}{H \times W}\sum_{i=1}^{H}\sum_{j=1}^{W} X_c(i,j))) \tag{11}$$

where σ is the sigmoid activation function, and δ is the ReLu activation function. W_1 and W_2 are fully connected operations.

g. Part G is a part of CBAM [25] weights. This is a mixed attention module and the feature map is as follows.

$$\begin{aligned} F_g &= \sigma(MLP(MaxPool(X(C,H,W))) + MLP(AvgPool(X(C,H,W)))) \\ &= \sigma(W_1(W_0(X^C_{\max})) + W_1(W_0(X^C_{avg}))) \end{aligned} \tag{12}$$

where σ denotes the sigmoid activation function, $W_0 \in R^{C/r \times C}$, and $W_1 \in R^{C \times C/r}$. Note that the MLP weights, W_0 and W_1, are shared for both inputs, and the ReLu activation function is followed by W_0.

h. Part H is a new significance attention module, and the detailed processing is shown in Figure 5. Firstly, we obtained the average pooling and maximum pooling of the feature map, then calculated the difference between the maximum pooling and the average pooling, and then combined the three parts together. Finally, the results were added to the activation function.

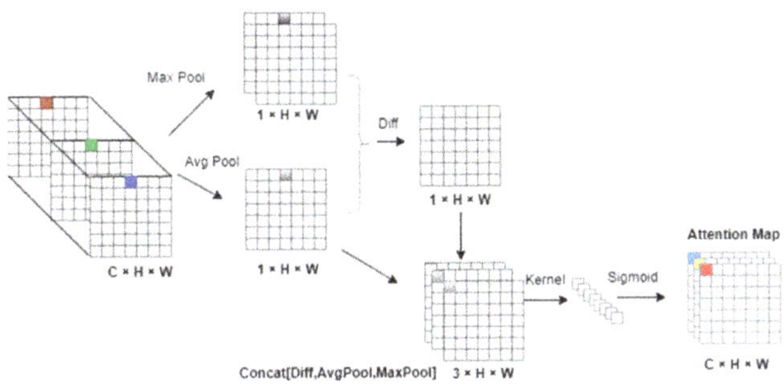

Figure 5. The processing of a new significance attention module.

The feature map F_h is as follows.

$$\begin{aligned} F_h &= \sigma(MLP(MaxPool(X(C,KH,KW)) - AvgPool(X(C,KH,KW))) + MLP(AvgPool(X)) + MLP(MaxPool(X(C,KH,KW)))) \\ &= \sigma(W_1(W_0(X^C_{\max - avg})) + W_1(W_0(X^C_{avg})) + W_1(W_0(X^C_{\max}))) \end{aligned} \tag{13}$$

where σ denotes the sigmoid activation function, $W_0 \in R^{C/r \times C}$, and $W_1 \in R^{C \times C/r}$. Note that the MLP weights, W_0 and W_1, are shared for both inputs, and the ReLu activation function is followed by W_0.

3. Experiments

3.1. Data Collect and Preparation

We used underwater forward-looking multi-beam sonar technology to capture enough sonar samples of drainage pipeline faults. The following Figure 6 demonstrates the collection equipment and operation scenario.

As shown in Figure 6c, the structure of the drainage pipeline sonar equipment includes (1) underwater ROVs (a), (2) transmission cables, (3) depth encoders, (4) integrated control boxes, and (5) computers, sometimes in conjunction with testing platforms such as boats or cars.

A large number of sample video files were obtained by selecting appropriate frame rates to convert the videos into images, in which the frame per second (fps) was 50 and the interval sampling frame was 10 and then labeling them to obtain a training sample set. The sample set was divided into training, validation, and testing sets in a ratio of

8:1:1. Referring to the urban drainage pipeline detection regulations, we divided the defect samples into nine classes, including branch concealed joints (AJ), variable diameter (BJ), two pipe mouth significances (CK), residual walls and dam roots (CQ), foreign object insertion (CR), hidden wells (MJ), pipeline rupture (PL), pipeline end disconnection (TJ), and obstacles (ZW). Figure 7 shows some sample images of pipeline sonar defects.

(**a**) (**b**) (**c**)

Figure 6. (**a**) The collection equipment. (**b**) Operation Scenario. (**c**) Structural schematic diagram.

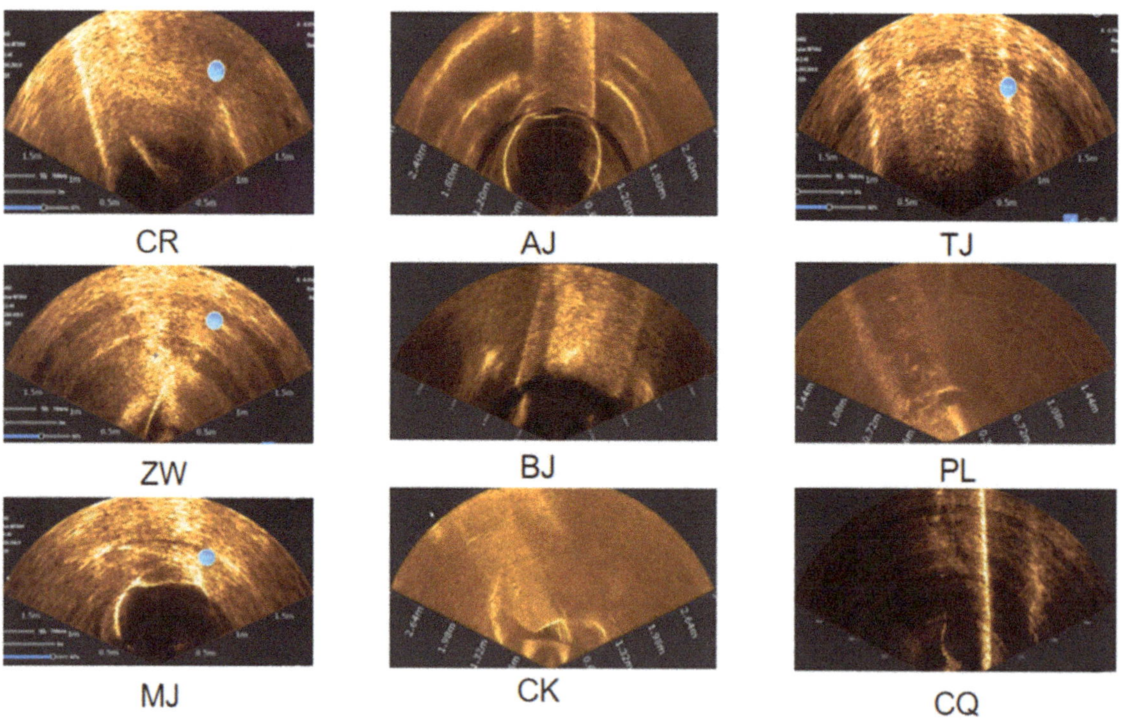

Figure 7. Sample images in our data sets.

At the same time, to increase sonar data richness, this paper adopted a hybrid data augmentation approach, including horizontal flipping, brightness, and saturation adjustment. The last dataset comprises a total of 2380 images, which are categorized into 3 distinct sets: a training set with 1904 images, a validation set with 238 images, and a test set with 238 images.

3.2. Evaluation Metrics

To comprehensively evaluate the performance of the model, precision (P), recall (R), average precision (AP), mean average AP (mAP), floating point operations (FLOPs), and the number of parameters (Params) are selected as the key evaluation metrics in this study.

Accuracy is the ratio of all predicted targets correctly identified by the model and is calculated as shown below.

$$P = \frac{TP}{TP + FP} \tag{14}$$

Recall is the ratio of all actual targets correctly recognized by the model and is calculated as shown in (15).

$$R = \frac{TP}{TP + FN} \tag{15}$$

where TP denotes the number of correctly identified sewer defects; FP denotes the number of incorrectly identified sewer defects, and FN denotes the number of unpredictable tags.

AP is equal to the area under the precision-recall curve with the formula shown in (16).

$$AP = \int_0^1 P(R)dR \tag{16}$$

The mAP is the average of the APs of all sample categories, which is the most commonly used evaluation metric in detection and intuitively reflects the performance of the current model, and is calculated as shown in (17).

$$mAP = \frac{1}{N}\int_0^1 P(R)dR \tag{17}$$

3.3. Training and Platform Setting

The hardware and software settings are shown in Table 1.

Table 1. Experimentation Platform and Settings.

Items	Value
Operating system	Windows 10
CPU	Intel(R)-Core(TM) i7-13700k
GPU	NVIDIA GeForce RTX 4090D
RAM	64 GB
CUDA	CUDAv12.1
Python	3.10.13
batch size	32
epochs	3000
Input_size	(640, 640)
Weight_decay	0.0005
Initial rate (lr_0)	0.01 (SGD)
Cyclical rate (lr_f)	0.1 (Cosine annealing)
momentum	0.937
Input size	640 × 640

After 348 epochs used 0.786 h, we obtained the detailed results shown below. The network converges well and the training is successful.

The R curve of all targets is shown in Figure 8. The results demonstrated that the detection precision of AJ, MJ, and CR is much better than the other classes, and all classes' average MAP@50 is 90.9%.

The detection and segmentation effects on the validation set are shown in Figure 9. It is evident from the figure that the segmented defect results are of defect types PL, BJ, CQ, ZW, and TJ.

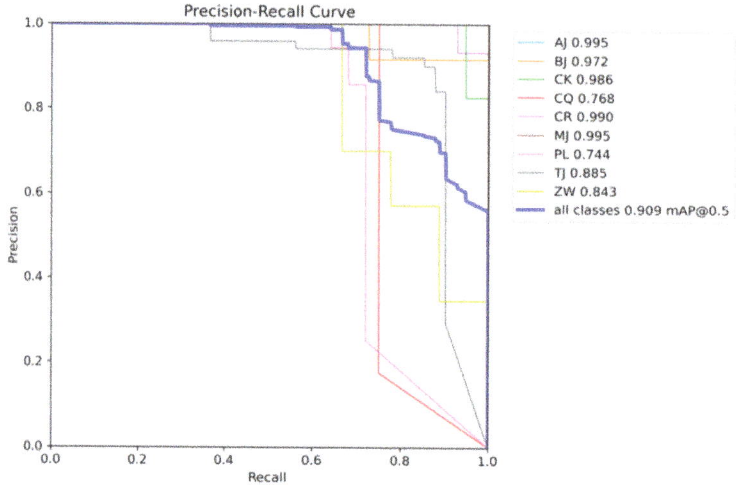

Figure 8. Precision–recall curve.

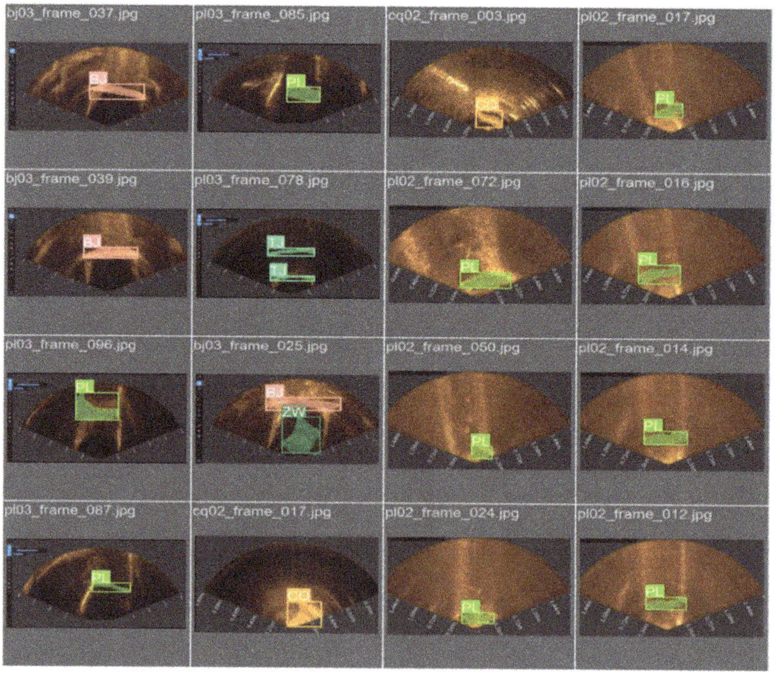

Figure 9. Segmentation results of original sonar images of drainage pipeline defects.

3.4. Comparison of the Other Attention Modules

To further illustrate the advantages of our proposed ENAM attention network module in object detection, we replaced four ENAM modules in the backbone network architecture in Figure 1 with V8n-SEG, CBAM [24], CA [23], RFA [22], EMA [26] attention module and YOLOV9 [14] segmentation module respectively. The Final evaluation results are shown in Table 2.

Table 2. Comparison with the other attention modules.

Model	Weight (M)	P^{mask}	MAP_{50}^{mask}	$MAP_{50:95}^{mask}$
V8n-SEG	6.8	88.6	84.1	44.2
SE	6.8	87.1	87.0	$46.2_{+2.0}$
CBAM	6.9	90.0	85.0	$46.9_{+2.7}$
CA	8.4	88.2	83.6	$45.4_{+1.2}$
EMA	6.8	86.7	83.9	$43.9_{-0.3}$
RFA	8.5	90.9	89.0	$47.1_{+2.9}$
ENCM (ours)	8.6	90.8	89.5	$49.3_{+5.1}$
Gelan-CSEG(V9)	53.1	90.0	89.9	$45.9_{+1.7}$
Gelan-C-DSEG(V9)	55.8	87.9	88.2	$46.5_{+2.3}$
ENAM (ours)	7.0	96.0	90.9	$47.9_{+3.7}$

From the results of Table 2, it is obvious that compared with the other attention mechanisms, the ENAM model we proposed obtains the best accuracy of 96.0% and MAP@0.5 of 90.9%, which is 7.3% higher than the highest MAP@0.5 of the CA model, and the precision is also 5.1% higher than the best RFA module.

3.5. Ablation Experiment in Different Datasets

To further illustrate the advantages of our proposed ENAM attention network module in object detection and segmentation, we performed validation evaluations on the COCO128-seg datasets. Meanwhile, we replaced three ENAM modules in the backbone network architecture with CBAM [24], SE [25], CA [23], RFA [22], DCA, YOLOV11, ENCM (Ours), and ENAM (ours). The final evaluation results are shown in Table 3.

Table 3. Comparison results with different attention modules of Coco128-Seg datasets.

Model	Weight (M)	Flops (G)	MAP_{50}^{box} (%)	$MAP_{50:95}^{box}$ (%)	MAP_{50}^{mask} (%)	$MAP_{50:95}^{mask}$ (%)
V8n + SE	7.3	12.6	80.8	62.6	71.3	42.2
V8n + CA	7.3	12.7	80.7	62.4	70.5	41.9
V8n + CBAM	7.4	12.7	84.5	66.2	75.3	47.2
V8n + EMA	7.3	12.6	76.4	58.5	65.7	37.3
V8n + RFA	9.5	15.3	85.0	70.8	76.7	49.7
V8n + DCA	8.9	15.0	83.0	64.4	73.2	44.4
V8n + RFCA	10.7	26.5	86.2	72.2	79.8	53.3
V8n + ENAM	10.9	27.1	87.4	**73.9**	82.1	**56.5**
V8s + ENAM	37.0	99.8	91.0	79.9	87.7	66.5
V8m + ENAM	74.3	219.9	94.3	85.6	92.6	74.6
V8l + ENAM	114.4	391.5	94.8	87.2	92.4	76.6
V8x + ENAM	178.2	611.1	95.1	87.8	93.9	77.9
V11n	6.7	10.4	85.3	71.0	79.6	54.2
+RFA	6.5	10.5	91.9	$78.1_{+7.1}$	89.0	$66.1_{+11.9}$
+RFCA	6.6	10.6	94.1	$83.0_{+12.0}$	91.6	$72.1_{+17.9}$
+RFD	7.1	20.1	93.3	$81.5_{+10.5}$	90.1	$68.5_{+14.3}$
+ENCM	6.6	10.6	94.0	$83.4_{+12.4}$	91.4	$71.0_{+16.8}$
+ENAM	7.3	20.2	93.8	$84.0_{\mathbf{+13.0}}$	91.0	$71.0_{+16.8}$

DCA is the combination of significance and coordinate attention, and RFDA is the combination of significance and receptive-filed attention. It is obvious from Table 3 that compared to SE, CA, RFA, and RFCA, our ENAM performs slightly better than them. On the COCO128 dataset, the MAP@50 of the V8n + SE model is 80.8%, while the MAP@50 of the V8n + ENAM model is increased to 87.4%. This shows that the ENAM attention module is more effective than the SE module in improving the accuracy of the model.

The attention module had various effects on model performance. For example, the performance of the V8n + ENAM model on the MAP@50:95 index (73.9%) was significantly better than that of the V8n + SE model (62.6%), indicating that the ENAM module performs better when dealing with more stringent evaluation criteria. As the size of the model increases, the performance usually improves. From V8 n to V8 x, with the increase of model weight and calculation amount, the MAP index also increases accordingly. For example, the V8x + ENAM model performs highest on all listed metrics.

This article adopted the SKU110k dataset to further verify the effectiveness of object detection. The dataset includes 11,762 images captured in dense scenes, with over 1.7 million bounding box annotations and approximately 1,733,678 instances. This dataset includes 8233 training sets, 588 validation sets, and 2941 test sets. Table 4 shows the performance comparison of different attention models on SKU-110K datasets. The ENAM model achieved the highest score of 90.6% on MAP@50 and 58.6% on MAP@50:95, the highest among all listed models.

Table 4. Comparison results with different attention modules of Sku-110k datasets.

Model	Weight (M)	Flops (G)	P (%)	R (%)	MAP_{50}^{box} (%)	$MAP_{50:95}^{box}$ (%)
V8n	6.2	8.0	90.3	83.6	89.1	56.3
+SE	6.3	8.1	90.5	83.8	90.2	58.2
+CA	6.3	8.1	90.5	83.6	90.0	57.9
+CBAM	6.3	8.1	90.3	83.2	89.9	57.9
+CAF	9.5	22.6	90.7	84.0	90.4	58.4
+ENAM	9.6	22.5	91.0	84.4	90.6	58.6

We conducted some experiments to assess the efficacy of our proposed ENAM and ENCM attention enhancement models in the context of pose estimation, we employed the Tiger-Pose datasets. These datasets tailored for pose estimation tasks, comprise 263 images sourced from YouTube videos, with 210 images allocated for the training phase and 53 for the validation phase. From Table 5, the ENAM and ENCM models proposed in this article MAP@50:95 was 32.3% and 39.4% higher than Yolov11n in Pose assessment, and also MAP@50:95, the evaluation of object detection was 6.6% and 6.7% higher than the base model, respectively.

Table 5. Comparison results with different attention modules of Tiger-Pose datasets.

Model	Weight (M)	Flops (G)	MAP_{50}^{box} (%)	$MAP_{50:95}^{box}$ (%)	MAP_{50}^{Pose} (%)	$MAP_{50:95}^{Pose}$ (%)
V11n	5.48	6.8	99.5	92.3	99.5	58.2
+RFA	5.63	6.9	99.5	98.9	99.5	$86.5_{+28.3}$
+RFCA	5.59	7.0	99.5	95.9	99.5	$82.7_{+24.5}$
+RFD	6.28	16.4	99.5	99.0	99.5	$89.1_{+30.9}$
+ENAM	6.32	16.5	99.5	$98.9_{+6.6}$	99.5	$90.5_{+32.3}$
+ENCM	5.74	6.9	99.5	$99.0_{+6.7}$	99.5	$97.6_{+39.4}$

3.6. Visualization of Attention Feature Maps

To further demonstrate the significant advantages of our methods, we provide a more intuitive analysis with the other attention module. We use the Grad-CAM [27] algorithm of visualization. We adopted the Coco2017 (118,289 training images, 5000 validate images, and 40,670 test images) classification dataset for training and utilized its validation set for attention validation. This algorithm highlights distinct regions of interest for specific class objects in different networks.

From Figure 10, it is clear that our ENAM attention, compared to CA and CBAM, can help the network detect better and highlight the key regions of targets.

Figure 10. Visualization of different attention feature maps (Coco2017).

ENCM is a combination of RFA, DA, and CBAM, which represents the addition of regional feature attention based on CBAM. To compare the differences in attention mechanisms among different defects in underwater sonar pipeline images in our datasets, it can be seen from Figure 11 that our ENAM attention mechanism is more focused and complete on the feature maps of sonar defects.

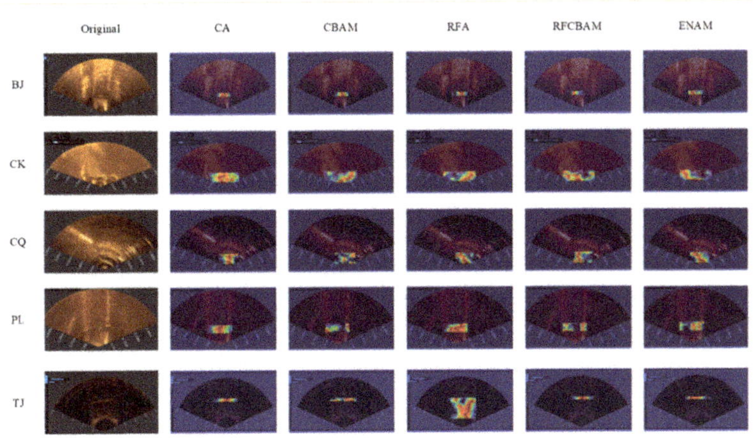

Figure 11. Different attention feature maps (Sonar Pipe defect).

4. Discussion and Conclusions

4.1. Quantitative Evaluation Analysis

Figure 12 indicates that the proposed methodology is capable of accurately evaluating the physical size features of underwater multi-defect. The geometric parameters of fractures can also be calculated when morphological data are computed. A is the defect's area.

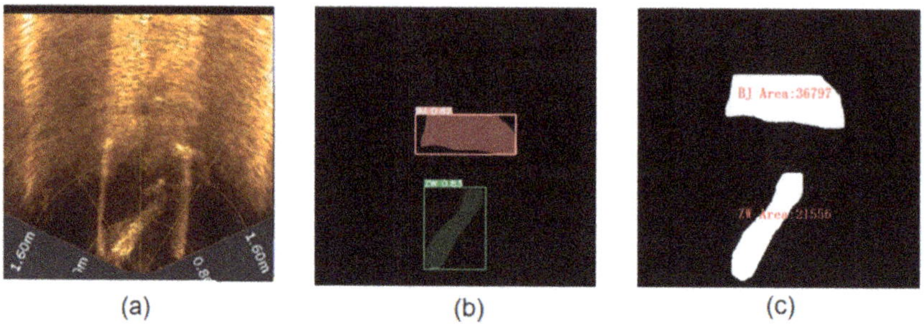

Figure 12. Quantification results for typical underwater defects. (**a**) Original image (**b**) Detection and segmentation prediction results (**c**) Defect type and pixel area.

Simultaneously, we have derived the current adjacent interval values from the original image via optical character recognition (OCR) technology. We have also calculated the scaling ratio of the present sonar system. By establishing a proportional relationship K between the actual environment and the pixel representation of the image, we have successfully determined the real defect area. OCR technology was applied to obtain the digital scale M in sonar images. Different scales indicated different depth information, which obtained the ratio K. K is as follows.

$$\mathrm{K} = k'M = k'\min((G(i/2) - G(i/2-1)), (G(i/2+1) - G(i/2)) \quad (18)$$

where $G(i)$ is an OCR-recognized array of numbers that have been removed from duplicates and sorted in ascending order, and i is the total number of digital scales after the above processing. In order to ensure accurate minimum interval scales, the minimum value between the middle and the difference between the front and back is taken. K′ is a fixed scaling factor provided by the device.

There are 4391 video frames in the underwater sonar video data utilized in this part for validation. It takes specialists roughly 54 s to locate and label the contour of the fault for each video frame. Extracting the defect feature information just by human observation takes 498 min, however, using the GUI software PySide6.0, detection time costs only 12.61 s.

4.2. Conclusions

We proposed a significance attention mechanism algorithm that combines other attention methods to form ENAM and ENCM multiple attention modules. We applied these algorithms to our network, and the network was implemented for multi-defect recognition and instance segmentation, and the experimental numerical calibration and analysis of multi-defect images were performed. Different algorithms were compared in this work, and the combination of significance attention learning and the other methods had higher recognition accuracy. The test and experimental results indicated that the model has a certain superiority. Through morphological processing, the area of the underwater multi-defect data was obtained, and the corresponding application software was developed. In the real-time underwater environment, this approach had a good detection and segmentation effect, and the network structure indicated a certain generalization ability. The MAP50:95 prediction results had higher accuracy than similar attention models, indicating that the multiple attention model proposed in this paper had some advantages in handling complex object detection and instance segmentation tasks.

Although the speed and accuracy of our network were improved to some extent by attention learning, the recognition network may not be able to cope with the challenge of highly complex underwater scenes, and may not achieve satisfactory recognition accuracy. In future work, we will further enrich our pipeline sonar defect datasets and improve recognition accuracy and visualization.

Author Contributions: Conceptualization, Q.H. and Q.J.; writing—original draft preparation, Q.J.; methodology, Q.J.; software, Q.J. writing—review and editing, Q.H. and L.S.; visualization, J.X.; resources, J.Q. and K.G. All authors have read and agreed to the published version of the manuscript.

Funding: This research was funded by the Natural Science foundation of China, grant number 12174085.

Institutional Review Board Statement: Not applicable.

Informed Consent Statement: Not applicable.

Data Availability Statement: The data presented in this study are available on request from the corresponding author.

Acknowledgments: Thanks for the equipment and data collection work provided by Auto Subsea Vehicles Inc., especially Jianhua Qian.

Conflicts of Interest: Author Jianhua Qian was employed by the company Auto Subsea Vehicles Inc. The remaining authors declare that the research was conducted in the absence of any commercial or financial relationships that could be construed as a potential conflict of interest.

References

1. Xu, W.; Yang, J.; Wei, H.; Lu, H.; Tian, X.; Li, X. A localization algorithm based on pose graph using Forward-looking sonar for deep-sea mining vehicle. *Ocean Eng.* **2023**, *284*, 114968. [CrossRef]
2. Cao, W.; Li, J. Detecting large-scale underwater cracks based on remote operated vehicle and graph convolutional neural network. *Front. Struct. Civ. Eng.* **2022**, *16*, 1378–1396. [CrossRef]
3. Sun, H.; Zhu, J. Nondestructive evaluation of steel-concrete composite structure using high-frequency ultrasonic guided wave. *Ultrasonics* **2020**, *103*, 106096. [CrossRef] [PubMed]
4. Wang, J.; Shan, T.; Chandrasekaran, M.; Osedach, T.; Englot, B. Deep learning for detection and tracking of underwater pipelinelines using multibeam imaging sonar. In Proceedings of the IEEE International Conference on Robotics and Automation Workshop, Montreal, Canada, 20–24 May 2019.
5. Sung, M.; Kim, J.; Lee, M.; Kim, B.; Kim, T.; Kim, J.; Yu, S.-C. Realistic Sonar Image Simulation Using Deep Learning for Underwater Object Detection. *Int. J. Control Autom. Syst.* **2020**, *18*, 523–534. [CrossRef]
6. Song, Y.; Liu, P. Segmentation of sonar images with intensity inhomogeneity based on improved MRF. *Appl. Acoust.* **2020**, *158*, 107051. [CrossRef]
7. Chen, Z.; Wang, Y.; Tian, W.; Liu, J.; Zhou, Y.; Shen, J. Underwater sonar image segmentation combining pixel-level and region-level information. *Comput. Electr. Eng.* **2022**, *100*, 107853. [CrossRef]
8. Tang, Y.; Jin, S.; Bian, G.; Zhang, Y. Shipwreck Target Recognition in Side-Scan Sonar Images by Improved YOLOv3 Model Based on Transfer Learning. *IEEE Access* **2020**, *8*, 173450–173460. [CrossRef]
9. Chen, L.; Zheng, M. Underwater Target Recognition Based on Improved YOLOv4 Neural Network. *Electronics* **2021**, *10*, 1634. [CrossRef]
10. Yu, Y.; Zhao, J.; Gong, Q.; Huang, C.; Zheng, G.; Ma, J. Real-Time Underwater Maritime Object Detection in Side-Scan Sonar Images Based on Transformer-YOLOv5. *Remote Sens.* **2021**, *13*, 3555. [CrossRef]
11. Li, S.; Li, C.; Yang, Y.; Zhang, Q.; Wang, Y.; Guo, Z. Underwater scallop recognition algorithm using improved YOLOv5. *Acquacult. Eng.* **2022**, *98*, 102273. [CrossRef]

12. Chen, R.; Zhan, S.; Chen, Y. Underwater Pipeline Target Detection Algorithm Based on YOLO and Swin Transformer for Sonar Images. In Proceedings of the OCEANS Conference 2022 Hampton Roads, Hampton Roads, VA, USA, 17–20 October 2022.
13. Zhen, Y.L. Side scan Drainage pipeline image target detection method based on improved YOLOv5 network. *Hydrogr. Surv. Charting* **2022**, *42*, 18–21, 26.
14. Yu, G.; Cai, R.; Su, J.; Hou, M.; Deng, R. U-YOLOv7: A network for underwater organism detection. *Ecol. Inform.* **2023**, *75*, 102108. [CrossRef]
15. Fuchs, L.R.; Noren, A.; Johansson, P. GAN-enhanced simulated sonar images for deep learning based detection and classification. In Proceedings of the Oceans 2022–Chennai Conference, Chennai, India, 21–24 February 2022. [CrossRef]
16. Chandrashekar, G.; Raaza, A.; Rajendran, V.; Ravikumar, D. Side scan sonar image augmentation for sediment classification using deep learning based transfer learning approach. *Mater. Today Proc.* **2021**, *80*, 3263–3273. [CrossRef]
17. Li, B.Q.; Huang, H.N.; Liu, J.Y.; Li, Y. Optical image-to-underwater small target synthetic aperture sonar image translation algorithm based on improved CycleGAN. *Acta Electron. Sin.* **2021**, *49*, 1746–1753.
18. Duan, Y.; Zhang, S.; Liu, Y.; Liu, J.; An, D.; Wei, Y. Boosting fish counting in sonar images with global attention and point supervision. *Eng. Appl. Artif. Intell.* **2023**, *126*, 107093. [CrossRef]
19. Cheng, K.; Yan, L.; Ding, Y.; Zhou, H.; Li, M.; Ghafoor, H.A. Sonar image garbage detection via global despeckling and dynamic attention graph optimization. *Neurocomputing* **2023**, *529*, 152–165. [CrossRef]
20. Zhao, D.; Ge, W.; Chen, P.; Hu, Y.; Dang, Y.; Liang, R.; Guo, X. Feature pyramid u-net with attention for semantic segmentation of forward-looking sonar images. *Sensors* **2022**, *22*, 8468. [CrossRef] [PubMed]
21. Huang, H.; Zuo, Z.; Sun, B.; Wu, P.; Zhang, J. DSA-SOLO: Double split attention solo for side-scan sonar target segmentation. *Appl. Sci.* **2022**, *12*, 9365. [CrossRef]
22. Zhang, X.; Liu, C.; Yang, D.; Song, T.; Ye, Y.; Li, K.; Song, Y. RFAConv: Innovating spatial attention and standard convolutional operation. *arXiv* **2023**, arXiv:2304.03198. [CrossRef]
23. Hou, Q.; Zhou, D.; Feng, J. Coordinate Attention for Efficient Mobile Network Design. In Proceedings of the 2021 IEEE/CVF Conference on Computer Vision and Pattern Recognition, Nashville, TN, USA, 20–25 June 2021; pp. 13713–13722.
24. Woo, S.; Park, J.; Lee, J.; Kweon, I. CBAM: Convolutional Block Attention Module. In *Computer Vision–ECCV 2018: 15th European Conference, Munich, Germany, September 8–14, 2018, Proceedings, Part VII*; Springer: Cham, Switzerland, 2018; pp. 3–19. [CrossRef]
25. Hu, J.; Shen, L.; Albanie, S.; Sun, G.; Wu, E. Squeeze-and-Excitation Networks. *IEEE Trans. Pattern Anal. Mach. Intell.* **2020**, *42*, 2011–2023. [CrossRef] [PubMed]
26. Ouyang, D.; He, S.; Zhang, G.; Luo, M.; Guo, H.; Zhan, J.; Huang, Z. Efficient Multi-Scale Attention Module with Cross-Spatial Learning. In Proceedings of the 2023 IEEE International Conference on Acoustics, Speech and Signal Processing (ICASSP), Rhodes Island, Greece, 4–10 June 2023; pp. 1–5.
27. Selvaraju, R.R.; Cogswell, M.; Das, A.; Vedantam, R.; Parikh, D.; Batra, D. Grad-cam: Visual explanations from deep networks via gradient-based localization. In Proceedings of the 2017 IEEE International Conference on Computer Vision, Venice, Italy, 22–29 October 2017; pp. 618–626.

applied sciences

MDPI

Article

Machine Learning-Based Lithium Battery State of Health Prediction Research

Kun Li and Xinling Chen *

School of Economics and Management, Tiangong University, Tianjin 300387, China; lk_neu@163.com
* Correspondence: cxl000308@163.com

Abstract: To address the problem of predicting the state of health (SOH) of lithium-ion batteries, this study develops three models optimized using the particle swarm optimization (PSO) algorithm, including the long short-term memory (LSTM) network, convolutional neural network (CNN), and support vector regression (SVR), for accurate SOH estimation. Key features were extracted by analyzing the temperature, voltage, and current curves of the battery, and health factors with high correlation to SOH were selected as model inputs using the Pearson correlation coefficient. The PSO algorithm was employed to optimize model parameters, resulting in the construction of three predictive models: PSO-LSTM, PSO-CNN, and PSO-SVR. The models were validated using the NASA PCoE battery aging datasets B0005, B0006, and B0007, with prediction accuracy evaluated based on Root Mean Square Error (RMSE), Mean Absolute Error (MAE), Mean Absolute Percentage Error (MAPE), and Coefficient of Determination (R^2). Results indicate that the optimized models achieved significant improvements in prediction accuracy, with RMSE and MAE reduced by over 0.5%, a minimum reduction of 38% in MAPE, and R^2 exceeding 0.8, demonstrating strong fitting capabilities and validating the effectiveness of the PSO strategy. Among the three models, PSO-LSTM exhibited the best predictive performance, achieving a minimum MAE of 0.67%, RMSE of 0.94%, MAPE of 45.82%, and R^2 as high as 0.9298 across the three datasets. These findings suggest that the PSO-LSTM model provides a robust reference for accurate SOH prediction of lithium-ion batteries and shows promising potential for practical applications.

Keywords: lithium-ion battery state of health; neural network model; particle swarm algorithm; support vector regression

1. Introduction

With the worsening global energy crisis and environmental pollution, electric vehicles have become a key industry that countries around the world are actively developing, as they help reduce dependence on traditional energy sources. Lithium-ion batteries, with their high energy density, long life, and low self-discharge rate, have emerged as a reliable power source for electric vehicles. However, as the number of charge and discharge cycles increases, lithium batteries gradually age. If their health status is not monitored in a timely manner, it may result in issues such as reduced driving range, decreased performance, and even safety hazards. Therefore, accurately assessing the state of health (SOH) of lithium batteries is crucial for the safe operation of electric vehicles [1]. The most common definition of SOH is the percentage of the battery's rated capacity that can be discharged from a fully charged state to the cut-off voltage under specific conditions [2].

Currently, SOH estimation methods for lithium-ion batteries can be broadly categorized into three types: direct measurement methods, model-based methods, and data-driven methods [3]. Direct measurement methods estimate SOH by directly measuring aging characteristic parameters such as battery capacity and internal resistance under laboratory conditions. For instance, Luo et al. [4] analyzed impedance data and established a mathematical model correlating impedance spectrum directional element parameters with SOH, achieving rapid SOH estimation with prediction errors within 2%. Similarly, Zhang et al. [5] combined the coulomb counting method with differential voltage analysis to estimate battery SOH, demonstrating high accuracy with a mean absolute error controlled within 0.4%, thereby enabling precise SOH prediction. However, direct measurement methods are often not prioritized for practical applications due to their high dependency on specialized testing equipment, time-consuming measurement processes, and limited applicability [6].

Model-based estimation methods typically require the establishment of an electrochemical model (EM) or an equivalent circuit model (ECM) to identify parameters related to battery performance degradation and failure [7]. These parameters are then updated using adaptive filtering techniques or nonlinear state observers to more accurately predict changes in the SOH of the battery. Xu et al. [8] utilized a simplified EM to describe the distribution of lithium ions within the battery and linked SOH to the irreversible loss of lithium ions. Recursive least squares was employed to online identify model parameters, while an unscented Kalman filter was used to estimate the SOH. The experimental data revealed that the method's average prediction inaccuracy was around 2%. Hosseininasab et al. [9] proposed a method based on a fractional-order battery EM combined with an iterative model observer technique to adaptively estimate both internal resistance and capacity degradation. This approach avoided the high computational cost caused by limited calibration parameters while maintaining high estimation accuracy. Similarly, Amir et al. [10] developed a 2-RC model to estimate SOH, where the open-circuit voltage was regarded as a function of the state of charge (SOC). Nonlinear least squares curve fitting was used to approximate the model parameters, enabling the model to adapt to and capture the dynamic changes in SOH. The results demonstrate that this method achieved SOH estimation with low RMSE. Zou et al. [11] applied recursive least squares to identify the parameters of a Thevenin model and implemented an extended Kalman filter to achieve joint estimation of SOC and SOH at different time scales. While model-based methods can achieve high estimation accuracy under stable external experimental conditions, they still face significant challenges in accurately modeling batteries under complex operating environments.

In contrast, data-driven SOH estimation methods do not require precise modeling of the complex electrochemical mechanisms inside the battery, nor do they necessitate identification of the battery's internal parameters [12]. Instead, these methods directly capture health indicators (HIs) related to aging from large amounts of historical data and utilize machine learning algorithms to establish a mapping relationship between HIs and SOH, thereby achieving SOH prediction. As a result, data-driven methods exhibit high transferability, robustness, and generalization capabilities. SVR and artificial neural networks (ANNs) are currently two widely used data-driven techniques.

ANNs are composed of a large number of interconnected neurons, following specific rules [13]. Through repeated training on experimental data, the model gradually optimizes the weight distributions and thresholds between neurons, thereby constructing accurate data models to predict system states. Yang et al. [14] used a three-layer BP neural network for modeling to estimate battery SOH. However, traditional BP neural networks are prone to local optima, leading to inaccurate SOH predictions. SVR, as a supervised learning algorithm, effectively addresses the issues of local optima and low convergence efficiency inherent in traditional neural networks. Wang et al. [15] reduced the dimensionality of battery current, voltage, and temperature data to obtain two His, operating temperature and energy efficiency, and combined these with the SVR algorithm to achieve high-accuracy SOH predictions. Similarly, Wei et al. [16] applied deep neural networks (DNNs) for SOH estimation, demonstrating that DNNs have stronger high-level feature extraction capabilities and higher prediction accuracy compared to BP neural networks. CNN, as feedforward neural networks, can efficiently extract local features of battery performance data with high computational efficiency. Sheng et al. [17] implemented online capacity estimation of lithium batteries using a deep CNN model. The model automatically learned features from charging data, achieving root mean square errors below 2.0% under varying conditions. Van et al. [18] argued that, compared to feedforward neural networks, LSTM networks are capable of capturing time-dependent information in feature sequences, providing higher accuracy. EL-DALAHMEH et al. [19] compared the SOH estimation performance of three data-driven algorithms: nonlinear autoregressive neural networks, CNN, and LSTM. Their experimental results showed that LSTM outperformed the other two algorithms in estimation accuracy under different operating conditions. However, both SVR and neural network methods often face challenges such as difficulty in determining hyperparameters and slow convergence, which directly affect prediction performance. To address these challenges, various intelligent optimization algorithms have been developed to enhance these models. For example, Li et al. [20] used an SVR model based on an improved ant lion optimization algorithm to estimate battery SOH, achieving a MAPE of 0.53%, significantly improving prediction accuracy. Qian et al. [21] addressed uncertainties in charge and discharge characteristics during practical applications by using a one-dimensional CNN model. They used data from multiple random time intervals of charging curves as input and optimized part of the model parameters with an improved PSO algorithm, thereby improving capacity prediction accuracy. Tong et al. [22] proposed an adaptive LSTM model for battery SOH estimation, which, combined with Bayesian optimization, required only 25% of aging data for training to achieve relatively accurate SOH estimation. Additionally, integrating various data-driven methods has further enhanced SOH prediction accuracy. For instance, Kong et al. [23] developed an online health prediction framework combining a deep convolutional neural network (DCNN) and a dual-layer LSTM, leveraging DCNN for spatial feature extraction and LSTM for time-series analysis to ensure high computational precision.

Based on previous studies, it can be observed that most researchers have primarily focused on using single models for lithium battery SOH prediction, with relatively few comparisons and evaluations across multiple models. Therefore, to comprehensively assess the applicability and performance of different methods in lithium battery SOH prediction, this study selects LSTM, CNN, and SVR for comparative analysis. By comparing the prediction performance of these models, this study aims to reveal the strengths and limitations of different methods under various conditions, providing more accurate and reliable decision support for lithium battery health state estimation. Furthermore, to improve prediction accuracy and reduce manual intervention during model training, this research employs the PSO algorithm to optimize the key parameters of the three models, further enhancing prediction performance and ensuring the accuracy and reliability of the results.

2. Review of Related Algorithms

2.1. Review of LSTM Algorithm

LSTM is a specialized type of recurrent neural network (RNN) that addresses the gradient vanishing and exploding issues encountered by traditional RNN when handling long-sequence data. As shown in Figure 1, each LSTM unit consists of a forget gate, an input gate, an output gate, and a cell state. These components work collaboratively to control the selective memory and forgetting of information, enabling the dynamic feature extraction of sequential data.

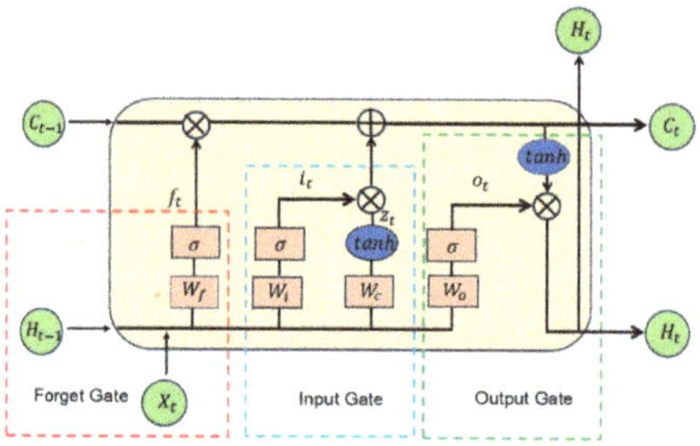

Figure 1. Basic structure of LSTM neural units.

1. Forget gate: Determines the portion of information in the current cell state that needs to be forgotten. The calculation formula is as follows:

$$f_t = \sigma(W_f \cdot [H_{t-1}, X_t] + b_f \tag{1}$$

where f_t is the output of the forget gate; W_f and b_f are the weights and biases of the forget gate; H_{t-1} is the hidden state of the previous moment; X_{t-1} is the input of the current moment; and σ is the Sigmoid function [24].

2. Input gate: Determines the new information that needs to be updated into the current memory. The calculation formula is as follows:

$$i_t = \sigma(W_i \cdot [H_{t-1}, X_t] + b_i \tag{2}$$

$$z_t = \tan \mathrm{h}(W_c \cdot [H_{t-1}, X_t] + b_c \tag{3}$$

where i_t is the output of the input gate and z_t is the candidate state.

3. Update cell state: Updates the memory unit based on the information from the forget gate and the input gate. The formula is as follows:

$$C_t = f_t \odot C_{t-1} + i_t \odot z_t \tag{4}$$

4. Output gate: Determines the hidden state H_t at the current time step, providing output information for the next layer.

$$o_t = \sigma(W_o \cdot [H_{t-1}, X_t] + b_o \tag{5}$$

$$H_t = o_t \odot \tan \mathrm{h}(C_t) \tag{6}$$

where o_t is the output of the output gate and H_t is the output at the current moment.

Through the above mechanisms, LSTM dynamically remembers and forgets information, enabling it to effectively capture both short-term and long-term dependencies in time-series tasks. This characteristic allows for LSTM to track the degradation patterns of battery performance over time, making it particularly advantageous in scenarios where the battery aging process spans a long period.

2.2. Review of the CNN Algorithm

CNN is developed from the neural perceptual machine concept, inspired by animal visual systems [25], and are a form of feedforward neural networks featuring convolutional operations within a deep structure [26], making CNN one of the representative algorithms in deep learning. The prediction principle of a CNN model lies in its layered structure comprising convolutional, pooling, and fully connected layers, which sequentially extract features from the data to make predictions.

(1) Convolutional layer: Extracts local features such as edges and textures by scanning the input data with convolutional kernels [27]. The kernels slide over the input data with a specified stride, performing dot products and weighted summations to compute the feature maps. The convolution operation is given by the following formula:

$$y_{i,j,k} = \sum_{m,n} x_i + m, j + n \cdot w_{m,n,k} + b_k \tag{7}$$

where x represents the input, w is the convolutional kernel, b is the bias, and y is the output feature map [28].

(2) Pooling layer: Reduces the dimensionality of the convolutional results, decreases the number of parameters, retains essential features, and enhances the model's computational efficiency and robustness. Common pooling methods include max pooling and average pooling [28].

(3) Fully connected layer: Combines the local features extracted by the previous layers into global information through fully connected operations, providing the final prediction output.

The CNN model leverages the convolutional layer for local feature extraction, the pooling layer for dimensionality reduction and noise filtering, and the fully connected layer for integrating global features. This structure offers significant advantages in predicting the SOH of lithium-ion batteries. Compared to other time-series prediction models, CNN demonstrates superior computational efficiency and the ability to capture local features, making it particularly suitable for analyzing the nonlinear changes and dynamic characteristics of battery health.

2.3. Review of the SVR Algorithm

SVR is an extension of the Support Vector Machine (SVM) and is specifically designed to address regression problems. It achieves data fitting by introducing the ε-insensitive loss function and utilizes kernel functions to map nonlinear problems into a high-dimensional feature space, thereby capturing complex nonlinear relationships.

The objective of SVR is to find a function $f(x) = w \cdot x + b$ that fits the data within an error margin of ϵ while maintaining a simple model with strong generalization capability. The key steps of SVR are as follows:

(1) ε-insensitive loss function: SVR introduces the ε-insensitive loss function to define an allowable error margin ϵ. If the error between the predicted value and the actual value exceeds ϵ, a loss is incurred; otherwise, the error is considered negligible.

(2) Optimization objective function: The goal of SVR is to balance the complexity of the model (achieved by regularizing the weights w) and the prediction error (measured by the loss function). The objective function is as follows:

$$\min\left\{\frac{1}{2}||w||^2 + c \cdot \sum_{i=1}^{s} (\xi_i + \xi_i^*)\right\} \quad (8)$$

$$s.t\begin{cases} w \cdot x_i + b - y_i \leq \varepsilon + \xi_i \\ y_i - w \cdot x_i - b \leq \varepsilon + \xi_i^* \\ \xi_i \geq 0, \xi_i^* \geq 0 \end{cases} \quad (9)$$

where $\frac{1}{2}||w||^2$ is the regularity term, which controls the model complexity; c is the penalty factor, which controls the trade-off between the error term and the model complexity; and ξ_i and ξ_i^* are the relaxation factors, which are used to deal with sample points with errors exceeding ϵ.

(3) Convex optimization: SVR transforms the regression problem into a convex optimization problem. By leveraging the dual form and Lagrange multipliers, SVR enhances its ability to fit nonlinear data effectively.

(4) Kernel function: To handle nonlinear problems, SVR employs kernel functions to map low-dimensional data into a high-dimensional feature space, where linear regression can be performed. Common kernel functions include the radial basis function (RBF), linear kernel, and polynomial kernel.

SVR aims to control prediction errors within an acceptable range, making it particularly suitable for precise modeling of battery SOH. Additionally, SVR exhibits strong generalization performance even under small sample conditions, effectively addressing the limitation of limited experimental data for lithium-ion batteries.

By summarizing relevant algorithms, it becomes evident that the LSTM model excels in handling time-series data, effectively capturing the complex temporal dynamics inherent in the data. Since SOH prediction for lithium-ion batteries is a typical time-series problem, LSTM leverages its memory capabilities to effectively track the long-term degradation of battery performance. The CNN model, on the other hand, is proficient in extracting local features efficiently, enabling it to identify critical characteristics from lithium battery parameter data while demonstrating strong noise tolerance. Lastly, the SVR model, through kernel functions (e.g., RBF kernel), effectively handles the complex nonlinear relationships in battery SOH and exhibits robust generalization capability with reduced risk of overfitting. Based on these considerations, this study selects LSTM, CNN, and SVR models to comprehensively address the challenges of lithium battery SOH prediction from three perspectives: long-term temporal dependency, local feature extraction, and nonlinear modeling.

2.4. Review of the PSO Algorithm

The PSO algorithm, proposed by Kennedy and Eberhart, was inspired by the foraging behavior of bird flocks and is a type of evolutionary computation technique based on swarm intelligence theory [29]. Due to its simplicity, few parameters, fast convergence, and ease of implementation, PSO has been widely applied in parameter optimization, path planning, production scheduling, and other fields. In a D-dimensional search space, a swarm is composed of N particles, each of which updates its velocity and position based on its individual experience and the collective experience of the swarm. The fitness function serves as the evaluation criterion, ultimately leading to the discovery of the global optimal solution. During the t-th iteration, the velocity vector of each particle is represented as $V_i^t = (v_{i1}^t, v_{i2}^t, \ldots, v_{id}^t)$,and the position vector is denoted as $X_i^t = (x_{i1}^t, x_{i2}^t, \ldots, x_{id}^t)$.

The formula for particle updating is expressed as follows:

$$v_{id}^{t+1} = \omega v_{id}^t + c_1 r_1 (p_{id}^t - x_{id}^t) + c_2 r_2 (p_{gd}^t - x_{id}^t) \tag{10}$$

$$x_{id}^{t+1} = x_{id}^t + v_{id}^{t+1} \tag{11}$$

where ω is the inertia weight, which characterizes the global search ability of the algorithm. c_1 and c_2 are acceleration factors that represent individual learning factors and social learning factors, respectively, and control the extent to which particles converge towards individual optimums and global optima. p_{id}^t is the individual optimal position, p_{gd}^t is the global optimal position. r_1 and r_2 are random numbers in the interval [0,1], which add randomness to the algorithm.

3. Data Analysis of Lithium Battery Aging and Extraction of Health Factors

3.1. Introduction to the Dataset

The dataset used in this study comes from the Predictive Center of Excellence (PCoE) of the National Aeronautics and Space Administration (NASA) [30]. The NASA PCoE laboratory conducted multiple charge–discharge cycle experiments on 2Ah 18,650 commercial lithium batteries under different ambient temperatures and charge–discharge modes. Key parameters characterizing the aging of lithium batteries, including capacity, voltage, resistance, and current, were obtained through charging, discharging, and impedance measurements. The charge–discharge cycle experiment process is divided into two stages: constant current–constant voltage (CC-CV) charging and constant current discharging [31]. First, the battery is charged at a constant current of 1.5 A until the voltage reaches 4.2 V and then continues to be charged at a constant voltage of 4.2 V until the current drops to 20 mA, at which point charging stops [32]. Finally, the battery is continuously discharged at a current of 2 A until the voltage drops to the set discharge cut-off voltage and discharging stops [33].

Considering the working conditions of electric vehicles during actual driving, this study selected the aging data of three batteries that were tested at a normal temperature of 24 °C for further research. The batteries were designated B0005, B0006, and B0007, respectively, and battery capacity was used as the main parameter to characterize the SOH of the battery. The specific charging and discharging experimental conditions are shown in Table 1.

Table 1. Specific experimental parameters for batteries.

Battery	Charge Current/A	Cutoff Current/mA	Discharge Current/A	Discharge Cut-Off Voltage/V
B0005	1.5	20	2	2.7
B0006	1.5	20	2	2.5
B0007	1.5	20	2	2.2

3.2. Extraction of Health Indicators

The selection of HIs is critical to the accuracy of SOH prediction, as appropriate HIs not only enhance model performance, but also reduce the time required for parameter adjustment [34]. Since battery SOH is influenced by various internal and external indicators and is closely related to the aging mechanism, it is difficult to obtain directly through measurement. However, studies have shown that HIs can be extracted from the patterns of external data such as voltage, current, and temperature generated during the aging process to indirectly predict SOH.

In this paper, we take the B0005 battery as an example for feature extraction. The graph in Figure 2a shows that the time to reach peak temperature shortens as the number of cycles increases, reflecting capacity decay. Therefore, this value can be used as a health indicator, denoted as HI1. The peak temperature under different cycle counts also varies, with the highest peak temperature during charging denoted as HI2. To fully utilize the existing temperature data, the average charging temperature is used as a health indicator, denoted as HI3. Figure 2b shows the discharge temperature curve, where the time to reach peak temperature shortens, the peak temperature increases, and the slope of the curve becomes steeper, indicating that the average discharge temperature is rising. Therefore, the peak time, peak temperature, and average discharge temperature are taken as HI4, HI5, and HI6, respectively.

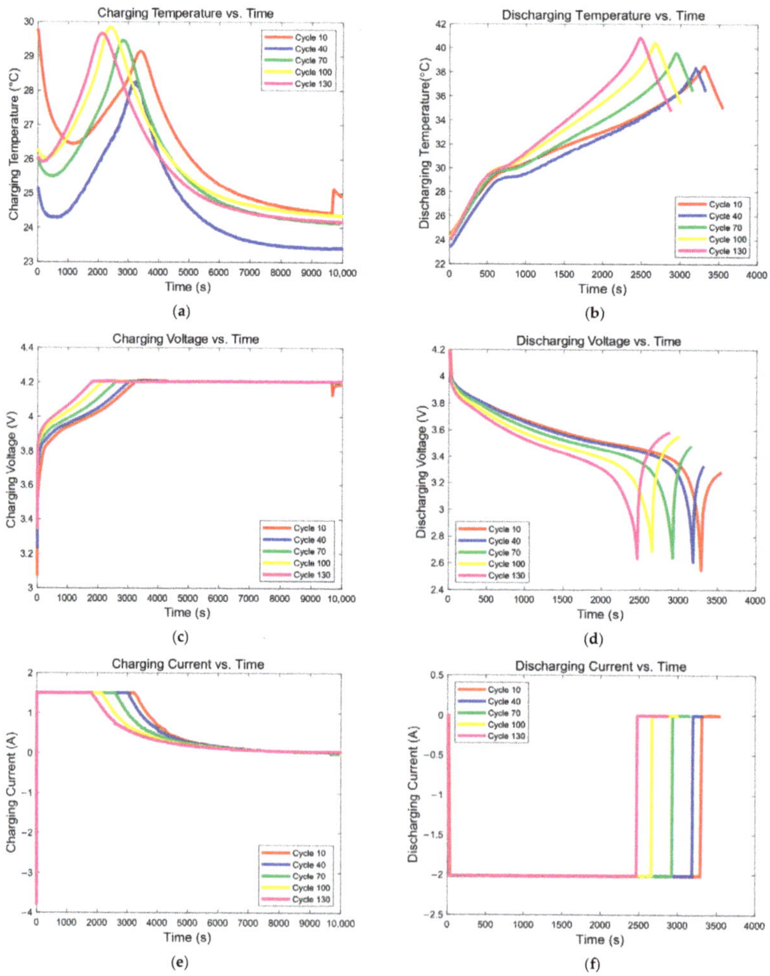

Figure 2. Temperature, voltage, and current versus time for different cycles of the battery. (**a**–**f**) represent the change curves of the battery charging temperature, discharging temperature, charging voltage, discharging voltage, charging current, and discharging current under different cycles.

From Figure 2c, it can be seen that the initial charging voltage decreases with the number of cycles, the difference between the charging voltage and cutoff voltage gradually decreases over the same time interval, and the area under the voltage curve during constant current charging also decreases. These changes reflect capacity decay, so the initial charging voltage, the isochronous average charging voltage difference, and the area under the voltage curve during constant current charging are denoted as HI7, HI8, and HI9, respectively. Figure 2d shows that over the same time interval, the rate of discharge voltage decrease increases, indicating battery capacity decay. Therefore, the isochronous discharge voltage drop is denoted as HI10, while the average discharge voltage is taken as HI11 due to varying working conditions during actual use.

From Figure 2e, it can be observed that the maximum slope of the current decreases with the number of cycles, especially between 0.5 A and 0.2 A, where the decline is more gradual, making it easier to analyze. Thus, the isochronous current drop time from 0.5 A to 0.2 A is denoted as HI12. Additionally, the difference between the charging current at the same moment and the constant current gradually increases, so the isochronous average charging current difference is denoted as HI13. The integral area of the current curve over time gradually decreases, indicating a decrease in charging power; thus, the area under the constant voltage charging current curve is taken as HI14. From Figure 2f, it is evident that the discharge current fluctuates significantly, indicating the complexity of the discharging process. Therefore, the average discharge current is chosen as HI15.

By analyzing the variations in the charging voltage and current curves, it can be observed that the charging time varies with the number of cycles. It takes approximately 3600 s for new batteries, while this time shortens to 1800 s for aging batteries. As a result, the charging energy is used as an indicator to characterize the health status of the battery, denoted as HI16.

A total of 16 health indicators were extracted from the temperature, current, and voltage profiles. However, the correlation between these indicators and battery capacity varies. Therefore, it is necessary to screen these indicators to remove invalid and redundant information, thereby reducing the computational complexity and training time of the network model. For this purpose, the Pearson correlation coefficient method was used to analyze the correlation of the 16 indicators, as shown in the following formula:

$$Pearson = \frac{\sum_{i=1}^{N}(x_i - \overline{x})(y_i - \overline{y})}{[\sum_{i=1}^{N}(x_i - \overline{x})^2 \sum_{i=1}^{N}(y_i - \overline{y})^2]^{\frac{1}{2}}} \tag{12}$$

The closer the absolute value of the coefficient obtained from the above equation is to 1, the stronger the correlation; the closer the value is to 0, the weaker the correlation [35]. In this study, health indicators with correlation coefficients less than 0.75 are eliminated, and the remaining indicators are used for model training and testing. The results of the correlation calculations are shown in Table 2.

Table 2. Table of health factor and capacity correlation coefficient values.

Indicator	B0005	B0006	B0007
HI1	0.927298	0.939146	0.892884
HI2	−0.570453	−0.245327	−0.492198
HI3	0.022756	0.266828	0.0600099
HI4	0.999935	0.999915	0.999725
HI5	−0.934350	−0.848772	−0.750104
HI6	−0.677412	−0.690275	−0.404310
HI7	−0.807104	−0.830185	−0.852246
HI8	0.944858	0.957171	0.890017
HI9	0.940103	0.955189	0.901295
HI10	−0.993799	−0.99226	−0.995370
HI11	0.992469	0.991659	0.9844268
HI12	−0.944819	−0.865739	−0.969182
HI13	−0.926612	−0.905479	−0.921218
HI14	−0.990445	−0.966795	−0.973317
HI15	0.321169	0.303079	0.303079
HI16	0.904601	0.925035	0.860496

From the results in Table 2, it can be observed that the correlation coefficients of four health indicators—HI2, HI3, HI6, and HI15—are lower than 0.75, and thus should be eliminated. Ultimately, the remaining 12 health indicators were used as the key indicators for battery SOH prediction.

4. Experimentation and Analysis of Results

4.1. Forecasting Modeling

To implement the proposed PSO-LSTM, PSO-CNN, and PSO-SVR models for state-of-health (SOH) prediction of lithium-ion batteries, aging datasets from three battery sets, B0005, B0006, and B0007, were employed for experimental validation. Considering that the selection of model parameters is very critical when predicting SOH of lithium batteries based on machine learning, the PSO algorithm is used to systematically optimize the key parameters of each model, minimize the need for manual adjustment, and improve the prediction accuracy and model stability. Based on parameter characteristics and model performance on the NASA dataset, the parameter optimization ranges were determined, as presented in Table 3.

Table 3. Optimization ranges for three model parameters.

Model	Parameter	Range
LSTM	Number of Hidden Neurons	range (10, 100)
	Time Steps	range (1, 10)
	Learning Rate	range (0.01, 0.3)
	Epochs	range (10, 100)
	Batch Size	range (10, 100)
CNN	Number of Filters	range (10, 100)
	Pooling Size and Strides	range (1, 10)
	Filter Size	range (1, 10)
	Learning_Rate	range (0.01, 0.3)
	Batch_Size	range (10, 100)
SVR	C	range (1, 100)
	Gamma	range (0.001, 0.1)
	Epsilon	range (0.001, 0.1)

The model optimization results are evaluated using Root Mean Square Error (RMSE), where an RMSE closer to zero indicates better parameter optimization. Figure 3 illustrates the optimization and prediction process for the three models: PSO-LSTM, PSO-CNN, and PSO-SVR.

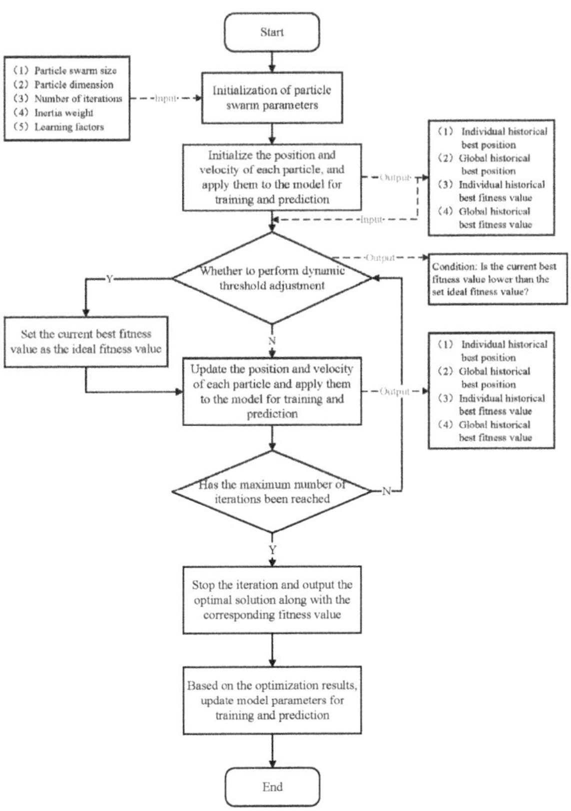

Figure 3. Prediction process of SOH of lithium battery by each model optimized based on PSO algorithm.

The detailed steps of PSO are as follows.

(1) Dataset division. For each set of aging dataset, the training and testing sets are divided in the ratio of 7:3, and 25% from the training set is divided as the validation set to enhance the generalization ability of the model.

(2) Data normalization. Min–max normalization is applied to all data, eliminating scale differences and ensuring uniform magnitude.

$$X_{norm} = \frac{X - X_{\min}}{X_{\max} - X_{\min}} \tag{13}$$

(3) PSO particle swarm parameter initialization. The number of iterations is set to 30, with a population size of 15; individual and social learning factors are both set to 1.7, and inertia weight to 0.6. Each population contains multiple particles, representing different hyperparameter combinations. For PSO-LSTM, the hyperparameters include Learning Rate, Number of Hidden Neurons, Time Step, Batch Size, and Epochs; for PSO-CNN, parameters include number of Filters, Pooling Size and Strides, Filter Size, Batch Size, and Learning Rate; and for PSO-SVR, parameters include Regularization Parameter C, Kernel Function Coefficient Gamma, and Epsilon.

(4) Initialize the position and velocity of each particle. Each particle corresponds to an LSTM, CNN, or SVR model. For each particle, the model was trained using the current combination of hyperparameters, and the prediction error RMSE of the test set was used as the fitness value of the particle. A lower fitness value represents a better prediction performance of the model for that hyperparameter combination. This gives the fitness value for each particle and the optimal fitness value in the population.

(5) Dynamic threshold adjustment. Unlike traditional PSO, this approach incorporates a dynamic threshold per iteration to decide if further optimization is required. If the current optimal fitness value is below this threshold, the threshold is updated.

(6) Position and velocity update. In each iteration, particles are updated based on individual and group optimal positions, speed, and current positions, and then fed back into the model for training and prediction.

(7) Continuous iteration search. The PSO algorithm iteratively performs "Dynamic Adjustment-Training Prediction-Update" steps to optimize hyperparameters across all iterations. Individual and group optimal positions are tracked and updated in each iteration, recording the population's optimal fitness values and corresponding hyperparameter combinations.

(8) Output of optimal parameters. After reaching the maximum iteration count, the algorithm outputs the optimal parameter combinations and corresponding fitness values.

(9) Final prediction. Each model's parameters are set to the optimal values, and SOH prediction is performed for the three datasets B0005, B0006, and B0007.

4.2. *Model Experiment Results*

4.2.1. Model Parameter Optimization Results

The PSO algorithm is applied to optimize the parameters of the LSTM, CNN, and SVR models for datasets B0005, B0006, and B0007. The results are presented in Figure 4 and Tables 4–6. Here, the optimization search results of the PSO algorithm on the B0005 dataset are specifically illustrated.

Table 4. PSO values for LSTM parameters.

Model	Dataset	RMSE	Learning Rate	Number of Hidden Neurons	Time Steps	Batch Size	Epochs
LSTM	B0005	0.008838	0.0192	11	1	51	50
	B0006	0.012059	0.033	16	1	61	74
	B0007	0.008049	0.0236	24	1	46	75

Table 5. PSO values for CNN parameters.

Model	Dataset	RMSE	Learning Rate	Number of Filters	Kernel Size	Pooling Size and Strides	Batch Size
CNN	B0005	0.012816	0.0119	29	1	4	53
	B0006	0.019277	0.0127	28	3	1	49
	B0007	0.010639	0.0392	41	1	5	79

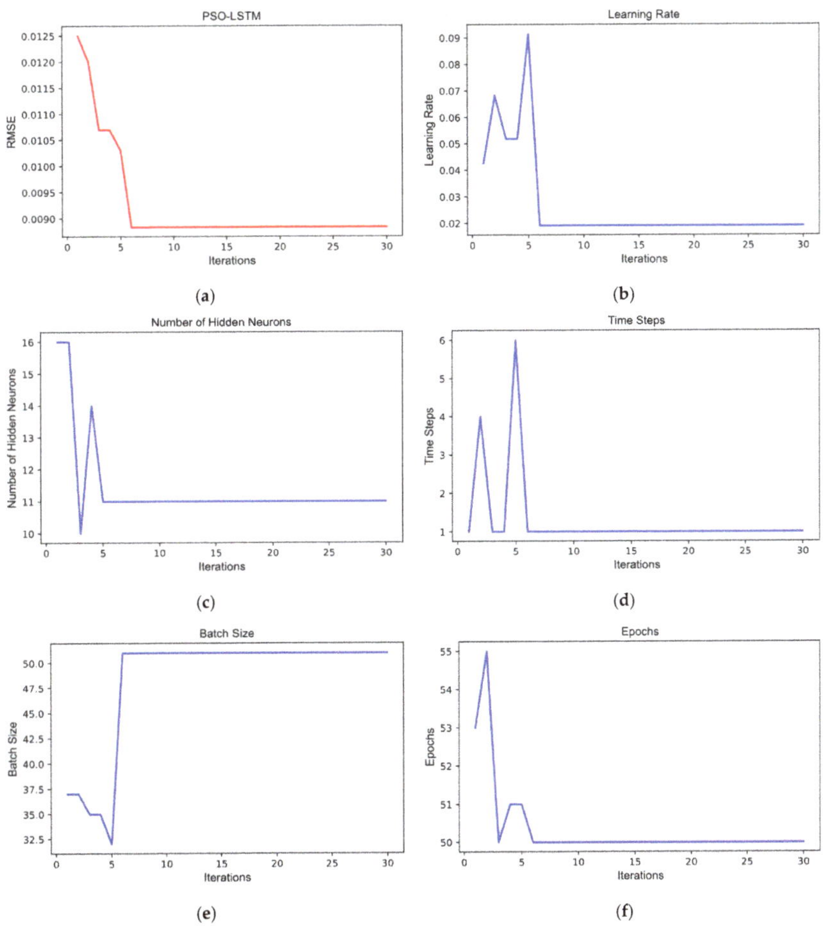

Figure 4. Optimization results of the PSO-LSTM model for the B0005 dataset. (**a**–**f**) represent the optimization results for the parameters Number of Hidden Neurons, Time Steps, Learning Rate, Epochs, and Batch Size.

Table 6. PSO values for SVR parameters.

Model	Dataset	RMSE	C	Gamma	Epsilon
SVR	B0005	0.013582	61	0.0195	0.0017
	B0006	0.020632	33	0.0062	0.0025
	B0007	0.017232	49	0.0017	0.0018

Figure 4 shows the optimization results of the PSO algorithm for the parameters of the LSTM model based on the B0005 dataset. It can be observed that after 30 iterations, the fitness value gradually decreases and eventually stabilizes. Simultaneously, each parameter being optimized also converges to a stable value during the iteration process. Therefore, the final stable values of these parameters after PSO can be considered to be the closest to the optimal solution for the dataset. Similarly, this method can be applied to obtain the optimization results for the parameters of the SVR and CNN models constructed from other datasets. The optimization results of the key parameters for each model are summarized in Tables 4–6.

4.2.2. Analysis of Experimental Results for Different Models

The PSO-LSTM, PSO-CNN, and PSO-SVR models were trained and tested using the parameter values obtained from the PSO described above. The predicted values of the PSO-LSTM, PSO-CNN, and PSO-SVR models on different datasets were then compared with the actual values, as illustrated in Figure 5.

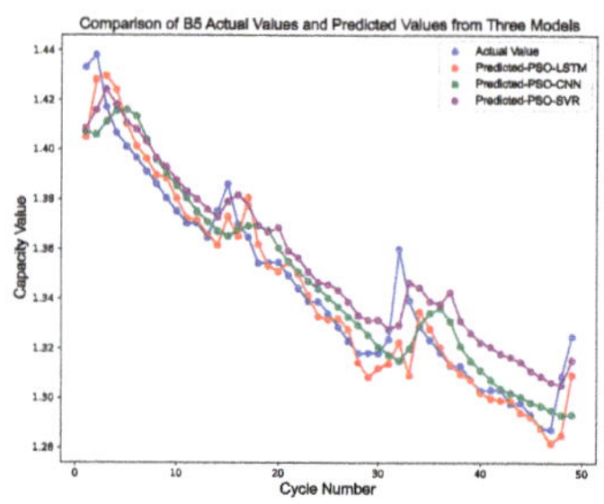

(a)

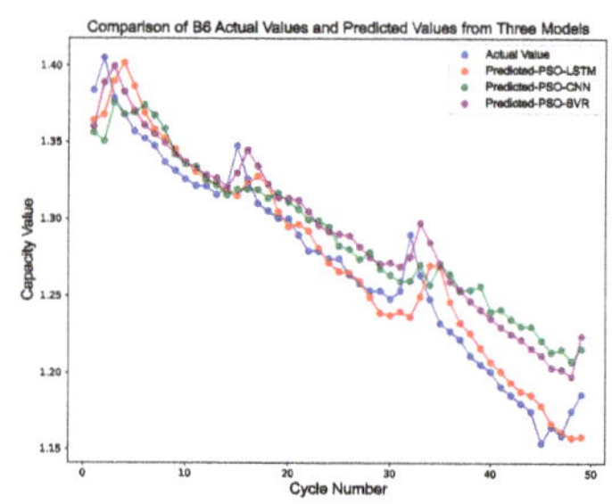

(b)

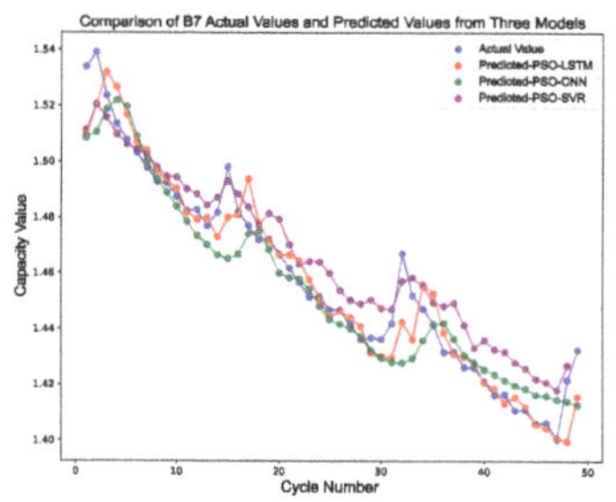

(c)

Figure 5. Prediction results of the three models on the three datasets. Among them, the plots in **(a–c)** represent the comparison between the predicted values and the true values of the B0005, B0006, and B0007 datasets on the three models, respectively.

As shown in the three graphs in Figure 5, the PSO-LSTM, PSO-CNN, and PSO-SVR models all demonstrate good prediction results for the SOH of cells B0005, B0006, and B0007, although there are differences in accuracy. The PSO-LSTM prediction curve is closest to the actual values, especially in most cycles, where it shows better tracking and smaller fluctuations, indicating its significant advantage in capturing the data's changing trends. Even in areas with large fluctuations, PSO-LSTM is relatively accurate in predicting the direction of data changes. The PSO-CNN model generally performs well in prediction, with its predicted values closely following the trend in actual values, particularly in areas where data changes are relatively stable. However, in regions with large data fluctuations, its prediction accuracy decreases, indicating certain limitations when dealing with complex patterns and sudden changes. Although the PSO-SVR model smoothly tracks real value changes in most data points, it exhibits more pronounced errors in areas with significant fluctuations, suggesting that while it performs well in capturing global trends, it is somewhat insufficient in regions with large data variations.

To comprehensively compare the predictive performance of the models, four commonly used evaluation metrics are employed: *RMSE*, *MAE*, *MAPE*, and the Coefficient of DR^2 [36]. The formulas for calculating each metric are as follows:

$$MAE = \frac{1}{N}\sum_{i=1}^{N}|\hat{y}_i - y_i| \tag{14}$$

$$RMSE = \sqrt{\frac{1}{N}\sum_{i=1}^{N}(\hat{y}_i - y_i)^2} \tag{15}$$

$$MAPE = \frac{100\%}{N}\sum_{i=1}^{N}\left|\frac{\hat{y}_i - y_i}{y_i}\right| \tag{16}$$

$$R^2 = 1 - \frac{\sum_i (\hat{y}_i - y_i)^2}{\sum_i (y_i - y)^2} \tag{17}$$

where N is the number of samples, $\hat{y}_i$ is the predicted battery capacity, and y_i is the actual battery capacity.

The model error comparison results are presented in Table 7. The results indicate that the PSO-LSTM model achieves the lowest MAE, RMSE, and MAPE values across all datasets, demonstrating its superior predictive performance. Additionally, the PSO-LSTM model exhibits the highest R^2 values, highlighting its excellent data-fitting capability, making it the most stable and optimal choice among the three methods. In contrast, the performance of the PSO-CNN model lies between that of PSO-LSTM and PSO-SVR. While its errors are slightly higher than those of PSO-LSTM, it significantly outperforms PSO-SVR, showcasing strong practicality and reliable fitting capability. On the other hand, the PSO-SVR model exhibits the weakest performance across all datasets and metrics, with larger errors and the lowest R^2 values, indicating its relatively poor predictive ability.

Table 7. Prediction errors of the dataset on different models.

Dataset	Prediction Model	MAE	RMSE	MAPE	R^2
B0005	PSO-LSTM	0.007413	0.010931	0.545550	0.921261
	PSO-CNN	0.010617	0.015898	0.930365	0.865900
	PSO-SVR	0.013960	0.015080	1.040813	0.845538
B0006	PSO-LSTM	0.012182	0.015197	0.961381	0.907357
	PSO-CNN	0.016638	0.018703	1.457222	0.833906
	PSO-SVR	0.018802	0.020316	1.835633	0.827241
B0007	PSO-LSTM	0.006739	0.009433	0.458213	0.929801
	PSO-CNN	0.009785	0.013172	0.899458	0.879730
	PSO-SVR	0.012121	0.014075	1.569721	0.839245

4.2.3. Analysis of the Effects Before and After Model Optimization

When the PSO algorithm is not used for optimization, the model parameters are set to default values for model building, training, and prediction. Figure 6 presents a comparison between the predicted and actual values for the same dataset before and after optimization for the three models.

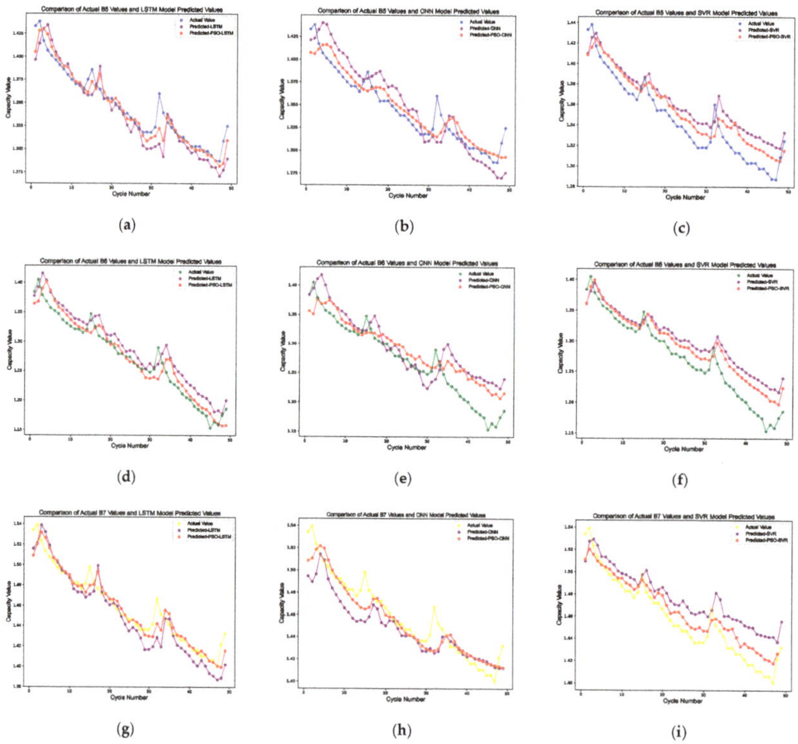

Figure 6. Comparison of the prediction results before and after model optimization on the same dataset. Among them, (**a**–**c**) represent the comparison between the predicted values and the actual values of the B0005 dataset before and after the optimization of the three models, respectively. (**d**–**f**) represent the comparison between the predicted values and the actual values of the B0006 dataset before and after the optimization of the three models, respectively. (**g**–**i**) represent the comparison between the predicted values and the actual values of the B0007 dataset before and after the optimization of the three models, respectively.

As shown in Figure 6, after optimization by the PSO algorithm, the three models exhibit a significant improvement in prediction accuracy for all three datasets. Compared to the unoptimized models, the optimized prediction curves are closer to the actual values, with smaller fluctuations, demonstrating higher accuracy and stability. This optimization effectively reduces errors and improves the model's fit to the actual values, verifying the effectiveness of the optimization strategy. Furthermore, a horizontal comparison of Figure 6 reveals that in subfigures (a) to (c), the purple curve (representing the unoptimized LSTM predictions) in subfigure (a) is closer to the blue curve (the actual values of battery B0005), indicating that the LSTM model outperforms the other two models even before optimization. Similar conclusions can be drawn from the prediction results of battery B0006 in subfigures (d) to (f) and battery B0007 in subfigures (g) to (i).

A comparison of the model performance evaluation metrics before and after optimization is presented in Table 8. The table shows that after optimization by the PSO algorithm, the error metrics for all three models improved significantly, with a notable decrease in MAE, MAPE, and RMSE, and an increase in R^2. The reduction in MAE and RMSE indicates enhanced prediction accuracy and precision, while the significant decrease in MAPE confirms increased reliability in predicting different capacity ranges. The increase in R^2 reflects improvements in the model's fitting ability, stability, and generalization. Furthermore, a comparison of the results before optimization reveals that the LSTM model outperforms the CNN and SVR models, achieving the smallest RMSE, MAE, and MAPE values, along with the highest R^2, which exceeds 0.8. This highlights its superior predictive capability and validates the earlier conclusion that, prior to optimization, the LSTM model demonstrated the best predictive performance among the three models.

Table 8. Comparison of prediction errors before and after model optimization.

	Dataset	MAE	RMSE	MAPE	R^2
Before LSTM optimization	B0005	0.012879	0.017659	1.059591	0.814509
	B0006	0.018616	0.020898	1.343736	0.805094
	B0007	0.011469	0.015693	0.952507	0.829709
After LSTM optimization	B0005	0.007413	0.010931	0.54555	0.921261
	B0006	0.012182	0.015197	0.961381	0.907357
	B0007	0.006739	0.009433	0.458213	0.929801
Before CNN optimization	B0005	0.01898	0.02042	1.74478	0.71724
	B0006	0.02186	0.02314	2.12921	0.70574
	B0007	0.01665	0.01871	1.647	0.72381
After CNN optimization	B0005	0.01062	0.0159	0.93036	0.8659
	B0006	0.01664	0.0187	1.45722	0.83391
	B0007	0.00979	0.01317	0.89946	0.87973
Before SVR optimization	B0005	0.01995	0.02116	1.72267	0.71014
	B0006	0.02442	0.02724	1.9643	0.69423
	B0007	0.0289	0.03065	2.27799	0.68511
After SVR optimization	B0005	0.01396	0.01508	1.04081	0.84554
	B0006	0.0188	0.02032	1.83563	0.82724
	B0007	0.01212	0.01408	1.56972	0.83925

5. Conclusions and Outlook

With the rapid development of the new energy industry, lithium-ion batteries are being increasingly used in electric vehicles and energy storage systems [37]. Health prediction technology plays a crucial role in promoting the efficient use of lithium-ion batteries in these fields and in supporting the transition to sustainable energy. This study utilized battery aging data from the NASA PCoE laboratory (B0005, B0006, and B0007). By analyzing battery temperature, voltage, and current curves, 16 features related to battery capacity degradation were extracted, and 12 highly correlated health indicators were selected. Subsequently, the PSO algorithm was used to optimize the parameters of three models: LSTM, CNN, and SVR. The models' prediction performance was evaluated using RMSE, MAE, MAPE, and R^2. The following conclusions were drawn:

(1) Effectiveness of the health factor. The health factor extracted by temperature, voltage, and current features can effectively characterize the battery capacity decline, which verifies the accuracy of the method.

(2) The PSO strategy can improve the model prediction performance. The PSO algorithm significantly improves the prediction effect of the three models, verifying the effectiveness of its optimization parameters, especially in improving the model stability and accuracy.

(3) The PSO-LSTM model has the best prediction effect. In the SOH prediction of B0005, B0006, and B0007 batteries, the PSO-LSTM model has the best performance in all evaluation indexes, especially in capturing the complex timing data trend and reducing the error, which provides an important reference for the accurate prediction of the SOH of batteries.

Due to the limitations of research time and experimental conditions, this study still has some deficiencies. Future research can be carried out in the following aspects:

(1) The dataset used in this paper is based on a single battery and is experimented on in an ideal environment, while the actual electric vehicle battery exists in the form of a battery pack which is affected by various factors. Therefore, charging and discharging data under actual working conditions can be collected for modeling studies in the future.

(2) In this paper, only three models, LSTM, CNN, and SVR, are used, and there may exist machine learning algorithms that are more suitable for SOH prediction that have not been tried. In addition, there are some limitations in optimizing the performance of a single model, and the prediction performance can be improved using model fusion in the future.

(3) Battery SOH estimation is only a part of the battery management system, and the estimation of SOP, SOF, and SOC of the battery all provide correlation information to each other and affect the accuracy of each other. In the future, SOH can be estimated jointly with other states to further improve the overall performance of the battery management system.

Author Contributions: Conceptualization, K.L.; Methodology, X.C.; Data curation, X.C.; Writing—original draft, X.C.; Project administration, K.L. All authors have read and agreed to the published version of the manuscript.

Funding: This research is partly funded by the National Natural Science Foundation of China (No. 71602143), and the (61403277).

Institutional Review Board Statement: Not applicable.

Informed Consent Statement: Not applicable.

Data Availability Statement: Data are contained within the article.

Conflicts of Interest: The authors declare no conflicts of interest.

References

1. Zhou, J.; Wang, S.L.; Cao, W.; Xie, Y.; Fernandez, C. State of health prediction of lithium-ion batteries based on SSA optimized hybrid neural network model. *Electrochim. Acta* **2024**, *487*, 144146. [CrossRef]
2. Zhao, K.; Ming, Z.; Wu, J.; Sun, W. Health state estimation of lithium ion batteries based on electrochemical impedance spectroscopy. *J. Phys. Conf. Ser.* **2022**, *2366*, 012030. [CrossRef]

3. Ge, M.-F.; Liu, Y.; Jiang, X.; Liu, J. A review on state of health estimations and remaining useful life prognostics of lithium-ion batteries. *Measurement* **2021**, *174*, 109057. [CrossRef]
4. Luo, F.; Huang, H.; Ni, L.P.; Li, T. Rapid prediction of the state of health of retired power batteries based on electrochemical impedance spectroscopy. *J. Energy Storage* **2021**, *41*, 102866. [CrossRef]
5. Zhang, S.; Guo, X.; Dou, X.; Zhang, X. A rapid online calculation method for state of health of lithium-ion battery based on coulomb counting method and differential voltage analysis. *J. Power Sources* **2020**, *479*, 228740. [CrossRef]
6. Su, L.; Xu, Y.; Dong, Z. State-of-health estimation of lithium-ion batteries: A comprehensive literature review from cell to pack levels. *Energy Convers. Econ.* **2024**, *5*, 224–242. [CrossRef]
7. Lanubile, A.; Bosoni, P.; Pozzato, G.; Allam, A.; Acquareone, M.; Onori, S. Domain knowledge-guided machine learning framework for state of health estimation in Lithium-ion batteries. *Commun. Eng.* **2024**, *3*, 168. [CrossRef]
8. Xu, Z.; Wang, J.; Lund, P.D.; Zhang, Y. Co-estimating the state of charge and health of lithium batteries through combining a minimalist electrochemical model and an equivalent circuit model. *Energy* **2022**, *240*, 122815. [CrossRef]
9. Hosseininasab, S.; Lin, C.; Pischinger, S.; Stapelbroek, M.; Vagnoni, G. State-of-health estimation of lithium-ion batteries for electrified vehicles using a reduced-order electrochemical model. *J. Energy Storage* **2022**, *52*, 104684. [CrossRef]
10. Amir, S.; Gulzar, M.; Tarar, M.O.; Naqvi, I.H.; Zaffar, N.A.; Pecht, M.G. Dynamic Equivalent Circuit Model to Estimate State-of-Health of Lithium-Ion Batteries. *IEEE Access* **2022**, *10*, 18279–18288. [CrossRef]
11. Zou, Y.; Hu, X.; Ma, H.; Li, S. Combined State of Charge and State of Health estimation over lithium-ion battery cell cycle lifespan for electric vehicles. *J. Power Sources* **2015**, *273*, 793–803. [CrossRef]
12. Xu, N.; Xie, Y.; Liu, Q.; Yue, F.; Zhao, D. A Data-Driven Approach to State of Health Estimation and Prediction for a Lithium-Ion Battery Pack of Electric Buses Based on Real-World Data. *Sensors* **2022**, *22*, 5762. [CrossRef]
13. Chen, L.; Li, S.; Bai, Q.; Yang, J.; Jiang, S.; Miao, Y. Review of Image Classification Algorithms Based on Convolutional Neural Networks. *Remote Sens.* **2021**, *13*, 4712. [CrossRef]
14. Yang, D.; Wang, Y.; Pan, R.; Chen, R.; Chen, Z. A Neural Network Based State-of-Health Estimation of Lithium-Ion Battery in Electric Vehicles. *Energy Procedia* **2017**, *105*, 2059–2064. [CrossRef]
15. Wang, S.; Zhao, L.; Su, X.; Ma, P. Prognostics of Lithium-Ion Batteries Based on Battery Performance Analysis and Flexible Support Vector Regression. *Energies* **2014**, *7*, 6492–6508. [CrossRef]
16. Wei, Z.X.; Han, X.J.; Li, X. Battery Health State Evaluation for Second-life Applications Based on Deep Neural Networks. *Sol. Energy* **2022**, *43*, 518–524.
17. Shen, S.; Sadoughi, M.; Chen, X.; Hong, M.; Hu, C. A deep learning method for online capacity estimation of lithium-ion batteries. *J. Energy Storage* **2019**, *25*, 100817. [CrossRef]
18. Nguyen Van, C.; Quang, D.T. Estimation of SOH and internal resistances of Lithium ion battery based on LSTM network. *Int. J. Electrochem. Sci.* **2023**, *18*, 100166. [CrossRef]
19. El-Dalahmeh, M.; Lillystone, J.; Al-Greer, M. State of Health Estimation of Lithium-Ion Batteries Based on Data-Driven Techniques. In Proceedings of the 56th International Universities Power Engineering Conference (UPEC), Middlesbrough, UK, 31 August–3 September 2021; pp. 1–6.
20. Li, Q.; Li, D.; Zhao, K.; Wang, L.; Wang, K. State of health estimation of lithium-ion battery based on improved ant lion optimization and support vector regression. *J. Energy Storage* **2022**, *50*, 104215. [CrossRef]
21. Qian, C.; Xu, B.; Chang, L.; Sun, B.; Feng, Q.; Yang, D.; Ren, Y.; Wang, Z. Convolutional neural network based capacity estimation using random segments of the charging curves for lithium-ion batteries. *Energy* **2021**, *227*, 120333. [CrossRef]
22. Tong, Z.; Miao, J.; Tong, S.; Lu, Y. Early prediction of remaining useful life for Lithium-ion batteries based on a hybrid machine learning method. *J. Clean. Prod.* **2021**, *317*, 128265. [CrossRef]
23. Kong, D.; Wang, S.; Ping, P. State-of-health estimation and remaining useful life for lithium-ion battery based on deep learning with Bayesian hyperparameter optimization. *Int. J. Energy Res.* **2022**, *46*, 6081–6098. [CrossRef]
24. Huang, Y.; Li, J.; Hou, W.; Zhang, B.; Zhang, Y.; Li, Y.; Sun, L. Improved clustering and deep learning based short-term wind energy forecasting in large-scale wind farms. *J. Renew. Sustain. Energy* **2020**, *12*, 066101. [CrossRef]
25. Alhussein, M.; Aurangzeb, K.; Haider, S.I. Hybrid CNN-LSTM model for short-term individual household load forecasting. *IEEE Access* **2020**, *8*, 180544–180557. [CrossRef]
26. Park, K.; Choi, Y.; Choi, W.J.; Ryu, H.Y.; Kim, H. TM-Based Battery Remaining Useful Life Prediction With Multi-Channel Charging Profiles. *IEEE Access* **2020**, *8*, 20786–20798. [CrossRef]
27. Zhu, R.; Dai, L.Y.; Liu, J.; Guo, Y. Diagnostic Classification of Lung Cancer Using Deep Transfer Learning. In Proceedings of the 2019 Chinese Intelligent Systems Conference, Haikou, China, 26–27 October 2019; Springer Science and Business Media LLC: Berlin, Germany, 2020.

28. Dai, Q.; Chen, H.; Yi, W.-J.; Zhao, J.-N.; Zhang, W.; He, P.-A.; Liu, X.-Q.; Zheng, Y.-F.; Shi, Z.-X. Precision DNA Methylation Typing via Hierarchical Clustering of Nanopore Current Signals and Attention-Based Neural Network. *Brief. Bioinform.* **2024**, *25*, bbae596. [CrossRef]
29. Kennedy, J.; Eberhart, R. Particle Swarm Optimization. *Proc. Int. Conf. Neural Netw.* **1995**, *4*, 1942–1948.
30. National Aeronautics and Space Administration Prognostics Center of Excellence. PCoE Datasets. Available online: https://ti.arc.nasa.gov/tech/dash/groups/pcoe/prognostic-data-repository/#battery (accessed on 16 August 2023).
31. Fan, Y.; Qiu, J.; Wang, S.; Yang, X.; Liu, D.; Fernandez, C. Incremental Capacity Curve Health-Indicator Extraction Based on Gaussian Filter and Improved Relevance Vector Machine for Lithium–Ion Battery Remaining Useful Life Estimation. *Metals* **2022**, *12*, 1331. [CrossRef]
32. Ma, Y.; Li, J.; Gao, J.; Chen, H. State of health prediction of lithium-ion batteries under early partial data based on IWOA-BiLSTM with single feature. *Energy* **2024**, *295*, 131085. [CrossRef]
33. Duan, W.; Song, S.; Xiao, F.; Chen, Y.; Peng, S.; Song, C. Battery SOH estimation and RUL prediction framework based on variable forgetting factor online sequential extreme learning machine and particle filter. *J. Energy Storage* **2023**, *65*, 107322. [CrossRef]
34. Dineva, A. Evaluation of Advances in Battery Health Prediction for Electric Vehicles from Traditional Linear Filters to Latest Machine Learning Approaches. *Batteries* **2024**, *10*, 356. [CrossRef]
35. Cai, H.; Duan, W. Narratives of a lost spatial pattern: A spatial analysis of the distribution of Namo Daoguans in Guangzhou in Republican period. *J. Asian Archit. Build. Eng.* **2023**, *22*, 2541–2566. [CrossRef]
36. Li, X.; Yu, H.; Wang, J.; Xia, Y.; Zheng, H.; Song, H. Prediction of state-of-health of lithium-ion battery based on CEEMDAN-SG-LSTM combined model. *Mater. Today Sustain.* **2024**, *28*, 100999. [CrossRef]
37. Ujiie, S.; Yagasaki, E. Characterization of EP-ash, collected at fossil fuel power plants, as electrode material. *J. Power Sources* **1999**, *79*, 97–104. [CrossRef]

applied sciences

MDPI

Article

An Improved YOLOv8 Model for Strip Steel Surface Defect Detection

Jinwen Wang [1], Ting Chen [2,3,*], Xinke Xu [1,*], Longbiao Zhao [4], Dijian Yuan [1], Yu Du [1], Xiaowei Guo [2,3] and Ning Chen [2,3]

1 College of Metrology Measurement and Instrument, China Jiliang University, Hangzhou 310018, China; wjw9969@163.com
2 Zhejiang Institute of Quality Sciences, Hangzhou 310018, China
3 Key Laboratory of Digital Precision Measurement Technology of Zhejiang Province, Hangzhou 310018, China
4 Naval Research Institute, Beijing 100070, China
* Correspondence: chentingzqs@zuaa.zju.edu.cn (T.C.); 17a0202147@cjlu.edu.cn (X.X.); Tel.: +86-0571-8512-9818 (T.C.); +86-0571-8691-4435 (X.X.)

Abstract: In the process of steel strip production, the accuracy of defect detection remains a challenge due to the diversity of defect types, complex backgrounds, and noise interference. To improve the effectiveness of surface defect detection in steel strips, we propose an enhanced detection model known as YOLOv8-BSPB. First, we propose a novel pooling layer module, SCRD, which replaces max pooling with average pooling. This module introduces the receptive field block (RFB) and deformable convolutional network version 4 (DCNv4) to obtain learnable offsets, allowing convolutional kernels to flexibly move and deform on the input feature map, thus, more effectively extracting multi-scale features. Second, we integrate a polarized self-attention (PSA) mechanism to improve the model's feature representation and enhance its ability to focus on relevant information. Additionally, we incorporate the BAM attention mechanism after the C2f module to strengthen the model's feature selection capabilities. A bidirectional feature pyramid network is introduced at the neck of the model to improve feature transmission efficiency. Finally, the WIoU loss function is employed to accelerate the model's convergence speed and enhance regression accuracy. Experimental results on the NEU-DET dataset demonstrate that the improved model achieves a classification accuracy of 81.3%, an increase of 4.9% over the baseline, with a mean average precision of 86.9%. The model has a parameter count of 5.5 M and operates at 103.1 FPS. To validate the model's effectiveness, we conducted tests on the Kaggle steel strip dataset and our custom dataset, where the average accuracy improved by 2.3% and 5.5%, respectively. The experimental results indicate that the model meets the requirements for real-time, lightweight, and portable deployment.

Keywords: defect detection; PSA; image processing; SCRD; strip steel defects; YOLOv8

1. Introduction

The detection of surface defects in steel strips is of significant importance in steel manufacturing and quality control. It not only directly affects the final quality of the products but also plays a key role in the control of the production process and the efficient utilization of resources. Steel strip is a type of steel product characterized by a relatively narrow width and an extremely long length, typically produced through hot or cold rolling processes. It is widely used in various industries, including automotive [1], construction [2], and manufacturing [3], due to its versatility and strength. During its production, various surface defects may occur, such as plaques, scratches, inclusions, edge cracks, and dents. These defects not only affect the performance of the steel strip but also diminish the overall quality of the product [4]. Furthermore, surface defects can lead to more serious issues, such as belt breakage and accumulation, resulting in significant losses in steel production [5]. Therefore, rapid and accurate detection of surface defects in steel strips is crucial for improving product quality and production efficiency, making it an important area of research in related technologies [6]. With the continuous advancement of automation technology, traditional manual inspection methods have gradually been replaced by advanced computer vision technologies and machine learning (ML) algorithms. The application of machine learning, particularly in the fields of image processing and data analysis, has introduced novel solutions for the detection of surface defects in steel strips.

Traditional machine vision methods mainly extract texture features from a local or global perspective to distinguish defects from the background [7,8]. For example, Liu, K. [9] et al., whose research was based on full differential image decomposition to detect various defects on texture homogeneous surfaces for tiny defects under low contrast conditions. Vaidelienė, G. [10] et al. explored DWT attributes and evaluated the explicit definitions of the spectral coefficients of the DWT, carrying out task-oriented subset correlation statistical analyses to generate a defect detection criterion to evaluate the multifaceted nature of the defects in the texture image. The literature [11] proposes an improved SVD phase space reconstruction algorithm by combining the nonlinear dynamics feature recognition theory and evaluating and comparing the experimental results to prove the feasibility of the proposed method. However, these image-processing-based methods face significant noise interference in industrial applications, which may lead to poor adaptability and robustness.

In recent years, machine learning-based methods have been successfully applied to many automated defect detection tasks. For instance, Prasanna et al. [12] proposed a deep learning-based method for detecting surface defects on steel strips, which automatically classifies surface defects using convolutional neural networks (CNN), achieving high classification accuracy and robustness. Zhang et al. [13] proposed a surface defect detection method for steel strips based on deep convolutional neural networks (CNN). The study successfully detected various common defects by applying data augmentation to the steel strip surface images and combining CNN for feature extraction and classification. The literature [14] explores the use of convolutional neural networks (CNNs) for real-time detection of steel surface defects. The research demonstrates how deep learning technologies can replace traditional defect detection methods, leading to better efficiency and fewer false positives in industrial settings. In addition, transfer learning has been widely applied in the field of steel strip defect detection. Transfer learning leverages models trained on other domains, such as bridge defect detection [15], and can fully exploit their advantages in steel strip surface defect detection. This approach allows for the effective enhancement of model detection capabilities, even when the dataset is insufficient. For example, Wan X et al. [16] demonstrated that transfer learning could significantly improve the accuracy and generalization ability of steel strip surface defect detection models, especially in scenarios with limited data.

With the continuous development of science and technology, deep learning has achieved remarkable progress in strip surface defect detection, addressing the robustness challenges often encountered by traditional machine learning algorithms [17]. By leveraging their ability to capture complex patterns in large datasets, deep learning models have proven particularly effective for defect detection on steel strip surfaces. This advancement has significantly improved detection accuracy, robustness, and efficiency, making deep learning an essential tool in industrial applications for surface defect detection. The YOLO series, with its high efficiency in detecting the target, provides a more efficient and accurate detection technique for strip surface defect detection. Yan Lu et al. [18] proposed a YOLOv5s-CSG model outperformed the original YOLOv5s by 7.3%. Y. Du et al. [19] proposed an adaptive focusing function AFF algorithm on the NEU-DET data set, achieving an average accuracy of 83.5%. Gui et al. [20] improved a YOLOX-ADS model, introducing the spatial depth (SPD) layer to prevent the loss of fine-grained information for small target features, and the improved SPPFCSPC module enhances the model's fusion and representation ability. The mean precision and mean average precision of this method on the NEU-DET dataset were 81.4% and 79.9%. In [21], a slim-YOLOv8 model was proposed, then the online reparameterization method was introduced to reduce the number of model parameters and was finally tested on the NEU-DET dataset. Compared with YOLOv8, the accuracy of IoU 0.5 was increased by 8.3%, and the number of model parameters was reduced from 3.0 M to 2.7 M. Lu, J et al. [22] investigated an enhanced lightweight YOLOv7 model and obtained a 4.5% improvement in mAP50 over YOLOv7. In [23], GSConv was used to design an improved and efficient aggregation network by improving YOLOv7-tiny by adding SPDConv modules for low resolution and small defects between the prediction head and neck networks, with an average accuracy mean of up to 74.1%. In F. Ren et al. [24], an ECA-SimSPPF-SIoU-Yolov5-based algorithm achieved good detection results in mean average precision and recall. In [25], a FANet fusion attention network was proposed for detecting various steel surface defects by changing the backbone of FANet to ResNet-50, and a higher detection speed was achieved. In [26], a real-time detection network (RDN) was proposed that balanced speed and accuracy achieved 80.0% mAP, while C. Li et al. [27], based on the improved YOLOX algorithm, achieved 77% mAP on the NEU-DET dataset with a speed of 100 FPS. The above methods have been improved and designed based on the YOLO series.

Nevertheless, steel strip surface defect detection still faces several challenges. First, due to the wide variety of defects and their variations in different production environments, automating defect identification and classification remains a challenge, particularly with regard to data diversity and quality. Second, surface defects often resemble the surrounding environment and the texture of the metal, making it difficult for models to detect subtle defects and accurately differentiate between surface textures and defects. Lastly, ensuring high detection accuracy while improving the system's real-time performance and computational efficiency—especially in large-scale data processing and high-speed production environments—remains an important direction for future research.

In summary, steel strip surface defect detection, particularly automation methods based on machine learning and deep learning, has become an essential part of modern steel manufacturing. In the future, with continuous algorithm optimization and the expansion of datasets, steel strip defect detection will become even more intelligent and efficient, further advancing the automation and intelligence of the steel industry. Here, we propose a YOLO-BSPB model aimed at improving the precision of surface defect detection in steel strips. The contributions of this paper are as follows:

(1) To address the information loss caused by the SPPF pooling layer in YOLOv8 during the feature extraction process, we replaced it with an enhanced SPPCSPC [28] module. This improvement includes substituting max pooling with average pooling and integrating DCNv4 [29] to dynamically adjust convolution operations, enabling the capture of diverse feature information. Furthermore, the addition of the RFB [30] module allows the model to achieve a global perspective while maintaining fine-grained local adaptability.

(2) To enhance the detection sensitivity to small targets and mitigate the loss of detailed information in the C2f module, we incorporated the BAM [31] attention mechanism to strengthen the model's focus on important regions. Additionally, the PSA [32] mechanism was employed to separately capture information in both channel and spatial dimensions, thereby enabling a more comprehensive feature representation.
(3) We utilized a bilateral improved feature pyramid structure (BiFPN) [33] to effectively combine fine-grained low-level features with high-level semantic features, improving detection performance for defects of varying sizes. Moreover, the WIoU [34] loss function was adopted to dynamically adjust weights, reducing performance loss and accelerating the convergence of the model.

2. Improved YOLOv8 Detection Principle

2.1. Improved Overall Structure of Yolov8n

YOLOv8 is a significant upgrade of the YOLO series of real-time object detectors launched by Ultralytics, aimed at enhancing the accuracy and efficiency of object detection. Its structure, as shown in Figure 1, features numerous innovations in architectural design. YOLOv8 employs a C2f backbone, which improves the model's ability to recognize targets in complex environments. Additionally, to enhance detection performance, the decoupled head design allows for the independent processing of object classification and localization. The model also introduces an anchor-free approach, reducing reliance on manual anchor box configuration and ensuring robust generalization capabilities. This version offers five different model options—n, s, m, l, and x. Each model varies in parameter count, computational complexity, and performance, catering to the requirements of different hardware environments and application scenarios. These improvements have enabled YOLOv8 to achieve outstanding results across multiple standard datasets, further advancing the progress of real-time object detection technology.

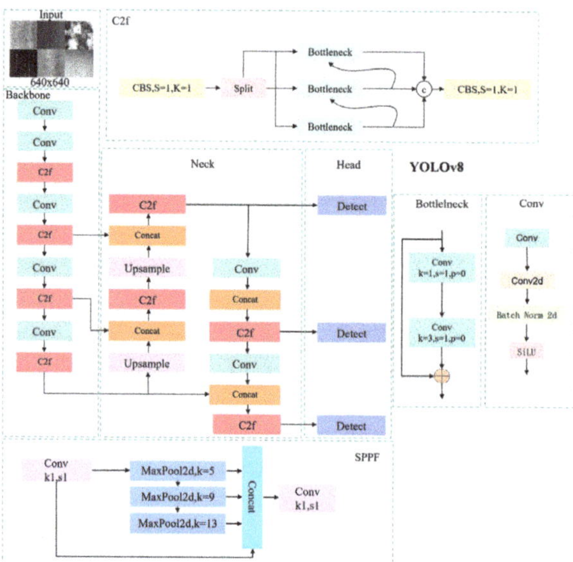

Figure 1. Structure diagram of YOLOv8.

The industrial requirements for defect detection are extremely stringent, and YOLOv8n offers a good balance between real-time performance and accuracy. As the smallest version in the YOLO series, YOLOv8n features a compact design that allows it to operate in resource-constrained environments, making it suitable for embedded systems and mobile devices. By adopting an optimized network architecture, it reduces computational load while maintaining good detection performance. This enables YOLOv8n to achieve high responsiveness when processing real-time data, allowing for timely identification and localization of surface defects. Therefore, this paper selects the smallest version of YOLOv8n as the basis for further improvement, proposing the YOLOv8-BSPB model. In this model, the SPPF pooling layer in the network structure is replaced with SCRD, the BAM attention mechanism is introduced into the C2f backbone, and the PSA mechanism is applied after the pooling layer. Additionally, the BiFPN is incorporated at the neck's connection layer to enhance model performance, and the loss function is replaced with the WIoU loss function to improve classification performance. The overall structure of the improved model is shown in Figure 2.

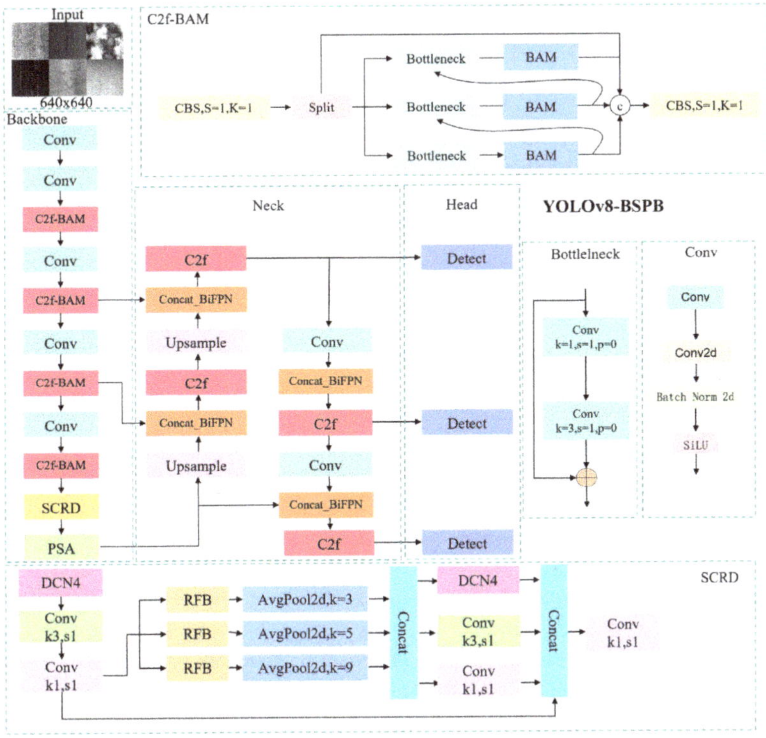

Figure 2. Structure diagram of YOLOv8-BSPB.

2.2. SCRD Improved Based on SPPCSPC

1. SPPCSPC

The multi-stage spatial pyramid pooling cross-stage partial concatenation (SPPCSPC) module, shown in Figure 3, combines the advantages of the SPP and CSPC modules. The SPP module enhances the detection of different defect scales by capturing more scales of features from the image through multi-scale pooling. The CSPC module introduces cross-stage partial concatenation, which improves the feature utilization efficiency and the gradient flow, and efficiently alleviates the gradient vanishing problem. By combining multi-scale feature extraction with efficient feature linking, the SPPCSPC module can better realize image classification and object detection and ultimately improve the overall performance.

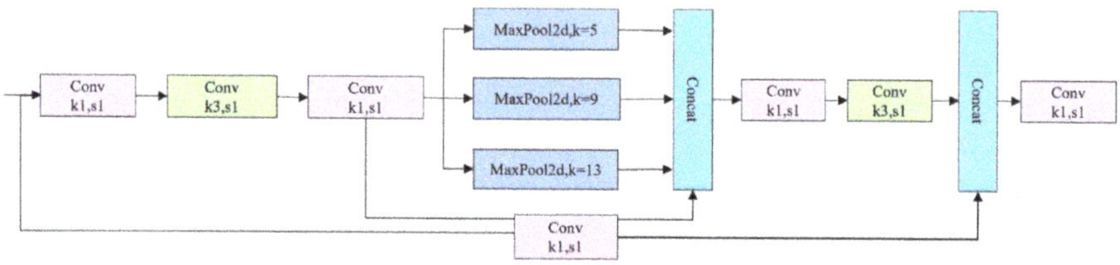

Figure 3. Structure diagram of SPPCSPC.

2. Variability Convolution DCNv4

DCNv4 integrates the sparse attention mechanism with convolution operations, enhancing dynamic attributes and representational capacity. By utilizing the sparse attention mechanism, DCNv4 adaptively adjusts the position and shape of the convolutional kernel, thereby improving the sampling mechanism. In DCNv4, a single thread processes multiple channels within the same group, allowing these channels to share sampling offsets and aggregated weights. This reduces workloads and enables the merging of multiple memory access instructions. Figure 4 illustrates the intra-group shared processing mechanism in DCNv4.

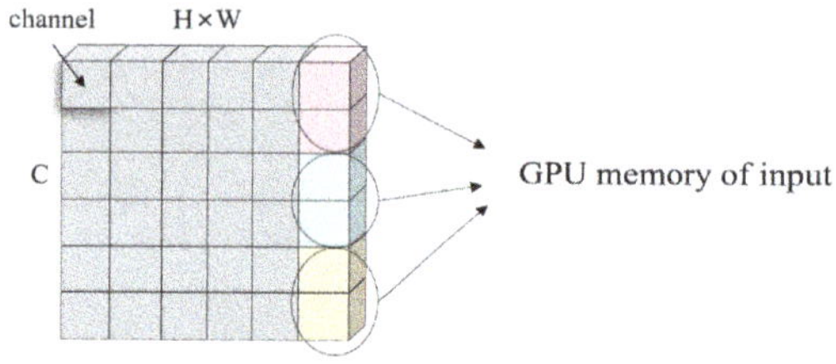

Figure 4. Intra-group sharing of DCNv4 processing.

When dealing with complex tasks, DCNv4 exhibits better adaptability. For the input feature map X, where the Δp_k grid samples the position pk offset, the convolution kernel is W_k, and the bias b and attention weight Δm_k are introduced. K represents the number of sampling points in the convolution kernel, p denotes the pixel coordinates of the image feature map, and p_k represents the relative position of the k-th sampling point in the convolution kernel. The output feature map Y obtained for each point p is

$$Y(\mathrm{p}) = \sum_{k=1}^{K} W_k \cdot X(p + p_k + \Delta p_k) \cdot \Delta m_k + b \quad (1)$$

3. Feature Extraction Module RFB

The receptive field block (RFB) module enhances traditional convolution operations through a series of convolution, normalization, and activation functions. As shown in Figure 5, the RFB module expands the receptive field, enabling the network to capture more comprehensive semantic information. Compared with traditional convolution operations, the RFB module significantly improves the perceptual capabilities of the network without increasing the number of parameters and computational costs. This improvement enables the model to better adapt to target objects of different sizes and shapes, thereby improving its performance in defect detection tasks.

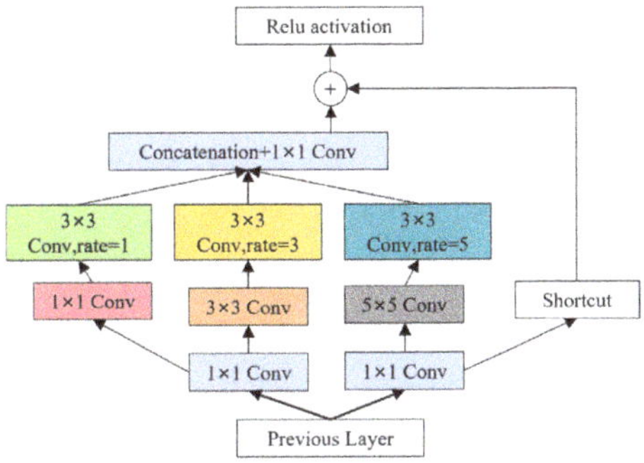

Figure 5. RFB structure diagram.

4. SCRD

To improve attention to the densely packed small target samples, the proposed SCRD structure employs the SPPCSPC module as its core component. To improve the SPPCSPC module to extract multi-scale features, previous studies have shown that modifications to the SPPCSPC module can significantly enhance detection accuracy in multi-scale object recognition tasks. For example, Zhao et al. [35] improved classification accuracy through SPPCSPC. Similarly, Wan et al. [36] replaced the SPPF module with an improved SPPCSPC to enhance feature representation capability and efficiently implement multi-scale feature fusion. Initially, dilated convolution and parallel convolution are integrated through the RFB module to enhance the model's feature extraction capabilities. Traditional max pooling, which selects only the maximum value as output, often overlooks detailed and local features, particularly in small target detection where such details are crucial. In contrast, average pooling retains more local detail information and is less sensitive to noise, making it more suitable for detecting small targets. Therefore, max pooling is replaced with average pooling, and the pooling kernel sizes are adjusted to 3, 5, and 9 to better preserve the local details of small objects. Additionally, variability convolution via DCNv4 is introduced in place of standard convolution layers, allowing for more flexible adaptation to target shapes and sizes and further enhancing feature extraction. The SCRD structure is illustrated in Figure 6.

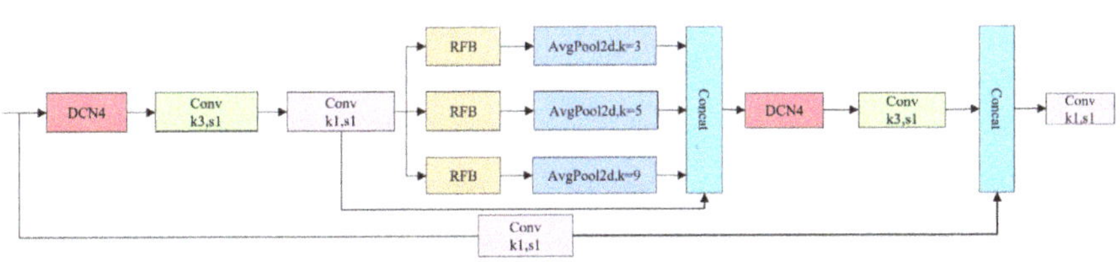

Figure 6. Structure diagram of SCRD.

2.3. Polarized Self-Attention Mechanism

PSA combines polarization filtering and enhancement design to improve pixel-wise regression, enabling the model to handle defect information more accurately. The PSA mechanism was introduced to optimize information transfer and suppress background noise, aiming to enhance the model's robustness in complex scenarios. Existing studies have shown that this mechanism can effectively improve detection accuracy. For instance, Cheng et al. [37] integrated the PSA mechanism into the YOLOv8 model for underwater target recognition, enhancing feature polarization across both channel and spatial dimensions, improving pixel-level prediction accuracy, and enabling lightweight edge deployment. Similarly, Lang and Lv [38] applied the polarized self-attention mechanism during feature extraction in a photovoltaic cell defect detection system, combining spatial and semantic features with the original input to enhance the network's feature representation capability. These studies provide strong support for the effectiveness of the PSA mechanism in improving model performance.

The structure of the PSA is shown in Figure 7. In the channel attention module, 1×1 convolution is used to convert feature X into Q and V. The Q channel is fully compressed, and the operator F_{SM} is used to enhance the Q information, while the V channel still maintains the number of $C/2$ channels. After matrix multiplication of Q and K, the dimension is raised to C by 1×1 convolution and normalization layer. Finally, the parameters are calculated by using the activation function. The channel branch weights $A^{ch}(X)$ are calculated as follows, where W_q, W_k, W_v are 1×1 convolutional layers, F_{GP} is the global pooling operation, σ_1, σ_2 are two tensor reshape operators, and $\times$ is the matrix dot-product operation:

$$A^{ch}(X) = F_{SG}\left\{W_{z|\theta_1}((\sigma_1(W_v(X)) \times F_{SM}(\sigma_2(W_q(X))))\right\} \tag{2}$$

where X represents the input features, F_{SG} denotes the nonlinear activation function, W_z is the learnable weight matrix, and θ_1 is the associated hyperparameter of this weight matrix. $W_v(X)$ represents a linear transformation of X, where W_v is the weight matrix. F_{SM} stands for the Softmax function, and $W_q(X)$ generates a representation of X that is related to the query.

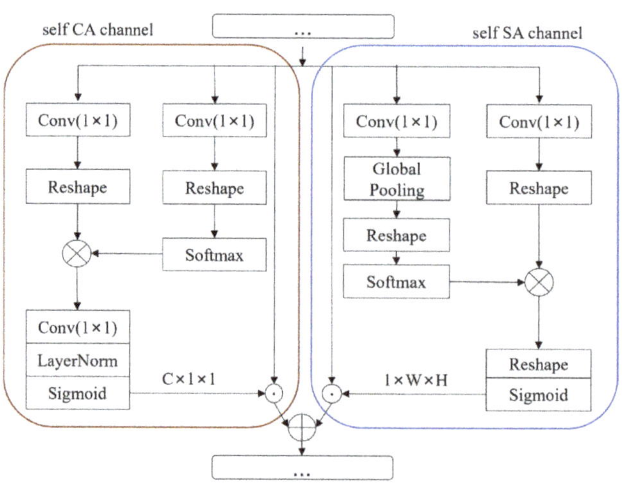

Figure 7. Structure diagram of PSA.

Similarly, in the spatial attention module, 1×1 convolution is used to transform features into Q and V. For Q, global pooling is used to compress the spatial dimension, and then, the Softmax operator is used to enhance Q, and V remains unchanged. The spatial branch computes the weight $A^{sp}(X)$ as follows, and $\sigma 3$ is the tensor reshape operator. F_{SG} represents the nonlinear activation function, σ_1, σ_2, and σ_3 are activation functions for processing linear transformations, F_{GP} denotes the global pooling operation, and $W_q(X)$ and $W_v(X)$ are linear transformation matrices:

$$A^{sp}(X) = F_{SG}\{\sigma_3(F_{SM}(\sigma_1(F_{GP}(W_q(X)))) \times \sigma_2(W_v(X)))\} \tag{3}$$

$PSA_p(X)$ is obtained by parallel fusion of the above two branch outputs, where *ch* is a channel-wise multiplication operator.

$$PSA_p(X) = A^{ch}(X) \odot^{ch} X + A^{sp}(X) \odot^{sp} X \tag{4}$$

2.4. C2f-BAM Attention Mechanisms

Due to the large difference in the target shape and size of the defects in the steel strip surface data set, it is difficult for the model to accurately identify and locate the defects, so the BAM attention mechanism is introduced. This mechanism improves the model's ability to focus on important features in complex environments, thus enhancing object detection performance. Hu et al. [39] effectively extracted the shape, contour, and structural features of steel strip defects by incorporating the boundary-aware module (BAM), which also accurately localized the defects. Zhao et al. [40] added a global attention mechanism module to C2f, focusing on defect areas and suppressing irrelevant regions, dynamically adjusting feature responses to more accurately recognize image features. These studies provide strong support for the effectiveness of the improvements to C2f in enhancing detection performance by increasing the focus on relevant features.

The structure of the BAM is shown in Figure 8. For the input feature map *F*, the channel dimension features are generated by using global average pooling in the channel attention module (CA), and then the input channel weights are learned through two fully connected layers *FC*, and finally, the $C \times 1 \times 1$ channel attention weights $M_c(F)$ are obtained.

$$M_c(F) = BN(W_1(W_0 AvgPool(F) + b_0) + b_1) \tag{5}$$

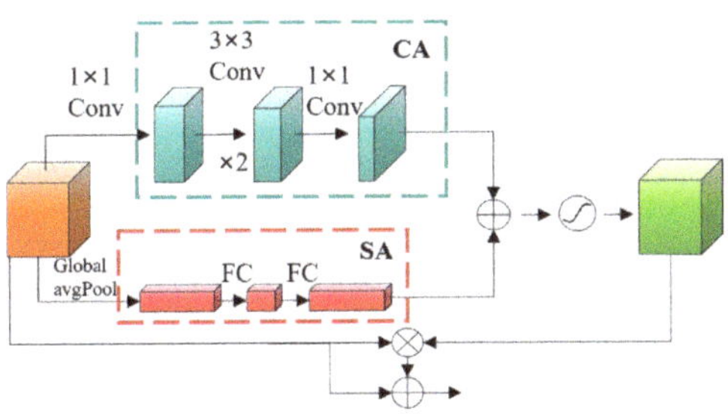

Figure 8. Structure diagram of BAM.

The spatial attention module (SA) uses 1×1 convolution for dimensionality reduction and then learns the spatial weight of the feature map by repeated dilated convolution operations. Finally, the feature weight is reduced to 1 by 1×1 convolution, and the spatial attention weight $M_s(F)$:

$$M_s(F) = BN(f_3^{1\times1}(f_2^{3\times3}(f_1^{3\times3}(f_0^{1\times1}(F))))) \tag{6}$$

The spatial and channel attention are summed element-by-element, and the original feature map is weighted by using the activation function and the final comprehensive attention weight to obtain F'.

$$F' = F + F \otimes (\sigma(M_c(F) + M_s(F))) \quad (7)$$

The neck attention module BAM is introduced at the bottleneck of C2f architecture to enhance the feature representation ability of the model. Through reasonable configuration and experimental verification, a good balance between performance and efficiency is achieved. Adding BAM directly to the bottleneck module can make full use of the power of the attention mechanism while maintaining the advantages of the original structure to improve the performance of the model in various tasks. The C2f-BAM structure diagram is shown in Figure 9.

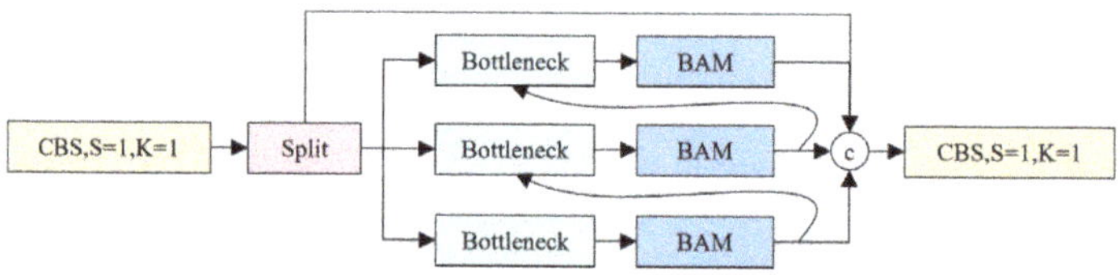

Figure 9. Structure diagram of C2F-BAM.

2.5. Bidirectional Feature Pyramid Network

The bidirectional feature pyramid network (BiFPN) enhances the traditional feature pyramid network (FPN), as illustrated in Figure 10a. By incorporating a weighted bidirectional feature fusion mechanism, BiFPN enables bidirectional information flow and multiple feature fusion, thereby improving both detection accuracy and speed. Kou et al. [41] adopted a de-weighted BiFPN structure to fully leverage feature information, enhancing feature fusion and reducing feature loss during the convolution process, which ultimately improved detection accuracy. Li et al. [42] introduced the de-weighted BiFPN structure at the network neck, balancing model size with detection accuracy and speed by adding step branches and reducing computation, leading to improved overall performance.

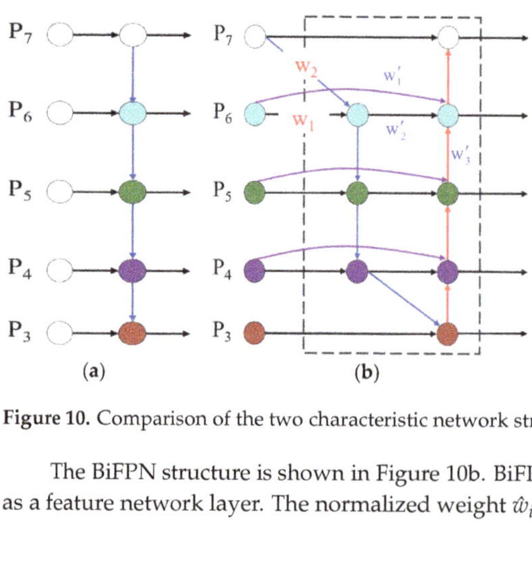

Figure 10. Comparison of the two characteristic network structures: (**a**) FPN, and (**b**) BiFPN.

The BiFPN structure is shown in Figure 10b. BiFPN considers each bidirectional path as a feature network layer. The normalized weight $\hat{w}_i$ is obtained for the *ith* input feature map weight parameter w_i. The normalized weights are used to perform weighted fusion on the i input feature maps xi, and the weighted fusion feature map x_f is obtained as

$$x_f = \sum_{i-1}^{N} \hat{w}_i \cdot x_i = \sum_{i-1}^{N} \frac{w_i}{\sum_{i-1}^{N} w_i + \varepsilon} \cdot x_i \tag{8}$$

Finally, the weighted fused feature maps are concatenated along the specified dimensions. For each weighted feature map, they are concatenated along the channel dimension to form a new feature map.

2.6. Loss Function WIoU

The WIoU loss function was introduced to reduce error propagation and improve model efficiency. It has been shown in the research that this loss function can effectively enhance accuracy in tasks requiring precise localization. For example, the WIoU bounding box loss function was used by Qin et al. [43] to address the class imbalance problem, improving the robustness of the model for steel strip surface defect detection. Similarly, the WIoU loss function was employed by Kong and Fan [44] to improve the localization accuracy of surface defects and enhance training efficiency for low-quality samples. Significant improvements in model performance were demonstrated by their experimental results.

The bounding box regression BBR loss directly determines the model localization performance. *IoU* is used to measure the degree of overlap between the predicted box and the true box.

Low-quality data annotations in the training process will interfere with the model convergence under the influence of efficient fitting loss function. In order to improve the generalization ability of the model, the loss function *WIoU* is introduced as the evaluation index.

IoU is used to measure the degree of overlap between the predicted box and the true box in the object detection task, L_{IoU} is an *IoU*-based loss function that adjusts the *IoU* loss weights in linear space:

$$L_{IoU} = 1 - IoU \tag{9}$$

WIoU, is a loss function with dynamic non-monotonic focused bounding box regression proposed based on *IoU*, which can better reflect the actual performance of object detection by introducing a better region alignment and weighting mechanism.

As shown in Figure 11, $C_B(x, y)$ represents the geometric center of the prediction box. As shown in Figure 11, $C_A(x_{gt}, y_{gt})$ represents the geometric center of the target box, W_g and Hg represent the width and height of the minimum outsourced rectangular box, preventing the anchor box loss R_{WIoU} from producing gradients that hinder convergence. "*" denotes a split operation. R_{WIoU}, a reweighted *IoU* loss, dynamically adjusts weights to address the adaptability issues for different types of samples in the loss function. By combining these two types of loss functions, the optimization of various aspects of the predicted bounding box performance is achieved, resulting in a more robust loss function, L_{WIoUv1}.

$$L_{WIoUv1} = R_{WIoU} L_{IoU} \tag{10}$$

$$R_{WIoU} = \exp\left(\frac{(x - x_{gt})^2 + (y - y_{gt})^2}{(W_g{}^2 + H_g{}^2)^*}\right) \tag{11}$$

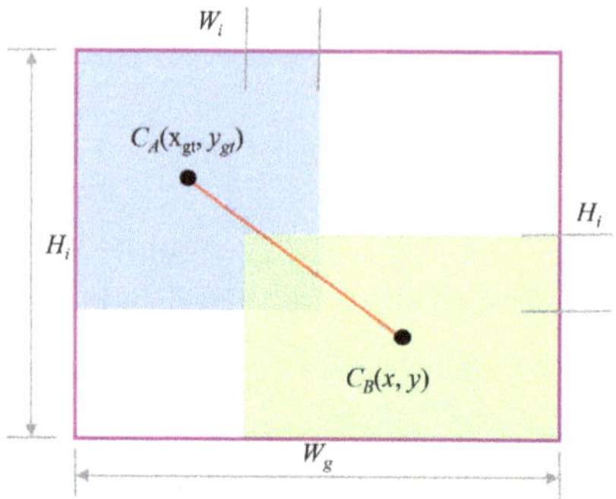

Figure 11. Geometric representation of WIoU.

In order to make BBR focus more on anchors of ordinary quality and improve its classification performance, a small gradient gain is assigned to anchor frames to effectively prevent large harmful gradients from occurring in low quality. A non-monotonic focusing coefficient L_{WIoUv3} is constructed by the anchor box outlier degree β, γ is the gradient gain, L_{IoU} represents an exponentially moving average with a momentum factor m, while L^{*}_{IoU} is an IoU loss function with a dynamic weighting mechanism that adjusts weights based on target attributes. Using β, a non-monotonic focusing coefficient L_{WIoUv3} is constructed for W_{IoUv1}, assigning smaller gradient gains to anchor boxes with larger outliers, and α and δ are the hyperparameters.

$$\beta = \frac{L^{*}_{IoU}}{L_{IoU}} \in [0, = \infty) \tag{12}$$

$$L_{WIoUv3} = \gamma L_{WIoUv1}, \gamma = \frac{\beta}{\delta\alpha^{\beta-\delta}} \tag{13}$$

3. Experiment and Analysis

3.1. Experimental Data Set

This study utilizes the publicly available NEU-DET dataset, which contains surface defect images of rolled steel strips.As shown in Figure 12, the dataset includes six defect types: rolled-in scale (RS), patches (Pa), cracks (Cr), pitted surface (PS), inclusions (In), and scratches (Sc), comprising a total of 1800 grayscale JPG images. Each defect type is represented by 300 images, each with a resolution of 200 × 200 pixels. The dataset is randomly split into a training set and a validation set in a 7:3 ratio, yielding 1260 images for training and 540 images for validation.

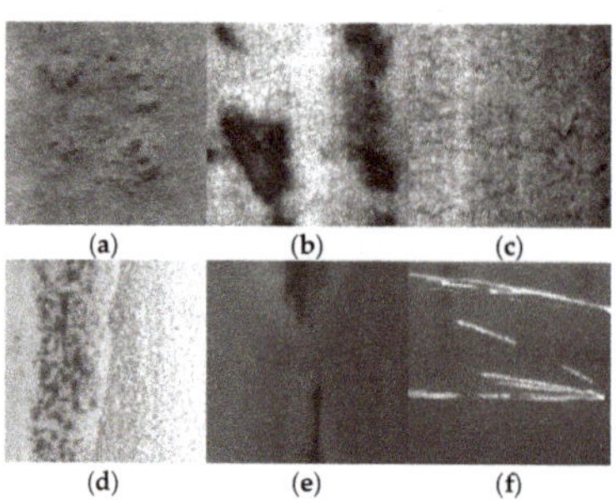

Figure 12. Strip surface defect categories in the NEU-DET dataset: (**a**) oxide skin; (**b**) patches; (**c**) cracks; (**d**) pitted surface; (**e**) inclusion; (**f**) scratches.

The experiments are conducted using a GPU in an Ubuntu 22.04.3 environment. The GPU utilized is an NVIDIA GeForce RTX 3090 with 24 GB of VRAM. The software environment includes CUDA 12.1 and the deep learning framework PyTorch version 2.1.1, with programming performed in Python 3.10.14. The model is trained for 120 epochs, with images resized to 640 × 640 pixels, a batch size of 16, an initial learning rate of 0.01, a weight decay of 0.0005, and a learning rate momentum of 0.937.

3.2. Evaluation Indicators

When evaluating the detection accuracy and speed of defect detection models, the commonly used evaluation indicators are precision P, recall R, average precision mean mAP, frames per second FPS, model calculation amount GFLOPs, and model parameters to measure the accuracy of detection. The formula is as follows Equations (14)–(17).

$$p = \frac{TP}{TP + FP} \tag{14}$$

$$R = \frac{TP}{TP + FN} \tag{15}$$

$$FPS = \frac{Framenum}{Elapsed - Time} \tag{16}$$

$$mAP = \frac{1}{n}\sum_{i=0}^{n} AP(i) \tag{17}$$

where TP, TN, FP, and FN represent the number of correctly detected defect targets, the number of correctly detected non-defect samples, the number of incorrectly detected defect targets, and the number of missed defect targets, respectively; Framenum is the total number of detected images; elapsed time is the total time spent in detection.

3.3. Performance Comparison of Feature Extraction Module

To assess the effectiveness of the proposed SCRD module, this study evaluates its performance in defect detection by comparing it with several feature extraction modules. Specifically, SPPF, SimSPPF, SPPELAN, and SPPCSPC are chosen for comparison. This comparison aims to illustrate the advantages of the SCRD module over existing methods in terms of detection accuracy and overall performance.

Table 1 shows that the precision of the proposed SCRD method is 3.9% higher than that of SPPCSPC. The recall rate of SCRD is 2.1% higher than that of SPPF, but 0.7% lower than that of SPPCSPC. However, the mean average precision of SCRD is 1.3% higher than that of SPPF and 0.6% higher than that of SPPCSPC. These improvements are mainly attributed to SCRD's ability to handle more complex defect patterns, making it more suitable for real-time detection under varying production conditions. Despite being computationally intensive, SCRD maintains relatively fast processing speeds, which makes it highly practical for steel strip defect detection.

Table 1. Comparison of feature extraction modules.

Model	P/%	R/%	mAP@50/%	GFLOPs	Paraments/M	FPS/s^{-1}
SPPF	76.4	76.0	82.9	8.1	2.87	454.5
SimSPPF	75.9	75.8	82.4	8.1	2.87	500.0
SPPELAN	73.4	73.7	80.8	8.0	2.72	625.0
SPPCSPC	74.4	78.8	83.6	9.5	4.40	454.5
SCRD	78.3	78.1	84.2	10.1	5.32	588.2

3.4. Experiments on Attention Mechanism

In order to verify the superiority of introducing C2f_BAM, the influence of different attention mechanisms on the experimental results was compared. In this paper, GAM, BoT3, MHSA, CBAM, MSDA, ShuffleAttention, and SENet are selected for comparison.

As shown in Table 2, after embedding GAM, BoT3, MHSA, CBAM, ShuffleAttention, MSDA, and SENet attention mechanisms, the precision, recall, and mean average precision have different degrees of decline. Compared with the baseline C2f, the mAP of the C2f_BAM attention mechanism is increased by 0.5%. The precision is improved by 0.6%, and the recall is improved by 0.8%. The slight decline in performance with the addition of other attention mechanisms may be attributed to their complexity, which could introduce unnecessary computational overhead or cause overfitting to certain features. This indicates that C2f_BAM strikes a better balance between model complexity and detection accuracy, highlighting its effectiveness in enhancing overall performance.

Table 2. Comparison of attention mechanisms.

Model	P/%	R/%	mAP@50/%	Paraments/M	GFLOPs/G
——	76.4	76.0	82.9	2.86	8.1
GAM	70.5	72.8	78.0	8.68	13.0
BoT3	68.3	69.4	75.4	2.64	7.9
MHSA	68.1	67.8	73.1	2.62	7.9
CBAM	69.7	69.9	75.7	2.43	7.8
ShuffleAttention	70.9	71.0	76.5	2.43	7.8
MSDA	73.6	74.7	82.1	4.41	9.4
SENet	66.9	71.6	75.9	2.56	7.9
C2f_BAM	77.6	76.8	83.4	2.62	7.9

3.5. Loss Function Comparison

In order to verify the usefulness of introducing the WIoU loss function reference, the commonly used loss functions GIoU, CIoU, EIoU, SioU, and DIoU were selected for comparison tests.

According to Table 3, compared with other loss functions, the introduction of the WIoU loss function can obtain higher total precision, recall rate, and mAP. For six kinds of defects, the mean average precision of Sc measurement by CIoU and WIoU both reached 97.9%. In addition, for other defect types, the detection results of WIoU were better than those of other loss functions. WIoU adopts a dynamic non-monotonic focusing mechanism, which enabled the model to focus more on the anchor box of common defect sample quality, thus ensuring the balance of model performance. This dynamic focusing mechanism allowed the model to prioritize the most relevant defect samples, improving detection accuracy. As a regression loss, WIoU loss improved the convergence speed of the model and then improved the prediction effect of the model. This faster convergence led to more efficient training, allowing the model to reach optimal performance quicker. Therefore, WIoU was selected as the loss function.

Table 3. Comparison of loss functions.

Model	mAP@50/%						P/%	R/%	mAP@50/%
	RS	Sc	Cr	Ps	Pa	In			
GIou	72.9	97.1	55.7	81.0	95.2	90.6	77.2	75.3	82.1
CIou	75.0	97.9	56.3	81.2	95.2	91.5	76.4	76.0	82.9
EIou	68.2	96.9	52.9	81.1	94.8	88.7	76.0	73.1	80.4
SIou	73.6	96.7	31.8	81.3	95.4	91.3	78.0	74.5	82.3
DIou	73.7	97.3	56.7	82.1	96.1	91.0	74.3	78.3	82.8
WIou	88.1	97.9	68.6	84.5	96.9	92.9	79.4 (↑)	79.2	85.5 (↑)

3.6. Ablation Experiments

In order to prove the effectiveness of the model, based on YOLOv8n, a comparative analysis of ablation experiments was designed on the NEU-DET public data set of Northeastern University, and all parameters were kept consistent during the experiment.

As shown in Table 4, 1 is the yolov8 benchmark model, 2 is the SCRD module, 3 is the PSA module, 4 is the C2f-BAM module, 5 is the BiFPN module, and 6 is the WIoU loss function. 1 + 2 refers to the introduction of the SCRD module alone, 1 + 3 refers to the introduction of the PSA module alone, 1 + 4 refers to the introduction of the C2f-BAM module alone, 1 + 5 refers to the introduction of the BiFPN module alone, and 1 + 6 refers to the introduction of WIoU loss function alone. The results after the introduction of SCRD and PSA modules are represented by 1 + 2 + 3; the overall precision was increased by 1.7%, the recall rate was increased by 2.6%, and the mean average precision was increased by 2.0%; 1 +...+ 4 indicated the results after adding SCRD, PSA, and C2f-BAM, with an improvement of 2.3%, 4.3%, and 2.9% in precision, recall, and average precision, respectively; 1 +...+ 5 represents the results after adding SCRD, PSA, C2f-BAM and BiFPN, accuracy and average precision compared to 1 +...+ 4 improved by 0.6% and 0.6%, but reduced recall by 0.5%. 1 +...+ 6 is for the introduction of 1+. The results after the +4 module and WIoU had the highest overall accuracy, and the mean average precision of the six defects detection was higher than that of the baseline model. The mean average precision of RS, Sc, Cr, Ps, Pa, and In were increased by 3.3%, 1.8%, 5.6%, 34.3%, 26.2%, and 3.4%, respectively.

Table 4. Ablation experiments.

Model	P/%	R/%	mAP@50/%	GFLOP/G	Parament/M	mAP@50/%					
						RS	Sc	Cr	Ps	Pa	In
1	76.4	76.0	82.9	8.1	2.87	75.6	96.8	58.6	53.0	72.2	90.9
1 + 2	78.3	78.1	84.2	10.1	5.32	74.1	97	60.9	83.5	97.5	92.2
1 + 3	77.4	74.4	82.9	8.4	3.03	74.0	97.3	56.1	82.4	95.6	92.0
1 + 4	77.6	76.8	83.4	8.1	2.62	75.7	97.0	58.1	80.9	97.3	91.6
1 + 5	79.4	77.4	85.0	8.1	2.87	78.0	97.0	62.0	82.3	96.8	93.7
1 + 6	79.4	79.2	85.5	8.1	2.87	73.7	97.9	63.9	84.5	96.9	92.9
1 + 2 + 3	78.1	78.6	84.9	10.4	5.48	74.4	98.6	61.7	83.7	98.0	92.8
1 +...+ 4	78.7	80.3	85.8	10.4	5.48	75.7	98.6	62.3	87.1	97.7	93.5
1 +...+ 5	79.3	79.8	86.4	10.4	5.48	78.2	97.6	64.2	87.1	98.3	94.3

To more effectively assess the performance of the YOLOv8-BSPB algorithm, a comparative analysis was conducted on the loss curves, as well as the precision and recall curves, of the YOLOv8 and YOLOv8-BSPB models during the training phase. As illustrated in Figure 13, the loss value for the YOLOv8-BSPB model decreases and stabilizes after 200 training epochs, demonstrating a superior convergence rate.

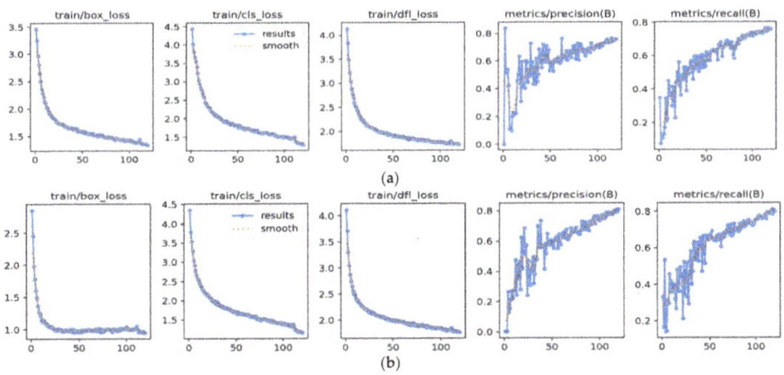

Figure 13. YOLOv8n-BSPB loss function curve: (**a**) YOLOv8; (**b**) YOLOv8-BSPB.

In addition, we evaluated the detection ability of YOLOv8N-BSPB using the confusion matrix, as shown in Figure 14. The false positive FP and false negative FN of Yolov8n in detecting most defect categories in the confusion matrix are comparable, indicating higher classification accuracy of the improved model.

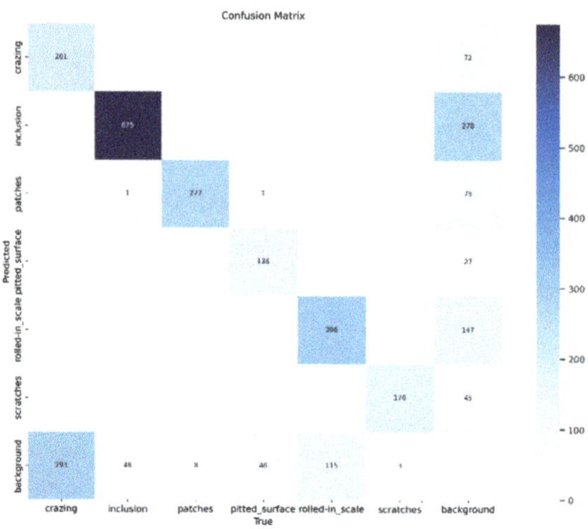

(a)

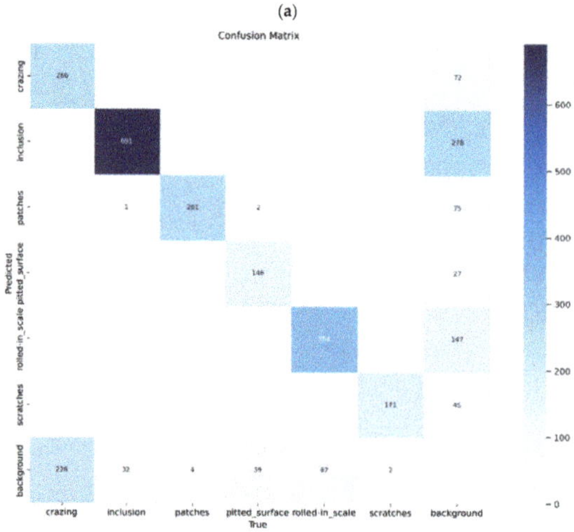

(b)

Figure 14. Confusion matrix for the validation set with defects of different classes: (**a**) confusion matrix of YOLOv8n; (**b**) confusion matrix of YOLOv8n-BSPB.

In order to visually compare the detection performance of the YOLOv8-BSPB proposed in this paper with that of the YOLOv8 model on surface defects of strip steel, the dataset is qualitatively analyzed, and the results are shown in Figure 15. From the Figure, it can be seen that the original model shows leakage and low detection accuracy, while the improved model improves in accuracy and is able to detect more targets, which verifies the improvement of the improved model in terms of defect sample detection.

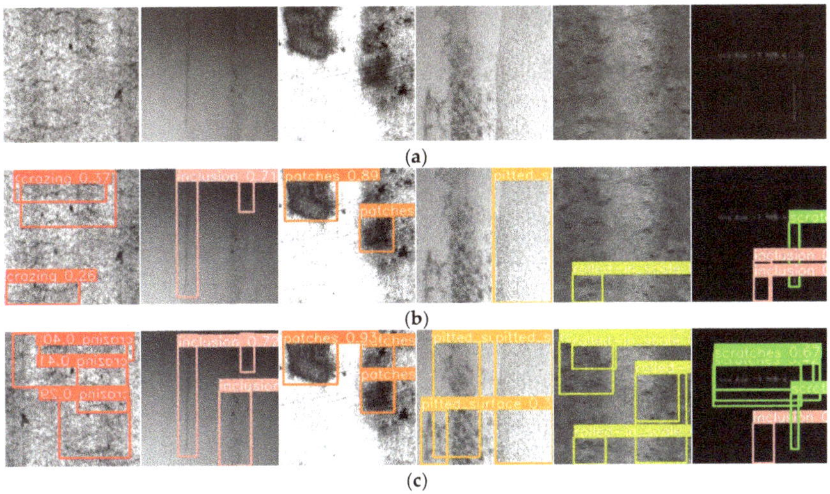

Figure 15. Visualization of detection results: (**a**) original images; (**b**) YOLOv8; (**c**) YOLOv8-BSPB.

3.7. Comparison Test of Different Models

To verify the performance of the improved model, it was compared with the current mainstream models YOLOv5n, YOLOv5s, YOLOX, YOLOv7, YOLOv7-tiny, Faster-RCNN [45], and SSD [46] algorithm, and the experimental results are shown in Table 5.

Table 5. Comparison of different models.

Model	P/%	R/%	mAP@50/%	Parameter/M	FPS/s^{-1}
Faster-RCNN	67.7	66.1	69.2	108.6	21.5
SSD	68.6	70.7	71.5	93.6	42.6
YOLOv5n	70.1	68.2	72.2	6.7	163.9
YOLOv5s	74.7	73.5	76.6	7.0	142.9
YOLOv6n	72.3	73.6	74.9	8.3	126.5
YOLOv7	70.4	73.7	75.5	36.5	64.9
YOLOX	77.0	72.5	79.8	9.0	67.1
YOLOv7-tiny	71.2	69.5	73.5	6.1	143.0
YOLOv8n	76.4	76.0	82.9	3.0	108.7
Our method	81.3 (↑)	80.3	86.9 (↑)	5.5	103.1

The table shows that Faste-RCNN and SSD have slow detection speeds and low accuracy, which makes it difficult to adapt to the current detection requirements. Compared with other methods, the proposed algorithm is better in precision, recall rate, and mean average precision. This indicates that the proposed algorithm addresses the limitations of previous methods, offering a more robust solution for defect detection. Compared with the YOLOv8n, the proposed algorithm is improved by 4.9%, 4.3%, and 4.0%, respectively, which proves the effectiveness of the improved algorithm. This improvement highlights the proposed algorithm's capability to achieve higher detection performance while maintaining efficient processing. Compared with other methods, the accuracy of YOLOv5n, YOLOv6n, and YOLOX was 11.3%, 9.0%, and 4.0% higher, and the parameter amount was reduced by 1.2 M, 2.8 M, and 3.5 M, and the model parameter amount is increased by 2.6 M. Although the model parameter increase may lead to a more complex structure, it contributes to enhanced performance. Due to the introduction of a complex module structure, the FPS of the algorithm in this paper is reduced compared with the detection frame number of

YOLOv8, but it is still suitable for real-time detection of strip surface defects. This trade-off between detection performance and FPS is a common characteristic in models that seek to balance accuracy with real-time processing capabilities. In summary, the improved YOLOv8-BSPB model has good detection performance in strip surface defect detection.

3.8. Experiments on Different Datasets

To validate the effectiveness of the model, the experimental environment was kept consistent, with the dataset randomly split into a training set and a validation set in a 7:3 ratio. The model was trained for 120 epochs, with an image resolution of 640 × 640 pixels, a batch size of 16, and an initial learning rate (lr0) of 0.01. The model was tested using both the Kaggle steel strip dataset and defect data collected from an actual factory. The Kaggle dataset contains four types of defects, as shown in Figure 16. The dataset includes a total of 6666 defect images, with 4666 images in the test set and 2000 images in the validation set.

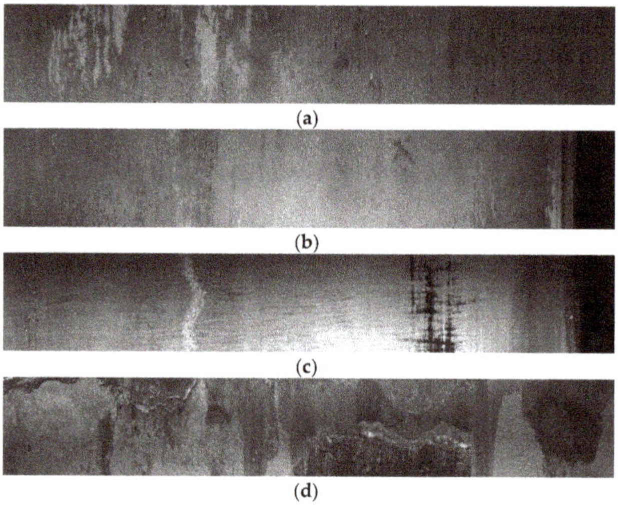

Figure 16. The four types of defects in the Kaggle dataset: (**a**) cracks; (**b**) inclusion; (**c**) patches; (**d**) scratches.

The defect dataset collected from the factory includes two types of defects were flaw (FL) and scratch (Sc), as shown in Figure 17. The dataset contains a total of 608 defect images, with 426 images in the test set and 182 images in the validation set.

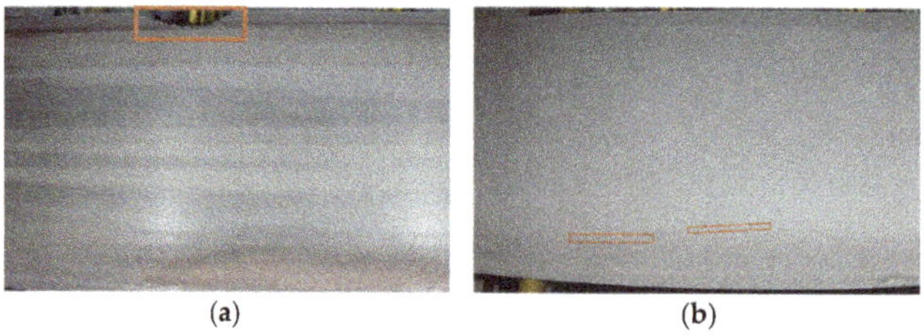

Figure 17. The two types of defects in the iron dataset: (**a**) flaw; (**b**) scratch.

The validation experiment results are shown in Tables 6 and 7, where the four defect types in the Kaggle dataset correspond to Cr, In, Pa, and Sc in Table 6, and the two defect types in the factory dataset correspond to FL and Sc in the table.

Table 6. Comparison results of Kaggle dataset.

Model	P/%	R/%	mAP@50/%	Parameter/M	mAP@50/%			
					Cr	In	Pa	Sc
YOLOv8	56.6	51.9	53.4	2.9	51.2	34.3	66.5	61.4
Our method	58.7	54.0	55.7 (↑)	6.6	55.9	38.3	67.4	68.6

Table 7. Comparison results of factory datasets.

Model	P/%	R/%	mAP@50/%	Parameter/M	mAP@50/%	
					FL	Sc
YOLOv8	57.6	54.6	50.1	2.9	25.7	25.5
Our method	67.6	56.5	55.6 (↑5.5)	6.6	38.7	28.9

As can be seen from Tables 6 and 7, for the Kaggle dataset, the improved Yolov8 model has the precision, recall rate, and mean average precision increased by 2.1%, 2.1%, and 2.3%, respectively, and the mean average precision of cracks, inclusions, patches, and scratch defects increased by 4.7%, 4.0%, 0.9%, and 7.2%, respectively. For the factory dataset, the mean precision, recall, and average precision of the improved model are increased by 10.0%, 1.9%, and 5.5%, and the mean average precision of scratch and flaw defects increased by 11.0% and 3.4%. In summary, the improved model has better generalization.

To more clearly and intuitively observe the performance improvement after mod-el enhancement, we provided visual images of YOLOv8-BSPB on two datasets. As shown in Figure 18, the improved YOLOv8-BSPB model demonstrates significantly better defect detection performance on both datasets compared to the original model. The enhanced model more accurately identifies and classifies defects in complex areas, particularly in cases where the original model missed or incorrectly detected defects. Additionally, the improved model shows fewer false positives in detecting scratches and edge losses. These enhancements not only increase the comprehensiveness and reliability of defect detection but also effectively reduce the false detection rate, validating the potential advantages of the improved model in practical applications.

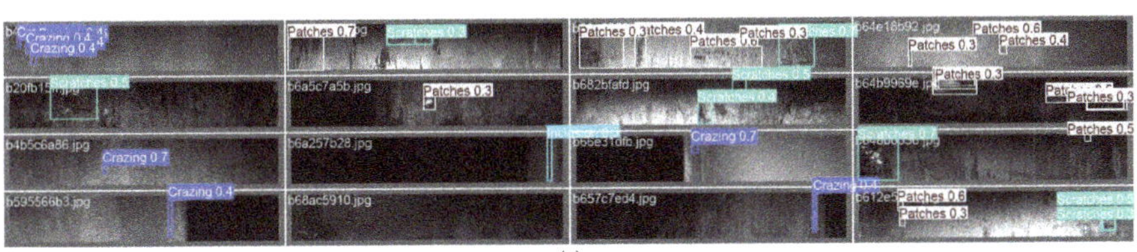

(a)

Figure 18. *Cont.*

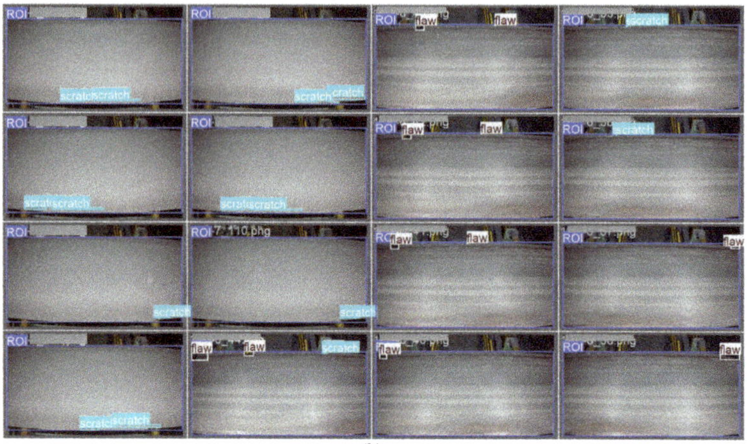

(b)

Figure 18. Two dataset visualizations: (**a**) Kaggle; (**b**) factory datasets.

4. Conclusions

Aiming at the problems existing in the actual detection of existing steel strips, yolov8n is improved, and the BAM attention mechanism is introduced into the backbone to strengthen the feature expression ability of the model. An improved spatial pyramid pooling feature extraction module SCRD was proposed to extract multi-scale features. The PSA polarized attention mechanism is introduced to optimize information transmission and suppress background interference. BiFPN bidirectional feature pyramid network is introduced in the neck to improve the detection speed of the model while maintaining accuracy. Finally, the WIoU loss function is introduced to reduce error propagation and improve the accuracy and efficiency of target detection. Through a large number of contrast tests and ablation experiments, the feasibility of the improved algorithm is verified. Compared to traditional methods like Faster R-CNN and SSD, the YOLOv8-BSPB model demonstrates significant advantages in both detection accuracy and real-time performance. Specifically, YOLOv8-BSPB improves detection accuracy while achieving real-time detection with lower computational cost, meeting the dual demands of speed and accuracy in industrial environments. Compared to YOLOv5n, YOLOv6n, and YOLOX, YOLOv8-BSPB optimizes the network structure, reducing computation while maintaining detection accuracy, which is more suitable for resource-limited environments. The YOLOv8-BSPB proposed in this paper has high detection accuracy and a small number of parameters, which has positive significance in the real-time detection of strip surface defects. In future work, we will continue to study how to lightweight the model while maintaining high accuracy so as to improve the detection speed while ensuring the detection accuracy.

Author Contributions: Conceptualization, J.W., X.X. and T.C.; methodology, J.W.; software, J.W.; validation, J.W., T.C. and X.X.; formal analysis, X.X.; investigation, X.X.; resources, T.C.; data curation, J.W.; writing—original draft preparation, J.W.; writing—review and editing, J.W.; visualization, J.W. and X.X.; supervision, T.C., X.X., L.Z., D.Y., X.G., N.C. and Y.D.; project administration, T.C. and X.X.; funding acquisition, X.X. and T.C.; All authors have read and agreed to the published version of the manuscript.

Funding: This research was funded by the Zhejiang Provincial Administration for Market Regulation Fund Project under Grant No. ZC2023010, and the Natural Science Foundation of Zhejiang province, grant number LY19F050008.

Institutional Review Board Statement: Not applicable.

Informed Consent Statement: Not applicable.

Data Availability Statement: The data presented in this study are available on request from the corresponding author.

Conflicts of Interest: The authors declare no conflicts of interest.

References

1. Lagneborg, R. New steels and steel applications for vehicles. *Mater. Des.* **1991**, *12*, 3–14. [CrossRef]
2. Cardellicchio, A.; Ruggieri, S.; Nettis, A.; Renò, V.; Uva, G. Physical interpretation of machine learning-based recognition of defects for the risk management of existing bridge heritage. *Eng. Fail. Anal.* **2023**, *149*, 107237. [CrossRef]
3. Sun, J.; Peng, W.; Ding, J.; Li, X.; Zhang, D. Key intelligent technology of steel strip production through process. *Metals* **2018**, *8*, 597. [CrossRef]
4. Chu, Y.; Yu, X.; Rong, X. A Lightweight Strip Steel Surface Defect Detection Network Based on Improved YOLOv8. *Sensors* **2024**, *24*, 6495. [CrossRef] [PubMed]
5. Chen, B.; Wei, M.; Liu, J.; Li, H.; Dai, C.; Liu, J.; Ji, Z. EFS-YOLO: A lightweight network based on steel strip surface defect detection. *Meas. Sci. Technol.* **2024**, *35*, 116003. [CrossRef]
6. Zhong, W.; Zhang, L.; Li, P.; Gui, W. Research on Steel Surface Defect Detection Algorithm Based on Improved YOLOv8n. *J. Physics Conf. Ser.* **2024**, *2832*, 012013. [CrossRef]
7. Luo, Q.; Fang, X.; Liu, L.; Yang, C.; Sun, Y. Automated Visual Defect Detection for Flat Steel Surface: A Survey. *IEEE Trans. Instrum. Meas.* **2020**, *69*, 626–644. [CrossRef]
8. Hao, B.; Fan, Y.; Song, Z. Crack defect detection using pulsed eddy current thermal imaging based on deep transfer learning. *Acta Opt. Sin.* **2023**, *43*, 146–154. [CrossRef]
9. Liu, K.; Luo, N.; Li, A.; Tian, Y.; Sajid, H.; Chen, H. A new self-reference image decomposition algorithm for strip steel surface defect detection. *IEEE Trans. Instrum. Meas.* **2019**, *69*, 4732–4741. [CrossRef]
10. Vaidelienė, G.; Valantinas, J.; Ražanskas, P. On the Use of Discrete Wavelets in Implementing Defect Detection System for. *Inf. Technol. Control* **2016**, *45*, 214–222. [CrossRef]
11. Chen, M.; He, Y. Wire Rope Weak Defect Signal Processing Methods Based on Improved SVD and Phase Space Reconstruction. *Inf. Technol. Control* **2021**, *50*, 752–768. [CrossRef]
12. Prasanna, S.; Manoharan, R.; Anandan, S. A Deep Learning-based Method for Steel Strip Surface Defect Detection Using Convolutional Neural Networks (CNNs). *J. Manuf. Process.* **2021**, *58*, 123–134.
13. Zhang, H.; Wang, L.; Li, Q.; Chen, Y.; Zhao, X.; Liu, J. Steel Strip Surface Defect Detection Using Deep Convolutional Neural Networks (CNNs). *IEEE Trans. Ind. Electron.* **2019**, *66*, 9554–9563.
14. Li, W.; Zhou, T.; Huang, Y.; Yang, X.; Chen, K. Real-Time Detection of Steel Surface Defects Using Convolutional Neural Networks and Data Augmentation. *J. Intell. Manuf.* **2020**, *31*, 1181–1193.
15. Di Mucci, V.M.; Cardellicchio, A.; Ruggieri, S.; Nettis, A.; Renò, V.; Uva, G. Artificial intelligence in structural health management of existing bridges. *Autom. Constr.* **2024**, *167*, 105719. [CrossRef]
16. Wan, X.; Zhang, X.; Liu, L. An improved VGG19 transfer learning strip steel surface defect recognition deep neural network based on few samples and imbalanced datasets. *Appl. Sci.* **2021**, *11*, 2606. [CrossRef]
17. Bhatt, P.M.; Malhan, R.K.; Rajendran, P.; Shah, B.C.; Thakar, S.; Yoon, Y.J.; Gupta, S.K. Image-based surface defect detection using deep learning: A review. *J. Comput. Inf. Sci. Eng.* **2021**, *21*, 040801. [CrossRef]
18. Lu, Y.; Huang, Z.-C.; Jiang, Y.-Q.; Liu, G.-R.; Wang, J.-X.; Ma, Q.-M. Lightweight-detection: The strip steel surface defect identification based on improved YOLOv5d. *Mater. Today Commun.* **2024**, *40*, 109814. [CrossRef]
19. Du, Y.; Chen, H.; Fu, Y.; Zhu, J.; Zeng, H. AFF-Net: A Strip Steel Surface Defect Detection Network via Adaptive Focusing Features. *IEEE Trans. Instrum. Meas.* **2024**, *73*, 2518514. [CrossRef]
20. Ribeiro, G.; Santos, V. Single and multi-tasked neural networks: Selection and deployment. In Proceedings of the 2024 IEEE International Conference on Autonomous Robot Systems and Competitions (ICARSC), Paredes de Coura, Portugal, 2–3 May 2024.
21. Zhao, J.; Liu, S.; Tao, H.; Liu, W. Slim-YOLOv8: A fast and accurate algorithm for surface defect detection of steel strips. *Ironmak. Steelmak.* **2024**. [CrossRef]
22. Lu, J.; Yu, M.; Liu, J. Lightweight strip steel defect detection algorithm based on improved YOLOv7. *Sci. Rep.* **2024**, *14*, 13267. [CrossRef] [PubMed]
23. Yang, L.; Li, M.; Hu, J.; Wang, D. Steel Strip Surface Defect Detection Algorithm Based on Improved Yolov7-tiny. *Comput. Eng.* **2024**, 1–9. [CrossRef]

24. Ren, F.; Fei, J.; Li, H.; Doma, B.T. Steel Surface Defect Detection Using Improved Deep Learning Algorithm: ECA-SimSPPF-SIoU-Yolov5. *IEEE Access* **2024**, *12*, 32545–32553. [CrossRef]
25. Yeung, C.-C.; Lam, K.-M. Efficient Fused-Attention Model for Steel Surface Defect Detection. *IEEE Trans. Instrum. Meas.* **2022**, *71*, 2510011. [CrossRef]
26. Wang, W.; Mi, C.; Wu, Z.; Lu, K.; Long, H.; Pan, B.; Li, D.; Zhang, J.; Chen, P.; Wang, B. A Real-Time Steel Surface Defect Detection Approach with High Accuracy. *IEEE Trans. Instrum. Meas.* **2022**, *71*, 941–953. [CrossRef]
27. Li, C.; Xu, A.; Zhang, Q.; Cai, Y. Steel Surface Defect Detection Method Based on Improved YOLOX. *IEEE Access* **2024**, *12*, 37643–37652. [CrossRef]
28. Wang, C.-Y.; Bochkovskiy, A.; Liao, H.-Y.M. YOLOv7: Trainable bag-of-freebies sets new state-of-the-art for real-time object detectors. In Proceedings of the 2023 IEEE/CVF Conference on Computer Vision and Pattern Recognition (CVPR), Vancouver, BC, Canada, 24 June 2023.
29. Xiong, Y.; Li, Z.; Chen, Y.; Wang, F.; Zhu, X.; Luo, J.; Wang, W.; Lu, T.; Li, H.; Qiao, Y.; et al. Efficient Deformable ConvNets: Rethinking Dynamic and Sparse Operator for Vision Applications. In Proceedings of the 2024 IEEE/CVF Conference on Computer Vision and Pattern Recognition (CVPR), Seattle, WA, USA, 17–21 July 2024; pp. 5652–5661.
30. Liu, S.; Huang, D. Receptive field block net for accurate and fast object detection. In Proceedings of the European Conference on Computer Vision (ECCV), Mico Milano, Italy, 29 September–4 October 2018; pp. 385–400. [CrossRef]
31. Park, J. BAM: Bottleneck Attention Module. *arXiv* **2016**, arXiv:1807.06514.
32. Liu, H.; Liu, F.; Fan, X.; Huang, D. Polarized self-attention: Towards high-quality pixel-wise mapping. *Neurocomputing* **2022**, *506*, 158–167. [CrossRef]
33. Tan, M.; Pang, R.; Le, Q. EfficientDet: Scalable and Efficient Object Detection. In Proceedings of the 2020 IEEE/CVF Conference on Computer Vision and Pattern Recognition (CVPR 2019), Seattle, WA, USA, 27 July 2020. [CrossRef]
34. Tong, Z.; Chen, Y.; Xu, Z.; Yu, Z. Wise-iou: Bounding box regression loss with dynamic focusing mechanism. *arXiv* **2023**, arXiv:2301.10051. [CrossRef]
35. Zhao, X.; He, Y.; Zhang, H.; Ding, Z.; Zhou, C.; Zhang, K. A quality grade classification method for fresh tea leaves based on an improved YOLOv8x-SPPCSPC-CBAM model. *Sci. Rep.* **2024**, *14*, 4166. [CrossRef]
36. Wan, W.; Wang, L.; Wang, B.; Yu, H.; Shi, K.; Liu, G. Space to depth convolution bundled with coordinate attention for detecting surface defects. *Signal Image Video Process.* **2024**, *18*, 4861–4874. [CrossRef]
37. Cheng, S.; Wang, Z.; Liu, S.; Han, Y.; Sun, P.; Li, J. Attention-Based Lightweight YOLOv8 Underwater Target Recognition Algorithm. *Sensors* **2024**, *24*, 7640. [CrossRef] [PubMed]
38. Lang, D.; Lv, Z. A PV cell defect detector combined with transformer and attention mechanism. *Sci. Rep.* **2024**, *14*, 20671. [CrossRef] [PubMed]
39. Hu, C.; Ma, R.; Du, X.; Gong, J. Boundary-aware residual network for defect detection in strip steel products. *Evol. Syst.* **2024**, *15*, 1649–1663. [CrossRef]
40. Zhao, C.; Fan, Y.; Tan, J.; Li, Q.; Lin, Z.; Luo, S.; Chen, X. FCS-YOLO: An efficient algorithm for detecting steel surface defects. *Meas. Sci. Technol.* **2024**, *35*, 086004. [CrossRef]
41. Kou, X.; Liu, S.; Cheng, K.; Qian, Y. Development of a YOLO-V3-based model for detecting defects on steel strip surface. *Measurement* **2021**, *182*, 109454. [CrossRef]
42. Li, Y.; Xu, S.; Zhu, Z.; Wang, P.; Li, K.; He, Q.; Zheng, Q. EFC-YOLO: An efficient surface-defect-detection algorithm for steel strips. *Sensors* **2023**, *23*, 7619. [CrossRef]
43. Qin, H.; Xiao, Y.; Wang, Y.; Chen, K.; Dong, F. Research on Surface Defect Detection of Steel Strip Based on Deep Learning. In Proceedings of the 2023 2nd International Conference on Computing, Communication, Perception and Quantum Technology (CCPQT), Xiamen, China, 4–7 August 2023; pp. 302–308.
44. Kong, X.; Fan, W. Surface Defect Detection of Steel Strips based on an Improved YOLOv8. In Proceedings of the 2024 5th International Seminar on Artificial Intelligence, Networking and Information Technology (AINIT), Nanjing, China, 29–31 March 2024; pp. 2325–2328.
45. Ren, S.; He, K.; Girshick, R. Sun Faster R-CNN: Towards real-time object detection with region proposal networks. *IEEE Trans. Pattern Anal. Mach. Intell.* **2017**, *39*, 1137–1149. [CrossRef]
46. Liu, W.; Anguelov, D.; Erhan, D.; Szegedy, C.; Reed, S.; Fu, C.-Y.; Berg, A.C. *SSD: Single Shot MultiBox Detector*; Springer: Cham, Switzerland, 2016. [CrossRef]

applied sciences

MDPI

Article

Analysis and Prediction of Wear in Interchangeable Milling Insert Tools Using Artificial Intelligence Techniques

Sonia Val *, María Pilar Lambán , Javier Lucia and Jesús Royo

Department of Design and Manufacturing Engineering, University of Zaragoza, 50009 Zaragoza, Spain; plamban@unizar.es (M.P.L.); javierluciamarco@gmail.com (J.L.); jaroyo@unizar.es (J.R.)

* Correspondence: sonia@unizar.es

Abstract: Milling machines remain relevant in modern manufacturing, with tool optimization being crucial for cost reduction. Inserts for compound cutting tools can reduce the cost of operations by optimizing their lifespan. This study analyzes the flank wear of cutting tools in milling machines, with an emphasis on evaluating different approaches to predict their lifespan. It compares three distinct modeling approaches for predicting tool lifespan using algorithms: traditional ensemble methods (Random Forest, Gradient Boosting) and a deep learning-based LSTM network. Each model is evaluated independently, and this comparative analysis aims to determine which modeling strategy best captures the intricate interactions between various process variables affecting tool wear. This method ensures greater efficiency and accuracy than conventional techniques, providing a scalable, resource-efficient solution for reliable and insightful tool wear predictions. The results obtained from the dataset of an insert tool can be extrapolated to other milling inserts and demonstrate the progression of tool wear over time under varying cutting parameters, providing critical insights for optimizing milling operations. The integration of uncertainty awareness in the predictive outputs is a unique feature of this research and enhances decision-making for smarter manufacturing. This proactive approach enhances operational efficiency and reduces overall production costs. Furthermore, the data-driven, AI-centric methodology developed in this study offers a transferable approach that can be adapted to other machining processes, advancing state-of-the-art tool wear prediction.

Keywords: artificial intelligence; cutting tools; machine learning; milling machines; predictive maintenance; tool wear

1. Introduction

Amid the Fourth Industrial Revolution, commonly referred to as Industry 4.0, significant technological changes are taking place. Industries must undergo modernization processes not only to sustain but also to compete in the current market, both locally and, thanks to globalization, all around the world. Material removal processes have had to optimize all aspects of their operations to enhance efficiency, and within this broad category, milling machines play a fundamental role in the sector, particularly in the production of high-precision parts with superior surface finishes.

Milling machines allow for the machining of surfaces and the creation of complex shapes in various materials, such as plastics and composites, though they are mostly used to work with metals. However, the significance of milling machines lies not only in their ability to perform precise and efficient cuts but also in their contribution to increasing productivity and reducing industrial costs [1]. Over time, these machines have evolved from simple manual devices to more sophisticated models managed by computer numerical control (CNC), thus improving precision and enhancing the repeatability of machining tasks to deliver unprecedented results.

The accuracy of cuts depends on the machine's rigidity, the quality of guides and spindles, and the operator's skill or the CNC program's accuracy, as well as the wearing of the cutting tool [2]. Shaping a workpiece requires coordinated movement of the worktable and the rotation of a cutting tool, called a milling cutter, which removes material at high speed, cutting small chips until achieving the final product [3]. Milling machines are versatile and capable of performing diverse operations such as milling, drilling, boring, and threading.

Cutting tools are essential for the machining process, enabling precise cutting, shaping, and finishing of parts. They are robust and efficient, capable of withstanding high forces and temperatures during milling. There are two main types: solid tools and interchangeable insert tools. Solid tools are made from a single piece of material, usually solid carbide or high-speed steel (HSS), known for their durability and precision; however, they require replacement or sharpening when worn, which can be highly costly and time-consuming. Interchangeable insert tools consist of a main body with replaceable cutting inserts, which can be changed when worn, reducing costs and minimizing machine downtime [4].

Inserts are small cutting pieces mounted on milling tools to perform precise cuts. They are utilized in disk cutters, slab mills, and grooving cutters for cutting flat surfaces, contours, and grooves, respectively. Choosing the right inserts and their proper use is crucial for optimal results in manufacturing and machining. Inserts are made from hard and wear-resistant materials, primarily tungsten carbide, ceramics, cermet (a mix of ceramic and metal), cubic boron nitride (CBN), and polycrystalline diamond (PCD). Tungsten carbide is the most widely used due to its high hardness and wear resistance, while ceramic inserts resist high temperatures, and cermet combines hardness and toughness. CBN and PCD inserts are used for specialized applications [5]. Inserts are classified based on shape, geometry, specific application, and the machined material. The shape determines the insert's cutting ability in different directions, with common shapes including square, round, triangular, rhomboidal, and hexagonal. Geometry influences cutting behavior, chip removal, and insert lifespan. Inserts are designed for various materials, such as steel, stainless steel, cast iron, non-ferrous metals, and composites, each requiring suitable properties to optimize cutting performance and minimize wear.

The cutting angles of inserts in milling machines are fundamental for tool performance and lifespan. The main angles to consider are the inclination angle and the clearance angle, which affect cutting efficiency, chip formation, and tool durability. Proper selection and adjustment of these angles, along with controlling chip thickness, optimize cutting conditions and improve productivity in milling.

Tool wear is an inevitable phenomenon affecting the efficiency, surface finish quality, and dimensional accuracy of machined parts. During milling, both the cutting tool and workpiece experience intense forces and frictions, resulting in various wear types. Optimizing cutting parameters like cutting speed (v_c [m/min]), feed rate (f_z [mm/rev]), and depth of cut (a_p [mm]) can minimize the forces and heat generated during the process. Understanding common wear types is vital for optimizing machining processes and selecting appropriate materials and cutting parameters.

The wear of cutting tools is a problem that can directly impact product quality, process efficiency, and, consequently, cost. Tool wear is an unavoidable phenomenon that primarily occurs due to friction forces, high temperatures, and adverse working conditions, eventually leading to a loss of precision and tool failure. This is why the prevention and management of tool wear in milling is a critical factor to consider in ensuring the continuity and quality of production. The study of cutting tool wear, particularly in milling machines, has been a constant point of interest in the manufacturing industry and has advanced significantly over time, encompassing everything from traditional monitoring and analysis methods to the application of cutting-edge AI techniques. Each stage has contributed to improving the accuracy and effectiveness of wear control.

Initially, cutting tool wear was assessed using rather rudimentary methods, such as visual inspection and direct measurement of deterioration on inserts using microscopes and high-resolution cameras [6]. While these techniques were useful for identifying evident damage to the tools, they had significant limitations. First, the process was labor-intensive, as it required the removal of the tools from the machine for detailed inspection, which disrupted production and increased downtime, thus reducing overall process efficiency. Additionally, the subjectivity tied to visual inspection could lead to inconsistent results, depending on the experience and judgment of the operator. In short, the need for constant monitoring and the lack of predictive capability limited their effectiveness in large-scale production scenarios. However, despite these limitations, these methods were fundamental in the initial phase of wear study, providing a foundation for the development of more advanced techniques that emerged over time.

As technology advanced and the need for improved production efficiency grew, more precise and less intrusive methods for wear evaluation were introduced. Among these, cutting force measurement became a popular technique, as it allowed for inference of tool wear conditions by monitoring the stresses experienced during the milling process. Similarly, vibration detection offered a way to identify anomalies in tool behavior that could indicate excessive wear or impending failure [6]. Monitoring the temperature in the cutting zone was also adopted as an indirect indicator of wear, given that an increase in temperature is usually associated with higher friction and, consequently, greater deterioration of cutting tools [7].

With the advent and improvement of advanced sensors and data acquisition systems, tool wear control became more precise and automated. Force, acoustic, thermal, and vibration sensors were integrated into milling machines to capture real-time data on operational conditions and the state of the tools [8]. These control systems, working alongside signal processing techniques, enabled earlier detection of wear and a better understanding of the mechanisms that cause it without the need to stop the machine for inspections. Despite these advances, the interpretation and analysis of the data remained significant challenges due to the complexity of wear patterns and the multiple variables involved.

However, these approaches had their limitations. Despite being more precise, they were still essentially reactive, meaning they detected wear only after it had occurred rather than predicting it before it manifested. Furthermore, they still required a high degree of supervision, albeit less than before, which could pose a problem in large-scale production scenarios where operating conditions vary considerably and resources for continuous monitoring may be limited.

The need to overcome these limitations led to the development of predictive and more integrated methods for wear evaluation, and this is where artificial intelligence (AI) and Machine Learning techniques played a critical role. These technologies enabled the development of models capable of predicting wear based on large volumes of real-time data [9]. These models use Machine Learning algorithms to identify complex patterns that would not be apparent through traditional methods. By analyzing data such as cutting forces, vibrations, temperature, and other parameters, they can anticipate wear before it occurs, allowing operators to take preventive measures to avoid failures and optimize the performance of milling machines.

As a result, AI models have been successfully applied to predict cutting tool wear, providing a more comprehensive and precise understanding of tool conditions to facilitate wear prevention in inserts. In recent years, various predictive models have been explored for cutting tool wear prediction, with approaches ranging from traditional Fuzzy Logic and Genetic Algorithms to more advanced Random Forest, Gradient Boosting, and Long Short-Term Memory networks. Each of these models offers distinct advantages and challenges, choosing models dependent on factors like data complexity, computation time, and the ability to predict wear patterns in real-time milling operations.

Fuzzy Logic has been a popular approach in systems where inputs are uncertain or imprecise. It relies on fuzzy set theory to deal with partial truth values, making it useful for decision-making in environments with vague or ambiguous data. However, despite its strengths in modeling uncertainty, Fuzzy Logic struggles with the complexity of modern predictive tasks. The challenge of dealing with high-dimensional, continuous data from milling processes makes fuzzy systems less effective in this context, as they do not naturally capture the nonlinear, time-dependent relationships involved in tool degradation [10]. This makes them less suitable for dynamic and real-time wear prediction in complex manufacturing settings.

Genetic Algorithms are another traditional approach used for optimization tasks, particularly in cases where the relationship between input variables is not easily defined. These models evolve solutions over generations through selection, crossover, and mutation. While they have proven effective for optimizing parameters like cutting speed and feed rate, they are computationally expensive and time-consuming, especially when applied to large datasets or real-time monitoring systems [11]. This limitation arises from the need for repeated evaluations of possible solutions, which can delay the prediction process, especially in environments where prompt wear detection is crucial.

Given the limitations of Fuzzy Logic and Genetic Algorithms, other Machine Learning models like Random Forest, Gradient Boosting, and LSTM have gained prominence due to their ability to handle complex, high-dimensional data, provide faster predictions, and achieve higher accuracy in predictive maintenance tasks.

Random Forest is an ensemble learning technique that constructs multiple decision trees and combines their outputs for robust predictions. It has become a widely used method in predictive maintenance because of its efficiency in handling large datasets with multiple variables [12]. The model's strength lies in its ability to manage missing data, provide feature importance, and avoid overfitting, making it particularly well-suited for real-time monitoring of tool wear where data may be noisy or incomplete.

Gradient Boosting is another powerful Machine Learning method that builds models sequentially, with each new tree attempting to correct the errors of the previous one. This method is known for its high predictive accuracy and ability to model complex relationships between variables [13]. Gradient Boosting is particularly useful when the data are noisy or contain outliers, as it reduces both bias and variance in predictions, making it a solid choice for tool wear forecasting where environmental conditions can vary.

Long Short-Term Memory networks are a type of recurrent neural network designed to handle time-series data, making them particularly well-suited for applications where wear evolves over time. Milling operations often involve continuous data streams, such as cutting forces and temperatures, that need to be analyzed over time to predict tool wear. The ability of LSTM to capture temporal dependencies is very relevant in tool wear prediction, where past operational conditions influence future wear patterns. Unlike traditional models, LSTM can detect subtle, time-dependent changes that may indicate impending tool failure, enabling proactive maintenance measures [14].

However, despite advancements in monitoring and analyzing wear using Machine Learning algorithms, challenges remain. For instance, the complexity of the generated data, the need for more robust and generalizable models, and the effective integration of these systems into real production environments are areas that require further research and development to achieve even greater precision and efficiency.

This research was performed with data from insert milling tools, which are widely used in industry today. With these approaches, more accurate and efficient results than in previous studies are expected, optimizing both the relevance of the analyzed parameters and the precision in wear prediction. An artificial intelligence pipeline was used to analyze key parameters such as cutting speed, feed per tooth, and cutting depth, as its models are widely accepted in the scientific field for their simplicity and effectiveness in translating various signals into reliable wear forecasts [6]. The results can improve knowledge about tool wear and lead to longer tool life, reducing downtime, tool costs, and, consequently, overall process costs.

2. Materials and Methods

2.1. Dataset

Data play a critical role in developing accurate models to predict tool wear and lifespan, as without data, models cannot be created. Generally, studies on tool wear in milling machines and other industrial equipment often rely on experimental data collected in controlled environments. These datasets may include measurements of forces, vibrations, and other operational parameters [15] that impact tool wear.

Researchers and professionals typically work with data obtained from internal experiments, which can be costly and labor-intensive to gather. However, today, the availability of public datasets provides a valuable alternative for research and development. These datasets allow researchers to access pre-collected data, facilitating experimentation and model development without the need to conduct new experiments from scratch. This enables those without access to physical machines to conduct experiments and compare results with other researchers.

In the search for a suitable dataset for predicting insert wear in milling machines, several relevant public datasets are available. Among them, a dataset from the Katulu project's GitHub repository [16] was selected as the origin of data for this research. A dataset called NUAA, extracted from NUAA Ideahouse, contains 24 variables and 234,902 rows of data from orthogonal milling experiments using solid carbide inserts and titanium parts (TC4).

Our research focuses on the NUAA dataset due to its greater number of variables, making it a more comprehensive option for a thorough analysis.

2.2. Method

To predict tool wear accurately, this study employed a systematic approach utilizing the NUAA dataset, which encompasses a wide range of inputs reflecting the complex relationships influencing tool wear. This comprehensive dataset facilitated the implementation of models specifically designed to learn and model nonlinear relationships between the input features and tool wear. The use of these models allowed us to capture intricate dynamics that would otherwise be overlooked, thus enhancing the accuracy of our predictions.

The analysis proceeded as follows:

1. **Data Selection and Preprocessing**: The dataset was carefully curated and preprocessed to ensure consistency and relevance. This included cleaning the data, handling missing values, and normalizing features to prepare them for analysis.
2. **Pipeline Implementation**: An AI pipeline consisting of ensemble models (Random Forest and Gradient Boost) and a Long Short-Term Memory network was used to compare the accuracy of tool wear prediction.
3. **Evaluation Metrics**: The models were evaluated using appropriate metrics, such as the mean squared error (MSE), Mean Absolute Error (MAE), and Coefficient of Determination (R^2 Score), to assess their predictive accuracy and generalization capabilities.
4. **Visual Analysis**: Different plots were generated to visually assess the distribution of predictions against actual values, providing insights into the best model's performance and calibration. Additionally, a correlation matrix was implemented to explore relationships between variables within the dataset. This algorithm analyzed how different input features correlate with tool wear, aiming to uncover patterns that can inform predictive modeling and enhance our understanding of tool wear mechanisms.

2.3. Data Analysis and Pipeline Creation

An excerpt of the code developed specifically for this task can be found in Appendix A.

To begin the analysis, the NUAA dataset was loaded. Then, it was cleaned and reorganized by removing irrelevant columns and rearranging the remaining ones. With the data prepared, an initial exploration was conducted to understand the structure of the dataset. Figure 1 offers descriptive statistics that are reviewed to gain a better understanding of the data characteristics, Figure 2a,b remain as support images to help illustrate a generic milling operation diagram to clarify milling parameters [17–19], and Figure 2c illustrates the different types of tool wear, out of which flank wear is the focus of this article.

234,902 rows

	timestamp	spindle_speed	axial_cutting_depth	feed_per_tooth	axial_force	bending_moment_x	bending_moment_y	torsion	spindle_power	spindle_current	force_z	vibration_x	vibration_y
count	234,902.0	234,902.0	234,902.0	234,902.0	234,902.0	234,753.0	234,753.0	234,753.0	234,902.0	234,902.0	234,902.0	234,902.0	234,902.0
mean	1168.5264	1788.1184	2.8477	0.0496	121.1849	0.8954	-6.5090	5.0905	2.1758	0.0470	121.1849	1.4919	2.0391
std	880.3653	37.9597	0.3753	0.0041	198.3735	2.2338	26.3179	26.2899	0.2684	1.3617	198.3735	4.9946	4.4314
min	0.0033	1750.0000	2.5000	0.0450	-904.3500	-9.2637	-101.4400	-88.7750	1.5137	-3.3671	-904.3500	-26.9570	-26.8360
25%	430.4998	1750.0000	2.5000	0.0450	-7.1774	-0.4503	-19.4960	-9.2486	1.9958	-1.4084	-7.1774	-1.1185	-0.9689
50%	956.8020	1800.0000	3.0000	0.0500	122.0200	1.0293	-1.1901	1.0847	2.1484	0.1755	122.0200	1.1392	2.2894
75%	1801.8899	1800.0000	3.0000	0.0550	258.3900	2.3802	7.2541	21.6370	2.3071	1.4257	258.3900	4.1996	4.9906
max	3323.7402	1850.0000	3.5000	0.0550	1055.1000	10.1000	89.4870	116.4600	3.4729	3.7875	1055.1000	74.6310	30.6270

Figure 1. Initial data exploration of the NUAA dataset.

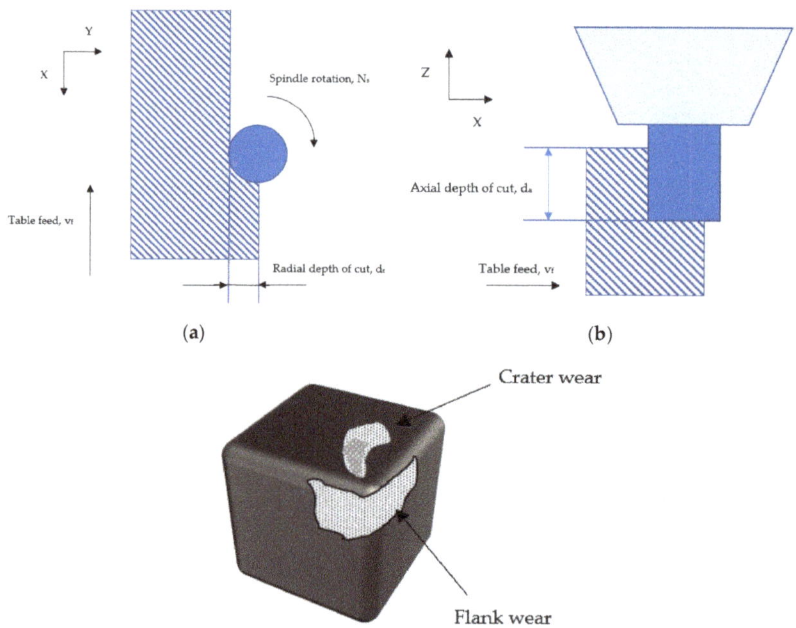

Figure 2. General milling operation sketches. (**a**) depicts a bending moment in one axis, (**b**) depicts general milling operation parameters, and (**c**) illustrates where flank wear occurs.

It can be seen (Figure 3) that out of the dataset, 149 rows are missing (null) values for bending moment x, bending moment y, and torsion, which could introduce errors during model training. To address this, we applied normalization to these rows instead of removing them.

To gain a deeper understanding of the data, histograms were generated from the initial data exploration of the dataset, illustrating the distribution of the different characteristics. These graphs helped identify patterns and anomalies in the data (Figure 4). In the *Y*-axis, the total count of data is plotted, whereas the X-axis shows the data range of each variable in its respective units.

```
Missing values in each column:
timestamp                0
spindle_speed            0
axial_cutting_depth      0
feed_per_tooth           0
axial_force              0
bending_moment_x       149
bending_moment_y       149
torsion                149
spindle_power            0
spindle_current          0
force_z                  0
vibration_x              0
vibration_y              0
dtype: int64
```

Figure 3. Missing values in each column.

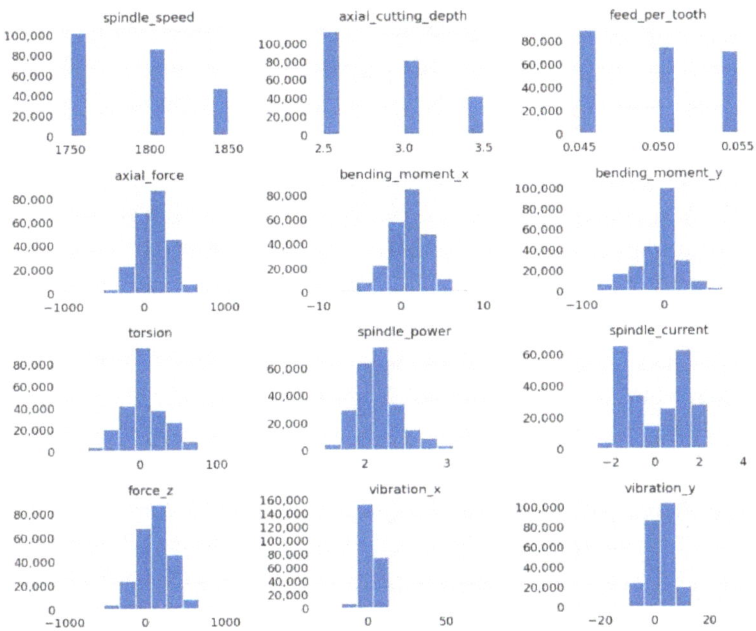

Figure 4. Graphical display of the data distribution from the NUAA dataset.

From the data distribution graphs, it can be observed that most values have a wide data range, except for the topmost three graphs, which have three columns, indicating that the owners of the dataset employed three constant values to obtain the others. Furthermore, while the Y-axis maintains consistent units across the graphs, the X-axis varies significantly. For example, in the feed per tooth graph, the values range from 0.04 to 0.06, whereas in the axial force graph, the range spans from -1000 to 1000. This disparity highlights the need for data normalization to ensure that the models can process the data effectively. It must be pointed out that those parameters can be negative, e.g., axial force can be physically negative because it represents a directional force along an axis. A negative axial force, such as -1000, indicates compression rather than tension, reflecting the direction in which the force is applied: positive forces align with the axis, while negative forces act in the opposite direction. In practical terms, a positive axial force pulls or stretches a structure, while a negative axial force pushes or compresses it.

To properly predict tool wear, a well-structured pipeline was leveraged to predict tool wear, employing a systematic approach to data preprocessing, model design, and validation. The process involved multiple steps, each critical to ensuring the models' accuracy and robustness. The steps can be summarized as follows:

1. Data Preparation and Preprocessing.

First, the dataset was processed to handle any potential inconsistencies. Any non-numeric values in the dataset were coerced into a numerical format, and any rows containing missing values were removed to ensure that the models only trained on clean and complete data, reducing the risk of errors during training.

Next, the dataset was split into features and the target variable. The features, which were various characteristics of the cutting tools, formed the input to the models, while the target variable was the tool wear, which the models were supposed to predict. Then, the data were divided into training and testing sets using an 80/20 ratio, meaning 80% of the data was used to train the models, while the remaining 20% was reserved for testing their performance.

Because different features can have different ranges and units, scaling was necessary. A standard scaler was used to normalize the feature values. The scaler transformed the data such that each feature had a mean of 0 and a standard deviation of 1. This step is critical for neural networks because it ensures that all inputs are treated equally by the model, helping to prevent any one feature from dominating the learning process.

2. Pipeline and Model Architecture Designs.

A comprehensive Machine Learning pipeline was developed incorporating ensemble models and an LSTM (Long Short-Term Memory) network to enhance the accuracy of tool wear prediction. The pipeline consisted of data preprocessing, ensemble modeling, temporal modeling with LSTM, and model evaluation.

Two ensemble models—Random Forest and Gradient Boosting—were employed to leverage the strength of multiple decision trees.

- **Random Forest:** Consisting of 100 decision trees, the model uses multiple random samples and averages predictions to reduce variance.
- **Gradient Boosting:** This model sequentially improves predictions by focusing on areas where previous models performed poorly, capturing complex data patterns.

Additionally, to capture temporal dependencies in the data, an LSTM network was implemented to model time-series behavior, which is useful for sequential dependencies like tool wear progression. A time window of 5 was set, allowing the model to predict tool wear based on the previous five time steps.

- **LSTM Architecture:** The LSTM layer, containing 64 units, captures sequential data patterns, while a dropout layer with a 20% rate helps prevent overfitting. The model's output layer consists of a single neuron, reflecting the regression task.
- **Early Stopping:** To avoid overfitting, early stopping was implemented, halting training once validation loss stopped improving.

3. Model Validation with Cross-Validation.

To evaluate the models' generalization performance, K-Fold cross-validation with five splits was used. This approach divided the training data into five equally sized subsets. The models were iteratively trained on four of these subsets and validated on the remaining one, cycling through each fold to ensure the models were tested on all portions of the data. By doing so, every data point in the dataset was used for both training and validation, providing a comprehensive assessment of the model's performance. This method reduces the variability that might result from relying on a single train–test split and ensures a more reliable estimation of the model's predictive capabilities.

The results from the cross-validation process were calculated using the MSE, which measures how well the models fit the training data. The average MSE across the folds was reported, along with the standard deviation, which gave an idea of how consistent the models' performances were across different subsets of the data.

4. Early Stopping Mechanism.

To prevent overfitting, an early stopping mechanism was implemented during training. Early stopping monitors the models' losses during training and halts the process if the losses fail to improve for a set number of epochs (in this case, 10). Additionally, the best weights of the models are restored once training stops, ensuring the best possible model is used for predictions.

5. Model Testing and Final Evaluation.

Once the models were trained and cross-validated, they were tested on the unseen test set. The test data were scaled using the same scaler fitted on the training data to ensure consistency. The model's performances on the test set were evaluated using the MSE, which provided an estimate of how well the models generalize to new, unseen data. This final MSE score gave a clear indication of the model's predictive accuracy.

3. Results

From the cross-validation training, the following cross-validated MSE values were obtained for the Random Forest, Gradient Boosting, and LSTE models, respectively: 2.05×10^{-8}, 8.14×10^{-5}, and 5.88×10^{-3} with a standard deviation of 2.18×10^{-8}, 4.56×10^{-6}, and 5.22×10^{-5}. Each fold of the training ran for a varying number of epochs, a variability that shows that early stopping occurred at different times across the folds to avoid overfitting.

After training on the complete training data, the following MSE values were obtained for the test set predictions, in the same order aforementioned: 4.7×10^{-9}, 7.4×10^{-5}, and 5.6×10^{-3}.

The Gradient Boosting model was chosen as the model to be implemented due to its ability to capture complex, nonlinear relationships and its adaptability to smaller datasets compared to Random Forest and LSTM. During cross-validation, Gradient Boosting demonstrated competitive performance, with its mean squared error being higher than Random Forest but still reasonable. Its ability to iteratively refine errors through its sequential learning process makes it more suited to datasets with intricate patterns. Furthermore, Gradient Boosting's stepwise construction of decision trees provides valuable insights into feature importance, improving the interpretability of its predictions. Although its standard deviation was higher than that of Random Forest, the model still exhibited strong generalization on the test set. This combination of performance, interpretability, and adaptability, particularly when hyperparameters are finely tuned, makes Gradient Boosting a robust choice. As such, it will be referred to as "the model" throughout the remainder of this study, representing the optimal approach for predictive tool wear analysis.

The first plot (Figure 5) compares actual and predicted tool wear values. This scatter plot, with a regression line, shows how closely the model's predictions align with the true values. Each point represents a prediction, and the red line represents perfect predictions, where actual equals predicted. Ideally, the points should be as close to this line as possible. In this case, the points generally follow the regression line, indicating a strong positive correlation between the actual and predicted values. However, the deviations observed at medium tool wear levels, often referred to as outliers, suggest that the model faces some difficulty in accurately predicting certain values. This does not necessarily indicate poor model performance. It is often a positive sign that the model is not overly precise, as perfect accuracy could imply overfitting. By not being excessively precise, the model demonstrates a level of flexibility, suggesting it will generalize better to new, unseen data. Balancing prediction accuracy with generalization is key to building a robust and reliable model.

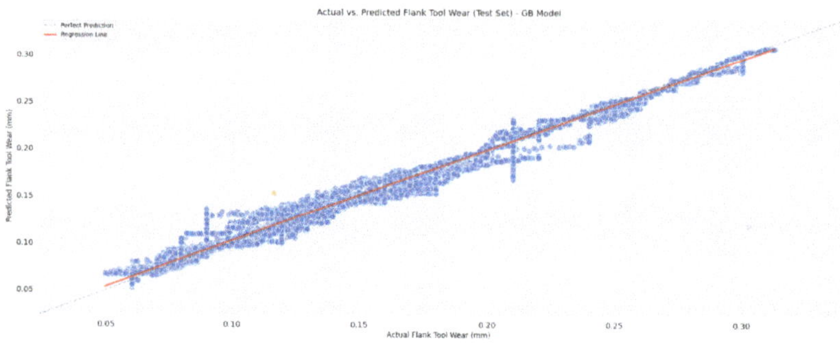

Figure 5. Actual vs. predicted flank tool wear of the test set.

The second plot (Figure 6) visualizes the error distribution, showing the difference between the actual and predicted values. The distribution should ideally be centered around zero, indicating that the model's errors are balanced and not consistently over- or underestimating tool wear. The histogram shows a relatively symmetric error distribution around zero, with a narrow spread. This suggests that most of the model's predictions are close to the actual values, with a few larger errors appearing in the tails of the distribution.

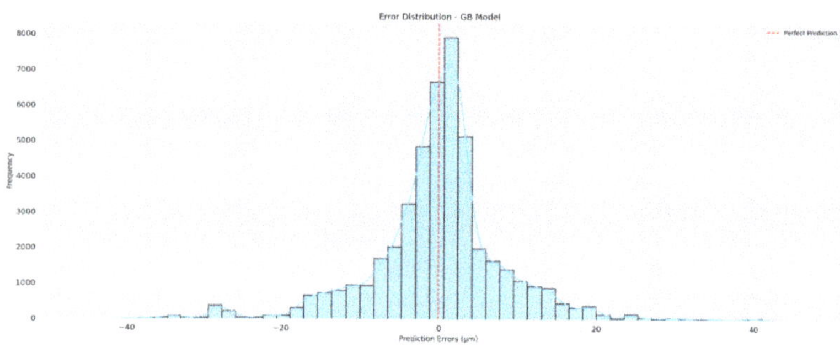

Figure 6. Error distribution of the test set.

The third plot (Figure 7) compares the overall distribution of the actual and predicted tool wear values using density plots. The blue curve represents the actual tool wear, while the orange curve represents the predicted values. The degree of overlap between these curves shows how well the model captures the general pattern of the data. Here, the predicted distribution follows the actual distribution quite well, especially for moderate tool wear values, although the model seems to underperform in predicting extreme tool wear, which leads to a slight mismatch at the higher values.

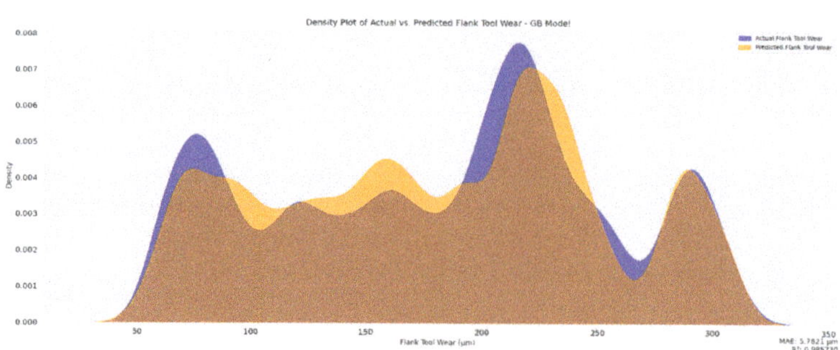

Figure 7. Density plot of actual vs. predicted flank tool wear in the test set.

The model's performance on the test set was evaluated using three metrics: MAE, MSE, and R^2 Score. MAE measures the average magnitude of the errors between predicted and actual values, indicating how much the predictions deviate from the actual values on average. MSE squares the errors, giving more weight to larger errors, and is useful for identifying significant prediction discrepancies. The R^2 score indicates how well the model explains the variance in the target variable. An R^2 score of 1 means the model perfectly predicts the target, whereas a value closer to 0 means the model performs no better than simply predicting the mean of the target variable. An acceptable R^2 for regression cases generally falls between 0.5 and 0.99, according to Ozili [20], with a strong R^2 score being above 0.90.

After predicting the tool wear values with our test set, the following parameters were obtained:

1. MAE: 5.7821×10^{-3} mm. This means, on average, the model's predictions are off by about that many units of tool wear.
2. MSE: 7.4263×10^{-5} mm. The low value indicates that the model makes a few large errors.
3. R^2 score: 0.9857. This suggests that the model explains 98.57% of the variance in the tool wear values.

3.1. Correlation Matrix

In the analysis of complex data, especially when dealing with multiple variables, it is important to identify patterns and relationships that may influence the desired outcome. In the case of this research, there is an interest in seeing how different variables affect tool wear. To facilitate this task, a correlation matrix can be used to quantify the linear relationship between each pair of variables in the dataset. This allows for the evaluation of how variables such as axial force, spindle speed, and vibrations correlate with tool wear. Each value in the matrix represents the Pearson correlation coefficient, which ranges from -1 to 1; values close to these limits indicate strong positive or negative correlations, respectively, while values close to 0 suggest a weak relationship. By visualizing the matrix (Figure 8) with a heat map, significant relationships and emerging patterns can be quickly identified, allowing the analysis and modeling to focus on the variables that have the greatest impact on wear [21] or even consider others that may have been overlooked.

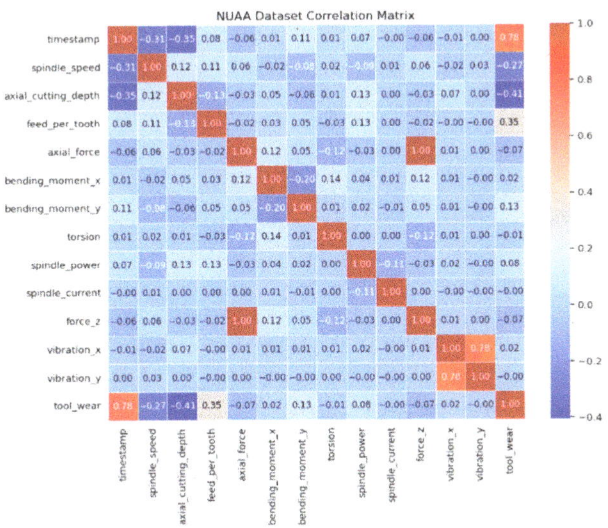

Figure 8. Correlation matrix of the NUAA dataset.

When analyzing the correlation matrix of the NUAA dataset (Figure 8), relationships emerge that align with the initial expectations of this study. In particular, the last row highlights a strong positive correlation between wear and time, which is consistent with the notion that wear tends to increase as time passes. Moreover, a moderate negative correlation is identified between wear and both cutting depth and spindle speed, suggesting that increasing these variables significantly reduces wear. The lack of data could imply that this might be due to factors such as energy dissipation or the distribution of stresses during the cutting process, though it could also be attributed to a decrease in cutting depth. A strong direct relationship is observed between vibrations on the *X*-axis and those on the *Y*-axis, as well as a total correlation between axial force and force *Z*, probably indicating that they have the same data with different variable names. Lastly, the feed per tooth shows a positive correlation with wear, although it is not very high, thus establishing that as the feed per tooth increases, wear also increases.

3.2. Remaining Useful Life (RUL) Prediction

RUL is a critical metric in predictive maintenance and reliability analysis, serving to estimate the time remaining before a component or system experiences failure or necessitates maintenance. In this analysis, predictive models are employed to forecast the future wear of a tool based on various input characteristics, thus indirectly calculating the RUL by predicting the expected wear that will occur over time. Rather than determining the RUL directly, the focus is on predicting future tool wear, which can effectively serve as a proxy for RUL if wear is recognized as an indicator of the time until failure.

To facilitate this analysis, the NUAA dataset is utilized due to its rich array of variables, which are essential for accurate predictions. In this approach, fixed values are applied, and a pre-trained Gradient Boosting model will assesses the tool's wear, ultimately allowing for the determination of the remaining useful life of the tool. This method harnesses the dataset's comprehensive features to enhance the reliability and accuracy of RUL predictions.

The graph presented in Figure 9 closely resembles the graph generated from Taylor's equation [22], with the *Y*-axis indicating the wear bandwidth as a function of time. It reveals three distinct zones, each representing specific wear levels and their corresponding severity for the cutting tool. The dataset demonstrates a wear range spanning from 0.049 mm to 0.31 mm, aligning well with the graph's scale and enabling the extraction of meaningful conclusions and physical interpretations from the mathematical results produced by the recently trained Gradient Boosting model.

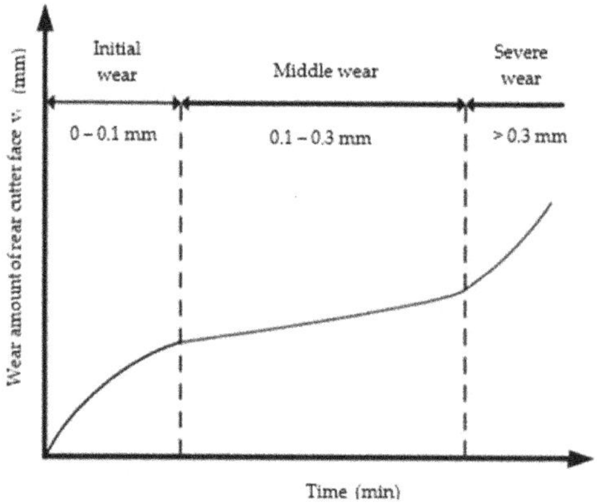

Figure 9. Wear bandwidth over time [14].

From the analysis of the correlation matrix, two parameters that exhibit a relatively somewhat strong relationship with tool wear are identified: spindle speed and feed per tooth. These parameters are input into the Gradient Boosting model, using mean values for the other variables, to calculate the time required to achieve the target wear levels. This analysis facilitates an approximate determination of the time it will take for the cutting tool to reach the specified wear thresholds corresponding to the three zones identified in the preceding graph (0.1 mm, 0.2 mm, and 0.3 mm).

Importantly, even with variations in parameters, the predictive model can be continuously updated with real-time data, allowing for adjustments to the expected time. Consequently, operators will maintain precise awareness of when the cutting tool is projected to enter the dangerous wear zone, with a recommendation to replace the insert when wear values fall between 0.25 mm and 0.3 mm to ensure a high-quality finish on the machined product.

While the NUAA dataset is derived from a specific insert and tool material, the findings can readily be extrapolated to other inserts if the materials are comparable. For instance, consider the insert model 345R-13T5E-ML S30T (Sandvik Coromant, Sandviken, Sweden) [23], which is coated with titanium and designed for use with solid carbide parts, consistent with the materials specified in the chosen dataset. From its manufacturer's datasheet, the following values are obtained:

- Feed per Tooth (f_z): 0.11 mm/rev; range: 0.07–0.20 mm/rev.
- Cutting Speed (v_c): 140 m/min; range: 120–155 m/min.

The value for f_z is directly incorporated into the code as a value. Subsequently, the cutting speed formula is applied, assuming a DC_{ap} (cutting diameter at the depth of cut) of 0.03 m to calculate the spindle speed, which, after rounding, yields a result of 1485 rpm. Upon inputting these parameters into the trained Gradient Boosting model, the following times are obtained:

- For a wear value of 0.1 mm: Time: 0:06:40.
- For a wear value of 0.2 mm: Time: 0:13:20.
- For a wear value of 0.3 mm: Time: 0:48:20.

To interpret these results, it is of great importance to reference Taylor's equation and analyze the preceding graph. In the initial phase (initial wear), which spans from 0 to 0.1 mm, the wear exhibits a nonlinear function, indicating rapid degradation based on the input parameters and the insert material. This phase is characterized by the tool's settling process, during which the cutting insert adjusts to the material being machined, effectively eliminating surface irregularities and micro-defects. Physically, this phase may reveal visible wear at the tool's cutting edges, a slight increase in surface roughness, and noticeable changes in cutting sounds and forces.

The second phase (middle wear), occurring within the wear range of 0.2 to 0.3 mm, presents a more stable and predictable wear rate. By this point, the insert has transitioned beyond the settling stage, resulting in more uniform wear. The relationship between time and wear becomes increasingly linear, enabling interpolation to predict wear at intermediate points, such as 0.25 mm. This threshold is critical, as it is the point at which tool replacement should be considered to prevent issues like diminished cut quality, increased cutting forces, and vibrations that could adversely impact the final product. During this phase, the tool's edge begins to round off more significantly, and while the cut quality starts to decline, it remains within acceptable limits.

As wear approaches 0.3 mm, operators enter a critical zone where wear accelerates again, marking the beginning of the third phase where severe wear occurs. At this stage, it is imperative to replace the tool, as prolonging its use could result in severe failures, such as tool breakage or damage to the workpiece. Operators may experience a marked increase in cutting forces, elevated temperatures, and alterations in sound, all indicating that the tool is under considerable stress. The quality of the machined surface may deteriorate rapidly, displaying visible signs of wear such as rough edges or even deformation of the workpiece. This phase is characterized by flank wear and potentially abrasive wear, making it essential not to exceed the recommended wear limits for inserts (between 0.25 and 0.3 mm).

4. Discussion

The objective of this investigation is to assess the tool wear of milling inserts through Machine Learning algorithms and data analysis. Numerous studies have explored the application of artificial intelligence techniques for monitoring wear in milling machines. According to a systematic review by Munaro et al. [6], there has been a notable increase in AI-related publications focused on machining operations since 2010, particularly concerning tool wear and the remaining service life of cutting tools. These studies highlight various Machine Learning algorithms, including artificial neural networks (ANNs), Fuzzy Logic algorithms (FLs), support vector machines (SVMs), decision trees and Random Forests (RFs), Bayesian networks (BNs), and convolutional neural networks (CNNs).

By turn, in their research, Kim et al. [24] proposed an innovative method for predicting tool wear using deep CNNs, known for their ability to learn hierarchical data representations through multiple convolutional layers. This CNN model architecture utilizes multi-scale convolutional kernels, which are crucial for capturing features of varying sizes in the input data. This capability is particularly relevant for predicting tool wear in machining processes, as it enables the model to recognize patterns across different scales, thus enhancing prediction accuracy. Additionally, Bayesian learning is introduced, incorporating principles of probability theory to manage uncertainty in predictions. By transforming the CNN model into a probabilistic framework, it becomes possible to estimate the uncertainty associated with wear predictions, which is vital for industrial applications. The integration of Bayesian learning allows the model to provide not only point predictions but also confidence estimates, making it especially valuable in processes like titanium alloy milling, where conditions can fluctuate and tool wear is unpredictable. While this model demonstrated a low mean square error and an R^2 score exceeding 0.9, our model achieved an R^2 value of 0.9857, surpassing the performance of the most advanced models, such as the Bayesian DMSCNN, which attained an average R^2 of 0.96. Notably, our model accomplished this with fewer cross-validation folds (5 vs. 10), fewer available data, and a broader range of variables, resulting in a very low mean MAE and MSE. This indicates that the model not only maintains competitive predictive accuracy but also demonstrates robust performance under challenging conditions.

Meng et al. [25] proposed an advanced approach for feature selection in predictive models by using singular value decomposition (SVD) combined with a voting scheme. In their study, the authors applied this technique to reduce the dimensionality of the dataset and then employed a voting scheme to select the most relevant features. This approach was complemented by three feature classification methods: Random Forest, Principal Component Analysis followed by Principal Component Regression (PCA + PCR), and Pearson Correlation Coefficient (PCC). The combination of SVD with PCA + PCR proved particularly effective, yielding the lowest mean square error [25]. This suggests that integrating advanced decomposition and feature selection techniques can significantly enhance the accuracy of predictive models by optimizing the relevance of the considered variables. Although our dataset was thoroughly cleaned and the features were handpicked to train the Gradient Boosting model, a careful examination of variables related to tool wear was conducted, adopting the findings of authors like Meng et al. Their research highlighted the significance of specific features in predicting tool wear, which can be corroborated by their results through a correlation matrix. This matrix effectively validated their hypothesis, demonstrating that the selected features have meaningful relationships with tool wear, which further strengthened the predictive capability of our model.

For its part, Random Forest regression effectively combines filter and wrapper methods to select features, providing a detailed analysis of each feature's importance in the model [26]. This not only evaluates the relevance of the features but also organizes them in descending order according to their impact on reducing model impurity. Feature selection in RFR utilizes a low-variance filter, Spearman correlation, and Random Forest regression [25]. The low-variance filter eliminates features with minimal variability, while Spearman correlation measures the monotonic relationship between variables, and RF regression analyzes how each feature contributes to the overall accuracy of the model. This allows the algorithm to manage complex data and provide a detailed assessment of each parameter's importance in the model's performance, thereby ensuring greater accuracy and generalization [25–28]. While Meng et al. achieved significant improvements with their methods, the approach presented in this article simplifies the selection process by training a Gradient Boosting model with all relevant variables that have a strong correlation with tool wear—as seen in the correlation matrix—while still yielding competitive performance.

Compared to neural network models, SVMs often exhibit superior generalization abilities, particularly when training data are limited. Gomes et al. [29] introduced a novel method for monitoring tool wear in the micro-milling process using an SVM artificial intelligence model combined with vibration and sound signals. This approach addresses the limitations of traditional methods by integrating multiple types of sensory data, significantly improving accuracy in detecting tool wear. The incorporation of vibration and sound signals resulted in a richer dataset, enabling the SVM model to discern patterns that may be obscured with a single type of signal. Building on this concept, Niu et al. [30] developed a multi-stage SVM to identify tool wear based on selected features, allowing for more precise identification by processing features through multiple analytical stages. This method proved effective in managing the complexity of and variability in wear data. However, the performance of the SVM model is heavily dependent on the selection of penalty factors and Kernel function parameters. In their study, the maximum prediction accuracies for BP neural networks, Bayesian networks, and SVM were 83%, 75%, and 72%, respectively—impressive performances, yet not the most precise. In contrast, our model was trained with the same (and more) variables—vibrations, for instance—and achieved a superior "accuracy" by obtaining lower MSE and MAE scores, thus improving the predictive capabilities of previous models.

Furthermore, the Gradient Boosting model of this study had an impressive performance given its relative simplicity, as shown by the obtained metrics; hence, there is no immediate need to enhance the model for this task. However, should the decision be made to pursue further improvements, several strategies could be considered to potentially enhance its efficacy. While the proposed models demonstrate promising performance, it is important to acknowledge the limitations associated with the dataset size. The training dataset, although robust, may not fully capture the wide range of variations in machining conditions and tool wear patterns encountered in real-world industrial settings. To mitigate this limitation, we employed K-Fold cross-validation with five splits to rigorously evaluate the generalization capacity of the models. While our current approach used standard K-Fold cross-validation, future research could benefit from implementing stratified K-Fold cross-validation to ensure a more consistent representation of tool wear characteristics across all folds. Future studies should also prioritize expanding the dataset by incorporating additional data points or generating new experimental data to further enhance the models' robustness and applicability to diverse scenarios.

In addition to the existing dataset limitations, future studies should prioritize the integration of environmental variables into the predictive modeling process. Factors such as temperature and humidity, which can significantly influence tool wear behavior, are not accounted for in the current dataset. To address this, we suggest the deployment of sensors capable of measuring these parameters during machining processes. The data collected from such sensors could serve as additional inputs for the predictive models, enabling them to capture a more nuanced understanding of the tool wear mechanisms under varying environmental conditions. Incorporating this layer of information would not only improve model precision but also enhance its applicability to real-world industrial settings where environmental fluctuations are prevalent.

Another limitation of this study lies in the reliance on pre-collected data from public datasets, which, while convenient and robust, may introduce inherent biases or omit key variables relevant to tool wear prediction. Factors such as variations in machining setups, operational conditions, or even data collection methods can affect the generalizability of the findings. To address this, future research should prioritize the creation of tailored experimental datasets that capture a broader range of tool wear scenarios, incorporating diverse materials, machining environments, and operating parameters. Additionally, leveraging advanced data preprocessing techniques, such as bias detection and correction methods, can help mitigate the influence of such limitations, ensuring that predictive models are both accurate and suited for various industrial applications.

Tool wear is a complex, dynamic process that is influenced by many variables, which are not only interconnected but also fluctuate over time, complicating the task of accurately modeling wear patterns. Current models, while valuable, are often limited in their ability to fully capture these dynamic interactions. As a result, future research should explore the application of advanced methodologies such as deep neural networks, which excel at modeling nonlinear and high-dimensional relationships. Furthermore, reinforcement learning offers the potential to continuously adapt to evolving wear conditions, providing a dynamic framework for more accurate predictions. These advanced approaches, coupled with transfer learning, could enable models to generalize more effectively across diverse machining environments, improving their applicability to real-world industrial settings.

Future research could also consider the development of hybrid or ensemble models as well that combine the strengths of different Machine Learning approaches, such as Gradient Boosting and LSTM. These models could harness the complementary benefits of ensemble techniques and sequential learning to improve predictive accuracy and robustness. Additionally, integrating these predictive models into existing manufacturing execution systems would facilitate their practical implementation in industrial settings. Such integration could enable real-time tool wear monitoring and adaptive decision-making.

Moreover, evaluating the model's performance on alternative test sets can yield valuable insights into its reliability across different scenarios. This practice is essential to ensure that the model is not overly fitted to a specific dataset, thus providing a more comprehensive understanding of its predictive performance.

Additionally, implementing a grid search for hyperparameter optimization could facilitate the identification of optimal model parameters. This methodical exploration of parameter combinations allows for fine-tuning, which can lead to significant improvements in accuracy and predictive capability. By rigorously refining the model in this manner, it can be ensured that it remains both robust and adaptable to diverse datasets and operational contexts. Such efforts would ultimately bolster the model's applicability and effectiveness in real-world applications.

While the NUAA dataset is based on solid carbide inserts and titanium parts, we have further analyzed the extent to which the proposed methodology can be applied to other materials and tools. By considering the fundamental properties that influence tool wear (e.g., hardness, thermal conductivity, and cutting force resistance), we have identified commonalities between the materials in the dataset and other industrial materials like high-speed steel and certain ceramic inserts. Future studies could validate these observations through experimental testing with diverse materials to confirm the robustness of the proposed models.

5. Conclusions

This research provides a comprehensive analysis of tool wear in milling machines, focusing on the precise prediction of remaining tool life through the implementation of artificial intelligence algorithms. Throughout the development, the purpose of milling machines and their cutting tools, particularly milling inserts, is thoroughly examined, highlighting the advantages they offer over solid tools. The phenomenon of tool wear is explained in detail, emphasizing its direct impact on product quality and process efficiency, culminating in a direct application of Machine Learning models to accurately predict wear in insert blades.

One of the key achievements of this work is the improvement in the accuracy of predictive models compared to what has been previously reported in the literature. The use of a simple Gradient Boosting model enables a more accurate prediction of wear compared to other models like neural networks and SVMs and is relatively close to that of more deep learning models that require extensive tuning and large datasets. Furthermore, it overcomes limitations observed in prior studies, where wear predictions relied on overly complex methods and a greater number of variables, obtaining an even lower R^2 score than our model.

Additionally, a proper relationship between variables is established, explaining the correlation between the different variables involved. Specifically, this work demonstrates that AI models, when properly tuned (cross-validation, train–test splitting, and regularization) and applied, can effectively capture the complex relationships among the involved variables, thereby enhancing the accuracy and reliability of their predictions.

The proposed methodology not only addresses the challenges of tool wear prediction but also provides valuable insights for future research in the field of machining. By employing a data-driven approach that integrates advanced AI techniques, this study lays the groundwork for developing more robust predictive models that can adapt to various machining conditions and contribute to solving potential limitations that could appear in these processes.

Moreover, the incorporation of uncertainty awareness in predictions paves the way for more informed decision-making processes within manufacturing environments. As industries increasingly shift towards smart manufacturing, the findings of this research can significantly enhance operational efficiency and product quality, ultimately leading to improved customer satisfaction as products will be affected by cost reduction.

The methodology employed in applying AI could be transferable to other manufacturing processes by adapting the algorithms.

This study underscores the importance of harnessing AI algorithms in machining processes to achieve reliable tool wear predictions. Future work could explore the applicability of these models in different manufacturing contexts and investigate the integration of real-time monitoring systems to further enhance predictive accuracy and reliability.

Finally, as previously noted, wear is a complex phenomenon influenced by numerous factors, including forces, vibrations, temperature, and material properties. While this model offers a reliable approximation, it could be significantly enhanced by integrating more relevant parameters related to the process. Using sensors and other techniques to collect real-time data thereby will improve input accuracy and ultimately enhance output predictions with artificial intelligence techniques.

Author Contributions: Conceptualization, S.V. and M.P.L.; methodology, S.V., M.P.L. and J.L.; software, J.L.; validation, S.V., M.P.L., J.R. and J.L.; formal analysis, S.V., M.P.L. and J.L. investigation, S.V., M.P.L., J.R. and J.L.; data curation, J.L.; writing—original draft preparation, S.V., M.P.L. and J.L.; writing—review and editing, S.V., M.P.L. and J.L. All authors have read and agreed to the published version of the manuscript.

Funding: This research received no external funding.

Institutional Review Board Statement: Not applicable.

Informed Consent Statement: Not applicable.

Data Availability Statement: The original contributions presented in this study are included in the article. Further inquiries can be directed to the corresponding author.

Conflicts of Interest: The authors declare no conflicts of interest.

Appendix A

Excerpt of the code used for the analysis.

```
# imports
import pandas as pd
import seaborn as sns
```

```
import matplotlib.pyplot as plt
import numpy as np
from sklearn.ensemble import RandomForestRegressor, GradientBoostingRegressor
from sklearn.model_selection import cross_val_score, GridSearchCV, KFold, train_test_split
from sklearn.preprocessing import StandardScaler
import tensorflow as tf
from tensorflow.keras.models import Sequential
from tensorflow.keras.layers import Dense, Dropout, LSTM
from tensorflow.keras.regularizers import l2
from tensorflow.keras.callbacks import EarlyStopping
from scikeras.wrappers import KerasRegressor
from sklearn.metrics import mean_squared_error, mean_absolute_error, r2_score
from datetime import datetime, timedelta

# Data Preparation
url = 'https://raw.githubusercontent.com/katulu-io/uniwear-dataset/main/data/nuaa_
orthogonal_bundle_high_resolution.csv'

df = pd.read_csv(url)
df = df.drop(columns=['Unnamed: 0', 'vibration1',
'vibration2', 'experiment_csv_n',
'dataset_tag', 'tool_wear_increment', 'wear_blade_1', 'wear_blade_2',
'wear_blade_3', 'wear_blade_4', 'experiment_tag'])

columns_order = ['timestamp', 'spindle_speed', 'axial_cutting_depth', 'feed_per_tooth',
'axial_force',
'bending_moment_x', 'bending_moment_y', 'torsion', 'spindle_power', 'spindle_current',
'force_z', 'vibration_x', 'vibration_y', 'tool_wear']
df = df[columns_order]
df2 = df.drop(columns=['tool_wear'])

num_rows = df.shape [0]
print("\n" + str(num_rows) + " rows\n")
print(df2.describe())
print("\n" + "Missing values in each column:")
print(df2.isnull().sum())

df.fillna({
'bending_moment_x': df['bending_moment_x'].mean(),
'bending_moment_y': df['bending_moment_y'].mean(),
'torsion': df['torsion'].mean()
}, inplace=True)

# Model Training
df = df.apply(pd.to_numeric, errors='coerce')
df = df.dropna()
X = df.drop(columns='tool_wear')
y = df['tool_wear']
scaler = StandardScaler()
X_train, X_test, y_train, y_test = train_test_split(X, y, test_size=0.2, random_state=42)
scaler = StandardScaler()
X_train_scaled = scaler.fit_transform(X_train)
X_test_scaled = scaler.transform(X_test)
```

```
rf = RandomForestRegressor(n_estimators=100, random_state=42)
gb = GradientBoostingRegressor(n_estimators=100, random_state=42)

kfold = KFold(n_splits=5, shuffle=True, random_state=42)
ensemble_scores = cross_val_score(rf, X_train_scaled, y_train, cv=kfold,
scoring='neg_mean_squared_error', verbose=3)
print(f"Cross-validated MSE (Random Forest): {-ensemble_scores.mean():.12f} (std:
{ensemble_scores.std():.12f})")
ensemble_scores = cross_val_score(gb, X_train_scaled, y_train, cv=kfold,
scoring='neg_mean_squared_error', verbose=3)
print(f"Cross-validated MSE (Gradient Boosting): {-ensemble_scores.mean():.12f} (std:
{ensemble_scores.std():.12f})")

def create_sequences(X, y, time_window):
X_time, y_time = [], []
if hasattr(y, 'values'):
y = y.values
if hasattr(X, 'values'):
X = X.values

for i in range(len(X) − time_window):
X_time.append(X[i:(i + time_window)])
y_time.append(y[i + time_window])
return np.array(X_time), np.array(y_time)

time_window = 10
X_train_time, y_train_time = create_sequences(X_train_scaled, y_train, time_window)
X_test_time, y_test_time = create_sequences(X_test_scaled, y_test, time_window)

def create_lstm_model(input_shape):
model = Sequential()
model.add(LSTM(64, input_shape=input_shape))
model.add(Dropout(0.2))
model.add(Dense(1))
model.compile(optimizer='adam', loss='mean_squared_error')
return model

def build_fn():
return create_lstm_model(input_shape=(time_window, X_train_scaled.shape [1]))

early_stopping = EarlyStopping(
monitor='val_loss',
patience=3,
min_delta=0.001,
restore_best_weights=True,
verbose=0
)

lstm_model = KerasRegressor(
model=build_fn,
epochs=100,
batch_size=32,
verbose=3,
validation_split=0.2
)
```

```
kfold = KFold(n_splits=5, shuffle=True, random_state=42)
lstm_scores = cross_val_score(
lstm_model,
X_train_time,
y_train_time,
cv=kfold,
scoring='neg_mean_squared_error',
fit_params={'callbacks': [early_stopping]}
)

print(f"Cross-validated MSE (LSTM Model): {-lstm_scores.mean():.12f}(std:
{lstm_scores.std():.12f})")

# Model Evaluation (Test Set)
rf.fit(X_train_scaled, y_train)
rf_pred = rf.predict(X_test_scaled)
rf_test_mse = mean_squared_error(y_test, rf_pred)
print(f"Test MSE with Random Forest: {rf_test_mse:.12f}")

gb.fit(X_train_scaled, y_train)
gb_pred = gb.predict(X_test_scaled)
gb_test_mse = mean_squared_error(y_test, gb_pred)
print(f"Test MSE with Gradient Boosting: {gb_test_mse:.12f}")

lstm_model.fit(
X_train_time,
y_train_time,
callbacks=[early_stopping],
validation_split=0.2,
epochs=100,
batch_size=32,
verbose=3
)

lstm_pred = lstm_model.predict(X_test_time)
lstm_test_mse = mean_squared_error(y_test_time, lstm_pred)
print(f"Test MSE with LSTM Model: {lstm_test_mse:.12f}")


# Flank Tool Wear Threshold Evaluations
model = gb

input_data = {
'timestamp': int(df['timestamp'].min()),
'spindle_speed': 1485,
'axial_cutting_depth': df['axial_cutting_depth'].mean(),
'feed_per_tooth': 0.11,
'axial_force': df['axial_force'].mean(),
'bending_moment_x': df['bending_moment_x'].mean(),
'bending_moment_y': df['bending_moment_y'].mean(),
'torsion': df['torsion'].mean(),
'spindle_power': df['spindle_power'].mean(),
'spindle_current': df['spindle_current'].mean(),
'force_z': df['force_z'].mean(),
```

```
'vibration_x': df['vibration_x'].mean(),
'vibration_y': df['vibration_y'].mean(),
}

target_wears = [0.1, 0.2, 0.3]
results = {target_wear: None for target_wear in target_wears}

timestamp_min = int(df['timestamp'].min())
timestamp_max = int(df['timestamp'].max())

timestamp_range = range(timestamp_min, timestamp_max, 100)

for timestamp in timestamp_range:
input_data['timestamp'] = timestamp
input_data_df = pd.DataFrame([input_data])

input_data_scaled = scaler.transform(input_data_df)
predicted_tool_wear = model.predict(input_data_scaled)
predicted_wear = predicted_tool_wear [0]

for target_wear in target_wears:
if predicted_wear >= target_wear and results[target_wear] is None:
results[target_wear] = (predicted_wear, timestamp)

for target_wear, result in results.items():
if result:
predicted_wear, timestamp = result
mean_time_predicted_wear = datetime.fromtimestamp(timestamp).strftime('%H:%M:%S')
print(f'For tool wear target {target_wear}mm:')
print(f'Predicted Tool Wear: {predicted_wear:.4f}mm')
print(f'Time when this wear occurs: {mean_time_predicted_wear}\n')
else:
print(f'For tool wear target {target_wear}mm:')
print('Target wear not reached within the given timestamp ange.\n')
```

References

1. Wen, J.; Xie, F.; Liu, X.; Yue, Y. Evolution and Development Trend Prospect of Metal Milling Equipment. *Chin. J. Mech. Eng.* **2023**, *36*, 33. [CrossRef]
2. Liu, C.; Ren, J.; Zhang, Y.; Shi, K. The Effect of Tool Structure and Milling Parameters on the Milling Quality of CFRP Based on 3D Surface Roughness. *Int. J. Precis. Eng. Manuf.* **2023**, *24*, 931–944. [CrossRef]
3. Son, N.H.; Hieu, T.T.; Thang, N.M.; Tan, H.N.; Can, N.T.; Thao, P.T.; Bao, N.C. Choosing the best machine tool in mechanical manufacturing. *EUREKA Phys. Eng.* **2023**, *2*, 97–109. [CrossRef]
4. Shirase, K.; Fujii, S. Machine Tool Automation. In *Springer Handbook of Automation*; Nof, S.Y., Ed.; Springer Handbooks: Cham, Switzerland, 2023. [CrossRef]
5. Jatakar, K.; Shah, V.; Binali, R.; Salur, E.; Sağlam, H.; Mikolajczyk, T.; Patange, A.D. Monitoring Built-Up Edge, Chipping, Thermal Cracking, and Plastic Deformation of Milling Cutter Inserts through Spindle Vibration Signals. *Machines* **2023**, *11*, 790. [CrossRef]
6. Munaro, R.; Attanasio, A.; Del Prete, A. Tool wear monitoring with artificial intelligence methods: A review. *J. Manuf. Mater.* **2023**, *7*, 129. [CrossRef]
7. de Oliveira Moreira, M.; Abrão, A.M.; Ferreira, R.A.M.; Porto, M.P. Temperature monitoring of milling processes using a directional-spectral thermal radiation heat transfer formulation and thermography. *Int. J. Heat Mass Transf.* **2021**, *171*, 121051. [CrossRef]
8. Mäkiaho, T.; Vainio, H.; Koskinen, K.T. Wear Parameter Diagnostics of Industrial Milling Machine with Support Vector Regression. *Machines* **2023**, *11*, 395. [CrossRef]
9. Soujanya, A.; Goud, O.S.C.; Prasad, K.S.; Reddy, G.P. Based pattern analysis using machine learning and artificial intelligence techniques for multiple featured dataset. *Mater. Today Proc.* **2017**, *4*, 9039–9048. [CrossRef]
10. Kim, D.; Jeon, D. Fuzzy-Logic Control of Cutting Forces in CNC Milling Processes Using Motor Currents as Indirect Force Sensors. *Precis. Eng.* **2011**, *35*, 143–152. [CrossRef]
11. Zhao, Y.; Cui, L.; Sivalingam, V.; Sun, J. Understanding Machining Process Parameters and Optimization of High-Speed Turning of NiTi SMA Using Response Surface Method (RSM) and Genetic Algorithm (GA). *Materials* **2023**, *16*, 5786. [CrossRef]

12. Shanmugasundar, G.; Vanitha, M.; Čep, R.; Kumar, V.; Kalita, K.; Ramachandran, M. A Comparative Study of Linear, Random Forest, and AdaBoost Regressions for Modeling Non-Traditional Machining. *Processes* **2021**, *9*, 2015. [CrossRef]
13. von Hahn, T.; Mechefske, C.K. Machine Learning in CNC Machining: Best Practices. *Machines* **2022**, *10*, 1233. [CrossRef]
14. Zhu, M.; Zhang, J.; Bu, L.; Nie, S.; Bai, Y.; Zhao, Y.; Mei, N. Methodology and Experimental Verification for Predicting the Remaining Useful Life of Milling Cutters Based on Hybrid CNN-LSTM-Attention-PSA. *Machines* **2024**, *12*, 752. [CrossRef]
15. Akbari, V.O.A.; Mohammadi, Y.; Kuffa, M.; Wegener, K. Identification of in-process machine tool dynamics using forced vibrations in milling process. *Int. J. Mech. Sci.* **2023**, *239*, 107887. [CrossRef]
16. Katulu-io. UniWear Dataset. Available online: https://github.com/katulu-io/uniwear-dataset (accessed on 17 March 2024).
17. Tu, J.F.; Corless, M. Review of Sensor-Based Approach to Reliable High-Speed Machining at Boeing—A Tribute to Jan Jeppsson. *High-Speed Mach.* **2014**, *1*, 1. [CrossRef]
18. Li, X. Detection of Tool Flute Breakage in End Milling Using Feed-Motor Current Signatures. *IEEE/ASME Trans. Mechatron.* **2001**, *6*, 491–498. [CrossRef]
19. Ito, Y.; Matsumura, T. Mechanisms for Metal Cutting and Grinding. In *Theory and Practice in Machining Systems*; Springer: Cham, Switzerland, 2017; pp. 1–9. [CrossRef]
20. Ozili, P.K. The Acceptable R-Square in Empirical Modelling for Social Science Research. MPRA Paper No. 115769. Available online: https://mpra.ub.uni-muenchen.de/115769/1/MPRA_paper_115769.pdf (accessed on 29 July 2024).
21. Zhang, T. *Mathematical Analysis of Machine Learning Algorithms*; Cambridge University Press: Cambridge, UK, 2023.
22. Taylor, F.W. On the Art of Cutting Metals. *J. Fluids Eng.* **1907**, *28*, 31–279.
23. Sandvik Coromant. *CoroMill*; Sandvik: Sandviken, Sweden, 2024.
24. Kim, G.; Yang, S.M.; Kim, D.M.; Kim, S.; Choi, J.G.; Ku, M.; Lim, S.; Park, H.W. Bayesian-based uncertainty-aware tool-wear prediction model in end-milling process of titanium alloy. *Appl. Soft Comput.* **2023**, *148*, 110922. [CrossRef]
25. Meng, X.; Zhang, J.; Xiao, G.; Chen, Z.; Yi, M.; Xu, C. Tool wear prediction in milling based on a GSA-BP model with a multisensor fusion method. *Int. J. Adv. Manuf. Technol.* **2021**, *114*, 3793–3802. [CrossRef]
26. Cheng, W.-N.; Cheng, C.-C.; Lei, Y.-H.; Tsai, P.-C. Feature selection for predicting tool wear of machine tools. *J. Adv. Manuf. Technol.* **2020**, *111*, 1483–1501. [CrossRef]
27. Kumar, S.; Kolekar, T.; Kotecha, K.; Patil, S.; Bongale, A. Performance evaluation for tool wear prediction based on bi-directional, encoder–decoder and hybrid long short-term memory models. *Int. J. Qual. Reliab. Manag.* **2022**, *7*, 1551–1576. [CrossRef]
28. Li, Z.; Zhong, W.; Liao, W.; Cai, Y.; Zhao, J.; Wang, G. A robust tool condition monitoring system based on cluster density under variable machining processes. *Appl. Sci.* **2023**, *13*, 7226. [CrossRef]
29. Gomes, M.C.; Brito, L.C.; da Silva, M.B.; Duarte, M.A.V. Tool wear monitoring in micromilling using support vector machine with vibration and sound sensors. *Precis. Eng.* **2021**, *67*, 137–151. [CrossRef]
30. Niu, B.; Sun, J.; Yang, B. Multisensory based tool wear monitoring for practical applications in milling of titanium alloy. *Mater. Today* **2020**, *22*, 1209–1217. [CrossRef]

applied sciences

MDPI

Article

Industrial Application of AI-Based Assistive Magnetic Particle Inspection

Julien Baumeyer [1,2,*], Hermine Chatoux [2], Arnaud Pelletier [1] and Patrick Marquié [2]

[1] CMPhy, 26 Rue Paul Sabatier, 71530 Crissey, France; arnaud.pelletier@cmphy.fr
[2] Laboratoire ImViA Dijon, 64 Rue Sully, 21000 Dijon, France; hermine.chatoux@u-bourgogne.fr (H.C.); patrick.marquie@u-bourgogne.fr (P.M.)
* Correspondence: julien.baumeyer@u-bourgogne.fr

Abstract: Magnetic Particle Inspection (MPI) is one of the most used methods in Non-Destructive Testing (NDT), allowing precise and robust defect detection on industrial-grade manufactured parts. However, human controllers perform this task in full black environments under UV-A lighting only (with safety glasses) and use chemical products in a confined environment. Those constraints tends to lower control performance and increase stress and fatigue. As a solution, we propose an AI-based assistive machine (called "PARADES") inside the hazardous environment, remotely manipulated by a human operator, outside of the confined area, in cleaner and safer conditions. This paper focuses on the development of a complete industrial-grade AI machine, both in terms of hardware and software. The result is a standalone assistive AI-based vision system, plug-and-play and controller-friendly, which only needs the usual power supply 230 V plug that detects defects and measures defect length. In conclusion, the PARADES machines address for the first time the problem of occupational health in MPI with an industrial standalone machine which can work on several parts and be integrated into current production lines. Providing cleaner and healthier working conditions for operators will invariably lead to increased quality of detection. These results suggest that it would be beneficial to spread this kind of AI-based assistive technology in NDT, in particular MPI, but also in Fluorescent Penetrant Testing (FPT) or in visual inspection.

Keywords: NDT; magnetic particle inspection; deep learning; object detection; dataset; vision system; quality control; hardware; software; transfer learning; faster R-CNN

1. Introduction

NDT is widely used in industries such as manufacturing, construction, aerospace, automotive and energy, among others. It helps ensure the quality [1] and safety of materials and structures by detecting flaws or defects early in the manufacturing or operational processes [2]. Some of the most used NDT methods are Magnetic Particle Inspection (MPI) and Fluorescent Penetrant Testing (FPT) because they are quick to implement, efficient and cost-effective. In this study, we will only explore MPI.

This method can suffer from lower control performance and increase stress and fatigue for human controllers due to the specific working condition, as depicted in Figure 1a. In order to overcome those drawbacks, a strong demand for a semi-automatic vision system has been expressed by our customers, leading to the design of the PARADES machine.

Customers' requirements concentrate on exceeding human controller performance in terms of percentage of defect detection, false detections and accuracy regarding defect length measurement. Parts' defect lengths and widths are summarized from several customer's requirements as well as the maximal system mechanical footprint. A summary of the important PARADES machine specifications is given in Table 1.

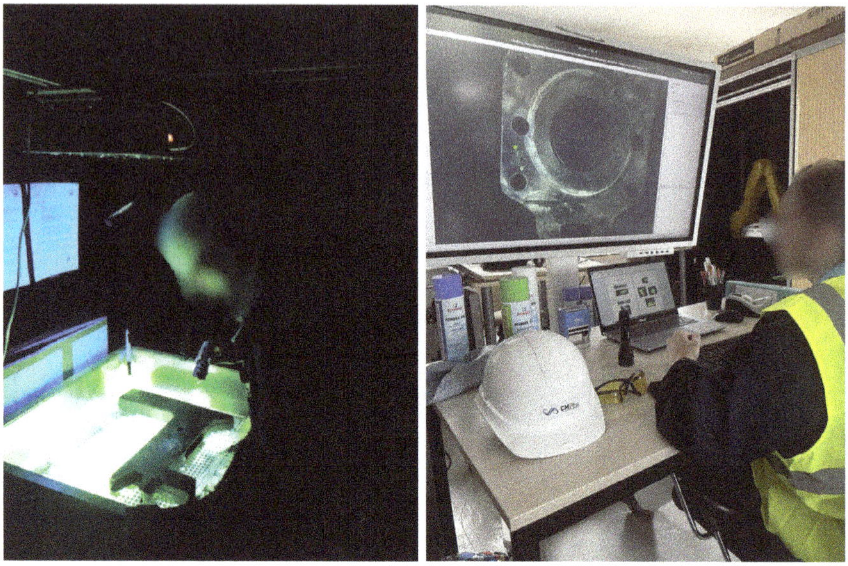

(a) (b)

Figure 1. MPI control without and with PARADES machine. (**a**) Human controller in a usual black, hazardous confined area. (**b**) Human controller remotely controlling defects with PARADES machine.

Table 1. PARADES machine important specifications.

Specification	Value
Percentage of Detection (PoD)	>90%
Max average FP per image	>1
MAE on defect length measurement	<3 mm
Scene area size	1200 × 800 mm
Minimal UV irradiance on scene area	1500 μW/cm^2
Maximal vision system size	700 × 400 × 150 mm
Min defect length	3 mm
Min defect width	50 μm

The assistive "PARADES" machine was briefly described in [3] and updated information can be found on the official company website (https://www.cmphy.fr/nos-produits/276/systeme-de-vision/, accessed on 7 February 2024). It consists of a standalone vision system with industrial-grade mechanical enclosure outer layer, powered from a standard 230 V socket. The operator can remotely use an external display as well as keyboard/mouse inputs as depicted in Figure 1b. A comparative control performed by a human operator with and without a PARADES machine is depicted in Figure 1.

Our main contributions are the following:

1. We developed an industrial machine for defect detection that can be integrated into every production line.
2. We proposed a way to design a customized lighting panel with specific simulation and optimization software.
3. We implemented a tool to measure defect length no matter the distance between the part and the PARADES machine.

The next sections will describe the related work (Section 2) and the PARADES machine, both in terms of hardware (Section 3) and software (Section 4). Hardware includes a custom UV-A lighting panel (Section 3.1), the RGB vision system (Section 3.3) as well as the depth sensor (Section 3.4) and an embedded AI-ready computer. The software section will detail the AI-based defect detection module (Section 4.2.1), the dedicated dataset in Section 4.1 (which was built using the PARADES machine itself) and the defect length measurement feature (Section 4.3). The results are then presented (Section 5), followed by the conclusion and future works.

2. Related Work

A substantial number of works have been found in the literature on the MPI topic. For instance, [4] shows an industrial application of a conventional computer vision algorithm on a rails wheelset. The non-standalone system built around a camera and a small lighting system must be coupled with an external PC for image processing, moreover, PoD results are not given and only qualitative examples are shown. Another example of industrial automatic defect detection in MPI is given by [5], showing material selection for specific image capturing on MPI. As stated in this article, optical filters are mandatory, and thus were used with the PARADES machine. A mobile machine based on a robot is depicted in [6] for use with MPI control on wide parts. It shows an interesting embedded magnetic excitation material as well as an MPI suspension dispenser. The defect-detection feature is not developed and shows the need for an external computer in order to process acquired images, due to the embedded Raspberry Pi Zero low computational performance.

The study in [7] is another example of an industrial machine used to detect defects with the MPI technique along with computer vision. It uses hardware as well as the usual computer vision algorithms to detect flaws. The paper focuses on detection speed rather than Percentage of Detection (PoD) and shows an insufficient detection accuracy (around 50%) for flaws less than 10 mm.

An experimental hardware setup is presented in [8] and aims to improve defect-detection precision using a conventional computer vision algorithm based on Hough transform. It visually shows how the false detections are removed but quantitative metrics are not given. Another approach was taken in [9], in order to find the 3D reconstruction of cracks on parts. The result is quite similar detection conditions as in PARADES, with defect size representing 1/4000 of the entire image with a few pixels for the defect width (7 pixels in their case). The scene area seems a lot smaller than ours, with a 7-pixel defect width at 2046×2040 full resolution compared to the PARADES high-definition > 96 Mp camera system.

An experimental platform is detailed in [10] and uses conventional image processing (i.e., a Canny operator); however, it cannot be used in a production line. Another example of an AI-based defect-detection system is presented in [11], which uses a CNN coupled with a set of cameras to find defects on a specific crankset. A 3000×4096 pixel camera is used to feed a ResNet34 CNN, which has been trained on 100×100 image patches. This study shows a very high score on classification and a 0.9 mAP@0.5 (mAP computed with IoU score over 0.5) for object detection. A segmentation AI-based method is proposed in [12].

Study [13] shows a good 96.5% recall with 91.7% precision on industrial bearing using a custom mobileNetV3 CNN but the hardware setup is not presented.

To summarize, a lot of work has been done on MPI using image processing and deep-learning methods, achieving better performance than human controllers while lowering risks and improving working conditions. However, most of the studies concentrate on developing the software side with a laboratory minimal hardware setup most of the time. The literature is missing industrial-grade production-ready hardware for MPI automatic defect detection. The goal of this paper is to present the PARADES machine, which is designed as a versatile platform with industrial production use in mind and with a focus on AI-ready image acquisition. The PARADES machine aims to achieve the best detection performance in MPI but might be sufficiently generic for a variety of industrial computer vision tasks.

3. Hardware

Fluorescent Magnetic Particle Testing uses 365 nm-based UV illumination. The PARADES machine is built around a custom designed lighting system that matches the international aeronautics standard limit of 1500 $\mu W/cm^2$. Indeed, international standard [14]

ISO 9934-1:2015 requires a minimum of 1000 μW/cm^2, but the PARADES machine will also be used by aeronautics manufacturers and so needs to reach those specific standards. Airbus Helicopter, Rolls-Royce, Turbomeca and other manufacturers have defined their own standards. For example, a minimum of 1500 μW/cm^2 is required (Airbus Helicopter EI070, Turbomeca TS-00616, Paris, France). In terms of spectrum, the requirement of the ISO standard [15] (ISO 3059:2012) matches aeronautics manufacturers and the most restrictive requirement is made by Airbus Helicopter. In this case, the UV peak must be within 365 and 370 nm. The FWHM's most restrictive criterion is given by Rolls-Royce RRES-90061 (Derby, UK) as a maximum of 20 nm. The PARADES specification is directly derived from those standards, so the minimal lighting irradiance will be 1500 μW/cm^2 with a peak between 365 and 370 nm and an FWHM < 20 nm, in order for PARADES to be able to be used in a variety of domains and by many customers. In the next section, we describe our custom lighting panel based on flood and narrow beam UV spots. As explained in the previous section, the illuminated scene size is 600 × 400 mm. First, we will test two versions of UV spots in order to characterize the relative emitted irradiance regarding angle. Then, multi-spot lighting simulation software which has been specifically developed is presented. Next, we propose to optimize UV spot location in order to fulfill the 1500 μW/cm^2 requirement on the 600 × 400 mm scene with a minimum number of spots, based on spot characterization and a simulation algorithm.

3.1. Wavelength Measurement

As explained in the previous section, the standards requirement, in terms of spectrum, is a peak between 365 and 370 nm with FWHM less than 20 nm. This validation was performed using a calibrated Hamamatsu mini spectrometer model TG-C9404CAH (Hamamatsu, Japan) with specifications given in Table 2.

Table 2. Spectrum measurement device Hamatsu TG-C9404CAH.

Feature	Value
Spectral response range	200 to 400 nm
Spectral resolution (FWHM) (typ.)	1 nm
Spectral resolution (FWHM) (max.)	2.2 nm

The measured spectrum is depicted in Figure 2 and shows a peak located at 369 nm with an FWHM of 11 nm. The results are summarized in Table 3.

Table 3. Spectrum data of narrow and flood UV spot compared to requirements.

Characteristic	Requirement Value	Ours
Main spectrum peak (nm)	365 < peak < 370 nm	369 nm
FWHM (nm)	<20 nm	11 nm

The characterized UV spots are within the standard requirement; they can then be used in the PARADES Machine.

3.2. Custom UV Lighting Panel

Industrial UV spots are used in order to light a surface of 600 × 400 mm with a minimum irradiance of 1500 μW/cm^2. A few 365 nm UV spots have been used in the PARADES machine. Two versions (flood and narrow beam) were used in a specific combination in order to reach the 1500 μW/cm^2 requirement on all of the 600 × 400 m scenes.

3.2.1. Characterization Setup

The setup consists of a tripod supporting the UV spot to be characterized, over a graduated line where a radiometer is moved along at a defined vertical distance of 68 cm above the scene area. The setup is depicted in Figure 3. A measurement is done using a radiometer every 1 cm, ranging from 0 to 40 cm. A horizontal radiometer position from spot beam axis is then converted to an angle using the distance between the spot and the scene surface.

The radiometer used to measure UV light irradiance is a calibrated Pfinder UVLuxCHECK.

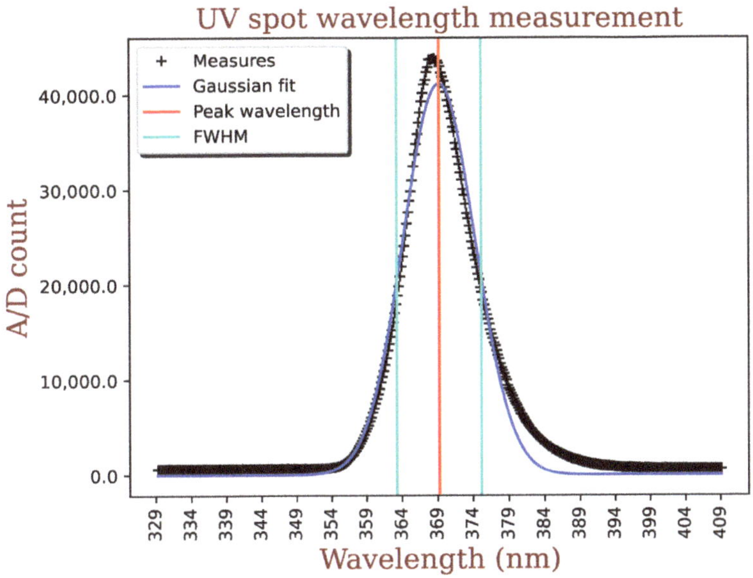

Figure 2. Measured spectrum of both narrow and flood UV spot used in the PARADES machine.

3.2.2. Measures and Characterization

Figure 4 give the measured and the interpolated data both with narrow (Figure 4a) and flood spots (Figure 4b). Those figures represent the relative irradiance versus the angle from vertical direction. As we can see, the flood spot (Figure 4b) has a wider emitting beam angle (35°) than the narrow beam spot (19°) shown in Figure 4a.

The flood spot emission feature is interpolated using a quadratic function clipped from 35°. The narrow spot is characterized using a SplineTransformer with 14 knots and degree 3. Despite the 68 cm distance between the radiometer and the spots, those measurements are then converted into an angle, allowing the use of those spots at any vertical distance.

The estimation error used is RMSE between measured and interpolated data. The narrow and flood spot RMSEs are given in Table 4 and show an error around 5 $\mu W/cm^2$ which is low compared to the requested 1500 $\mu W/cm^2$. Basically, an RMSE less than 10 $\mu W/cm^2$ is considered as acceptable.

Table 4. RMSE error for flood and narrow beam UV spot characterization.

Spot Type	RMSE ($\mu W/cm^2$)
Flood	5.25
Narrow	6.52

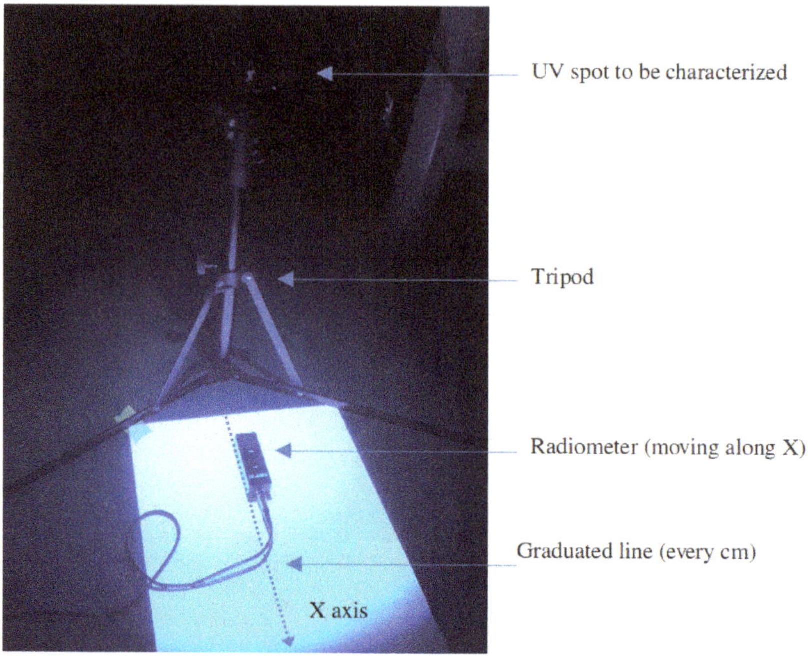

Figure 3. Hardware material setup to characterize UV spots.

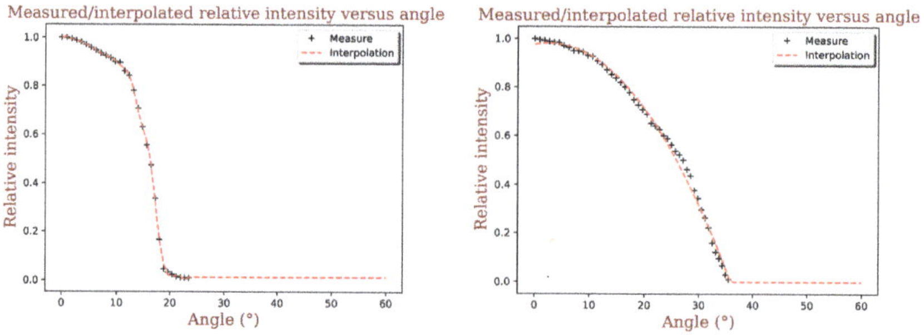

(**a**) (**b**)

Figure 4. Measured and interpolated narrow and flood spot characteristics. (**a**) Relative irradiance regarding angle for narrow beam UV spot. (**b**) Relative irradiance regarding angle for flood beam UV spot.

3.2.3. UV Lighting Simulation

Based on those results, we are able to estimate the amount of UV light received at any location (X and Y) of our studied surface (e.g., 600 × 400 mm) by either a narrow or a flood spot located in a parallel plane distant from a defined distance (50 cm in our case). This is given by the spot characterization performed in the previous section. The next step is to estimate the received amount of light when a given set of narrow or flood spots are used with their own spatial disposition. Please note that in our case, all spots are located on the same surface, parallel to the illuminated surface and 50 cm distant. Basically, the UV irradiance received by a X,Y point from several spots is the sum of the irradiance received by each spot at this location. This computation is done for every point of a 1 × 1 mm grid pattern of the 600 × 400 mm studied surface. In order to perform the computation, simulation software written in

Python has been specifically developed and uses the spots characterization. A screenshot of a simulation performed with our software is shown in Figure 5.

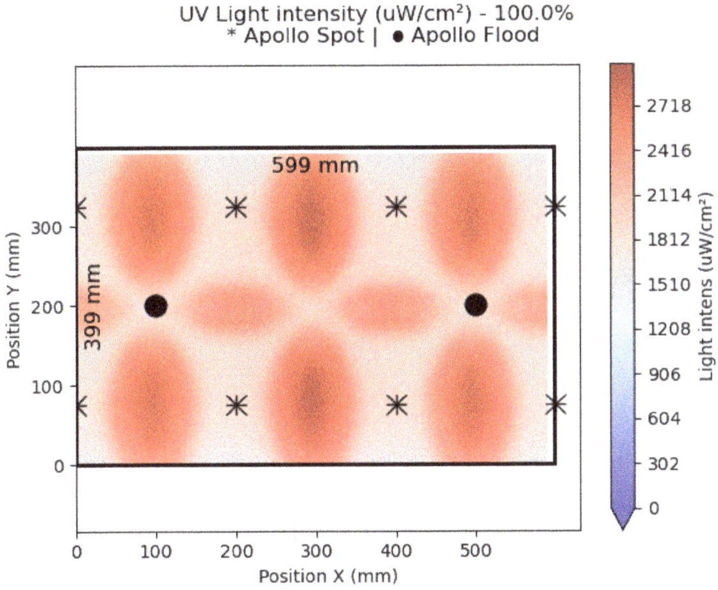

Figure 5. Two-dimensional visualization of UV light received.

3.2.4. UV Spots Disposition Automatic Optimization

Achieving 1500 µW/cm^2 on every point inside our studied surface of 600 × 400 mm is not an easy task. Trying and testing several spot dispositions in order to reach the minimum UV light specification will require a lot of tests with unlimited combinations of:

- Spot type (narrow, flood);
- Spot number;
- Spot disposition (X and Y).

To obtain a working disposition (all of the 600 × 400 mm surfaces above 1500 µW/cm^2), we used an optimization algorithm based on our light simulator results which gives us a suitable spots disposition (X and Y for each spots). Spot number and spot type will be tuned manually. Optimization is done using a combination of "Simulated annealing" and the "Nelder–Mead" algorithm, both implemented with Scipy library in Python. The optimized output variables are the (XY) location of each spot. All spots are considered on a fixed horizontal plane, on the same Z location as the camera system (e.g., 50 cm) above the illuminated surface.

In our tests, a minimum of 10 spots are required, with the following mix:

- 8 narrow spots;
- 2 flood spots.

The validation step was performed using the same Pfinder UVLuxCHECK radiometer, measuring the absolute received UV irradiance on a 10 cm × 10 cm grid pattern under the lighting system at 50 cm distance (Figure 6), covering a surface of 600 × 400 mm. The results show a perfect similarity with the simulated one, since no measured points were less than 1500 µW/cm^2.

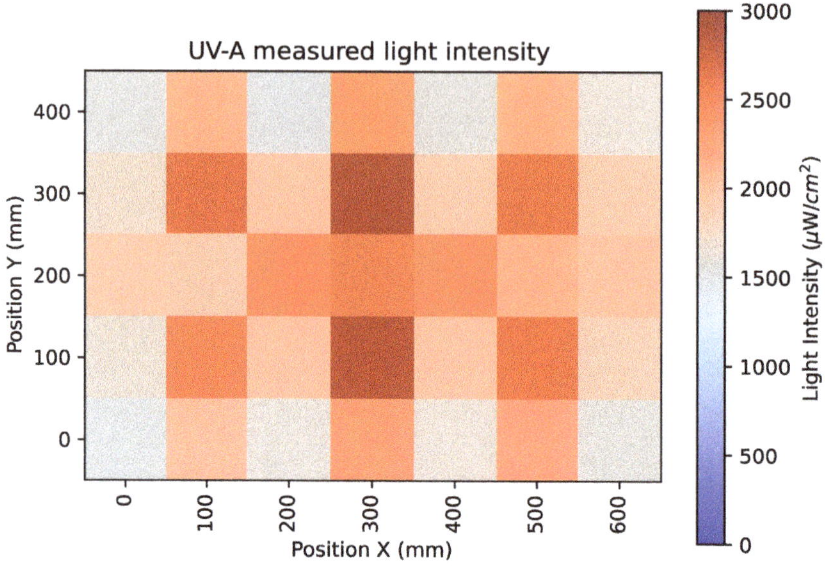

Figure 6. Measured UV light irradiance under the lighting system at 50 cm distance.

3.3. High-Definition Camera Imaging System

The machine must be able to detect defects with length above 3 mm with at least 50 μm width. In order to achieve this requirement, we state the hypothesis that we need at least one pixel representing the defect width (e.g., 50 μm).

Contrary to the initial scene size requirement (1200 × 800 mm), we decided to reduce this value to 600 × 400 mm due to hardware and performance limits. Considering we need at least 1 pixel on the defect width, this leads to:

- 50 μm on 400 mm → 8000 pixels;
- 50 μm on 600 mm → 12,000 pixels

We need at least a sensor with 8000 × 12,000 (96 M) pixels, so we choose a camera sensor with more than 100 M pixels fulfilling the requirement.

3.4. Depth Camera System

In addition to defect detection, a human controller needs to measure the defect length. Defects are located on a part's surface. Due to the complex and unpredictable 3D shape of the part, the distance between the defect and the 2D camera could be different from 50 cm, leading to an over or under length estimation, if length is computed from a 2D image only with a fixed pixel-to-distance ratio. In order to overcome this issue, a depth camera was used to obtain the previously unknown distance between the defect and the camera system, enabling an accurate defect length measurement. The integration of the depth sensor inside the vision system is depicted in Figure 7.

The depth camera is able to provide 2D images (Figure 8a) as well as a depth array (Figure 8b). Both data can be merged using the manufacturer SDK. A custom method to fill the missing depth point information has been implemented as shown in Figure 8c.

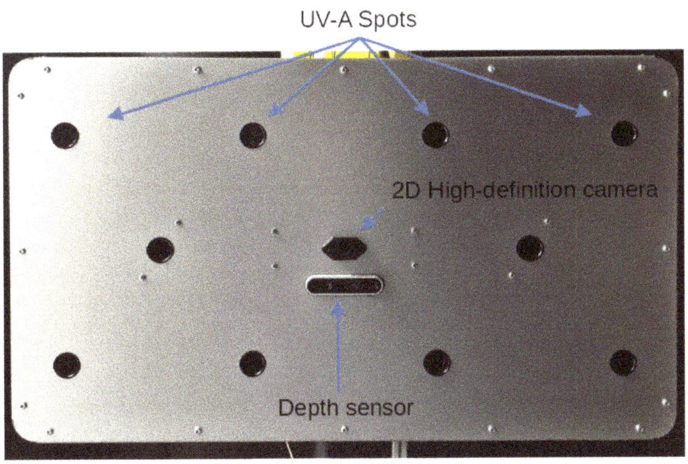

Figure 7. Integration of depth sensor along with 2D camera and UV-A spots in PARADES standalone vision system.

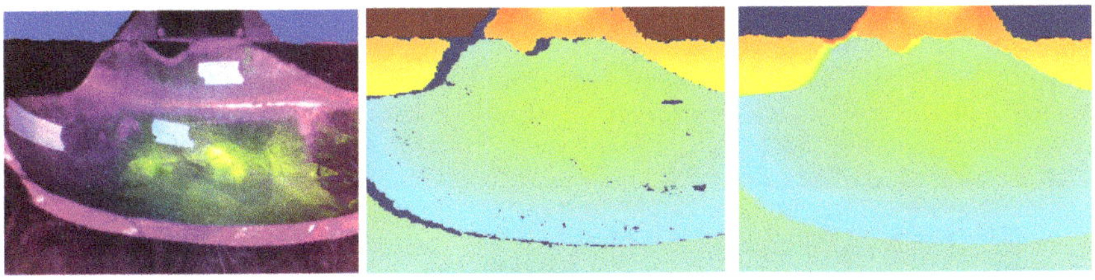

(**a**) (**b**) (**c**)

Figure 8. Output images from depth camera system. (**a**) 2D image. (**b**) Colorized depth image. (**c**) Processed depth image.

The complete hardware setup is depicted in Figure 9a,b.

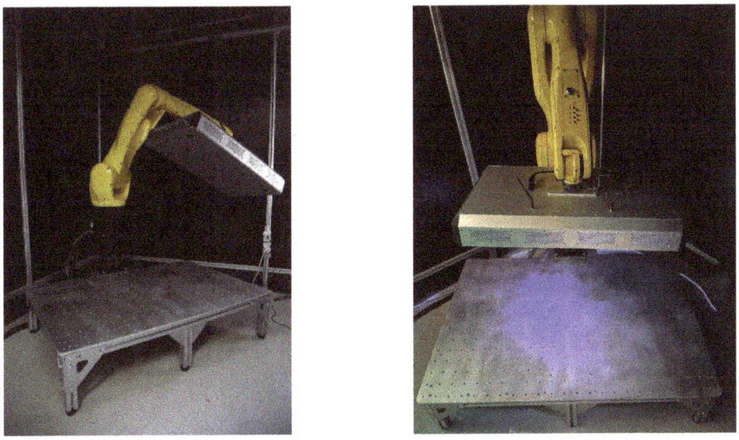

(**a**) (**b**)

Figure 9. Pictures of PARADES machine ready for production. (**a**) PARADES vision system. (**b**) PARADES vision system with UV lighting on.

4. Software

The hardware detailed in the previous section permits us to acquire MPI images in a controlled and repeatable way, thus enabling efficient AI-based defect detection. The images acquired by our system are >100 MP where defect length is about 3 $\rightarrow$ 20 mm with a minimal width of 50 μm. Only one class representing defects is used which changes from the usual paradigm in object detection where images are usually smaller (640 × 480 pixels in COCODataset) with more classes (80 classes in COCODataset). In this case, objects to be detected are only four or five times smaller than the image size. Those characteristics remain the same with the PascalVOC dataset with 207 images of size 500 × 375 pixels with 20 classes. With our hardware, in our use-case, a small defect (50 μm × 3 mm) is ~8000 times smaller than the image size, leading to a lot of missing detections if using current unchanged object-detection models. An example of an acquired MPI processed part is given in Figure 10. The part size is 80 × 80 mm in the 600 × 400 mm scene and as observed in this figure, a defect is present on the part with a size of ~20 × 20 mm. In addition, none of the trained object-detection models have labels for cracks and none of the image datasets used during training have fluorescent colorimetry, contrary to our acquired images. Indeed, usual crack datasets such as in [16] represent cracks in visible light but not using UV-based fluorescence. Moreover, cracks appear a lot bigger in the acquired image than defects seen with the PARADES machine. Based on those observations, a PARADES dedicated dataset is mandatory and so must be created. For that purpose, a custom UV-based MPI dataset is first built using PARADES hardware and will be detailed in the next section. Then, a specifically tuned object-detection module based on Faster R-CNN will be trained and presented. Finally, a defect length measurement feature is detailed.

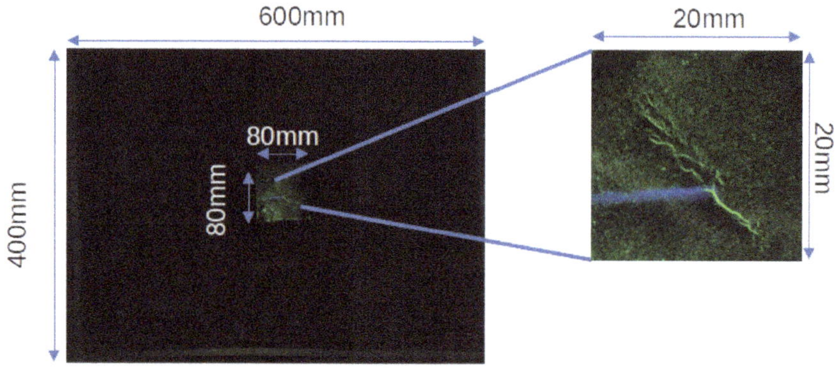

Figure 10. An example MPI result acquired with PARADES machine.

4.1. MPI Custom Dataset

A dataset of images similar to those acquired via the PARADES system is mandatory to train our model. The images in this dataset must match the distribution of images that will be acquired by controllers using the PARADES machine in the production line. Therefore, the images must be green fluorescent (linked to the MPI process) with defects (linear scratches) ranging from 2 to 25 mm length in a 600 × 400 mm field of view. Those images must show mechanical steel parts with granulated and smooth textures in order to represent forged, milled and foundry parts. In the literature, previous similar works in deep learning in MPI have been found [9,11–13]; authors had to build their own dataset because there is no such dataset available publicly on the market. Unfortunately, those authors did not give access to their datasets. On the other hand, similar dataset in other fields could have been used, datasets of cracks on concrete [17,18] representing scratches, but unfortunately, images in those datasets are not green fluorescent and furthermore cracks appear bigger in the images than in our case. Another approach could have been to use the NEU-DET dataset [19] or GC10-DET dataset [20], dedicated to steel surface defects. However the images are not fluorescent; images are small compared to our required 600 × 400 mm field of view. In addition, the images in those datasets show defects other than scratches (crazing, inclusions, patches, etc.) which are useless in our case of MPI, containing only linear scratch defects. Regarding those considerations, we decided to build our own dataset, with the PARADES machine. Existing customer (forgery, foundry, etc.) parts have been used, along with manually created defects on metal plates or tubes in order to increase the number of total defective parts.

Some image examples of our built dataset are given in Figure 11.

The complete dataset consists of 233 train images and 29 test images. Some continuous, discontinuous, curved or straight line defects are presents. Defect length varies from small (3 mm) to longer defects (>25 mm). Several types of manufactured parts compose the dataset, such as welded, milled and raw foundry parts. Defect types can be foundry defects (Figure 12) or defects induced by mechanical stress (Figure 13).

An example of a discontinuous foundry defect is shown in Figure 12.

For the trainset, some smartphone-based (not PARADES machine) images acquired by a human controller during control sessions have been also included. The testset only includes images from the PARADES machine with all the diversity explained previously. This dataset is still under construction and will continuously be completed. It contains some proprietary manufacturer parts and so cannot be publicly available at the time of writing.

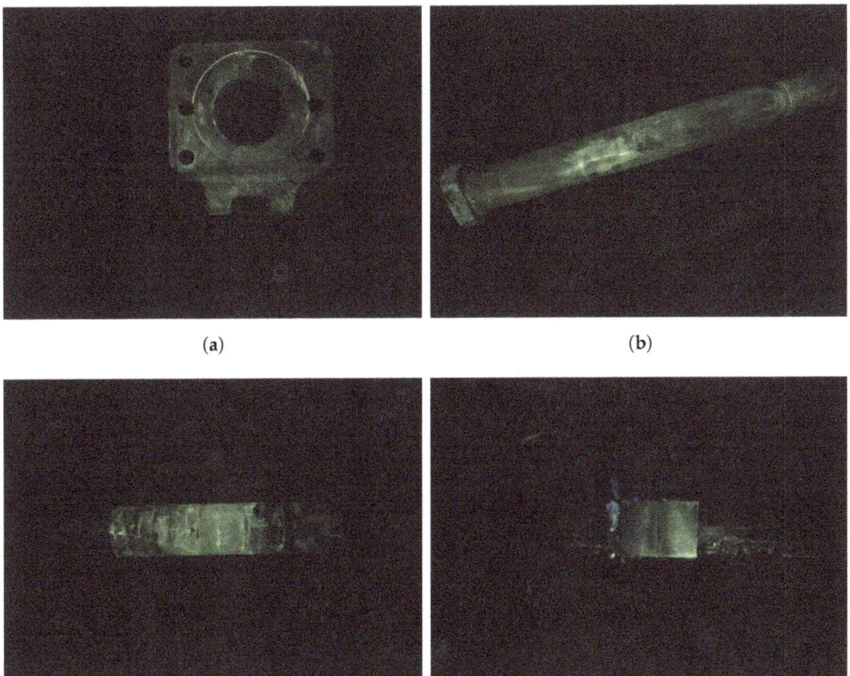

(**a**) (**b**) (**c**) (**d**)

Figure 11. Customer's part used in our dataset (1st row) and manually created parts (2nd row). (**a**) Forgery defect on customer's part. (**b**) Stress defect on a customer's screw. (**c**) Manually created defect on a metal tube. (**d**) Manually created defect on welded metal sheet.

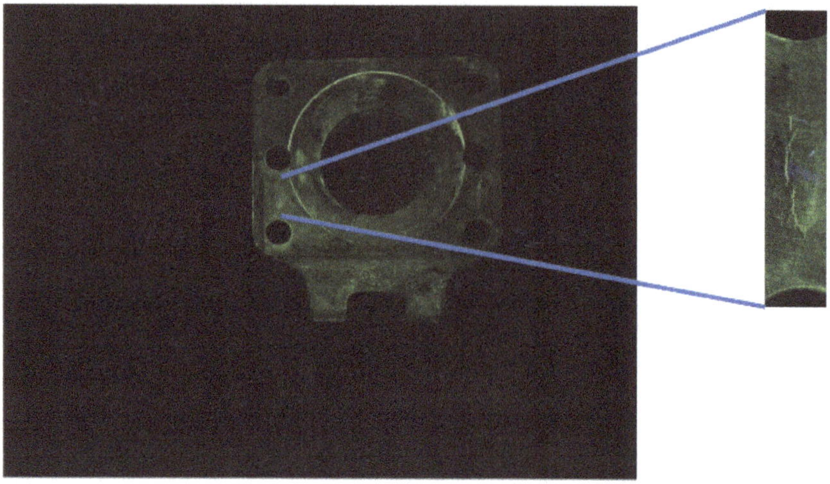

Figure 12. Example of a foundry defect on image acquired with the PARADES machine.

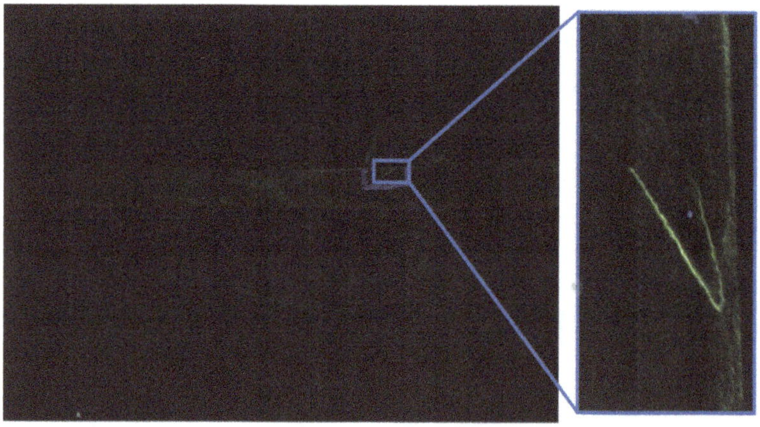

Figure 13. Example of stress-induced defect on a rod, acquired with the PARADES machine.

4.2. Defect Detection

4.2.1. Faster-RCNN Architecture

The usual available AI-based architectures for object detection are:

- Yolo [21];
- Faster R-CNN [22];
- Transformer-based (Deformable-DETR [23]).

Real-time detection is not a requirement for our project (the maximum allowed detection time is set to 60 s), but having a high PoD (which is similar to high recall) is mandatory. This eliminates Yolo-based architecture which is faster than the accurate sort due to its single stage architecture, contrary to Faster-RCNN [19,24] built around a two-stage architecture. Transformers (*-DETR) using ViT cannot be used because of the non-uniformity of the lighting. Indeed, the attention mechanism will detect non-uniformity as possible defect zones, leading to a high number of false detections.

Faster-RCNN is based on feature extraction (backbone), then a Region Proposal Network (RPN) followed at the end by the classification. Backbone based on convolution layers provides features to the RPN. RPN selects the best possible regions of interest (RoIs), to feed the classifier (Figure 14). The RPN part is relevant to our defect detection. Indeed, this defect represents 1 to 5 % of the whole image. Therefore, selecting RoI is more efficient. In this study, we used the official Pytorch implementation of Faster R-CNN.

During training, trainset images were compressed to be able to train with the CCUB (Centre de Calcul de l'Université de Bourgogne) cluster. The validation on the testset was carried out with the original image size. Data augmentation [9,11,13,20] was performed on the trainset only, with rotation as well as horizontal and vertical flips. This is motivated by the fact that during machine utilization defects can appear rotated and horizontally or vertically flipped. Data augmentation is implemented using a Python library called 'Albumentations'.

Transfer learning [11,25] using argument 'weights' set to 'IMAGENET1K_V2' was used with Pytorch implementation of Faster R-CNN to obtain the best possible recall with a limited number of training epochs. Indeed, only the fully connected layers used for classification are trained; the backbone remains frozen. This is useful for re-using the feature extraction capability of the backbone trained with a huge dataset, while specializing our network for our specific detections.

The PARADES machine is a commercial product; thus, the source code and implementation details cannot be released.

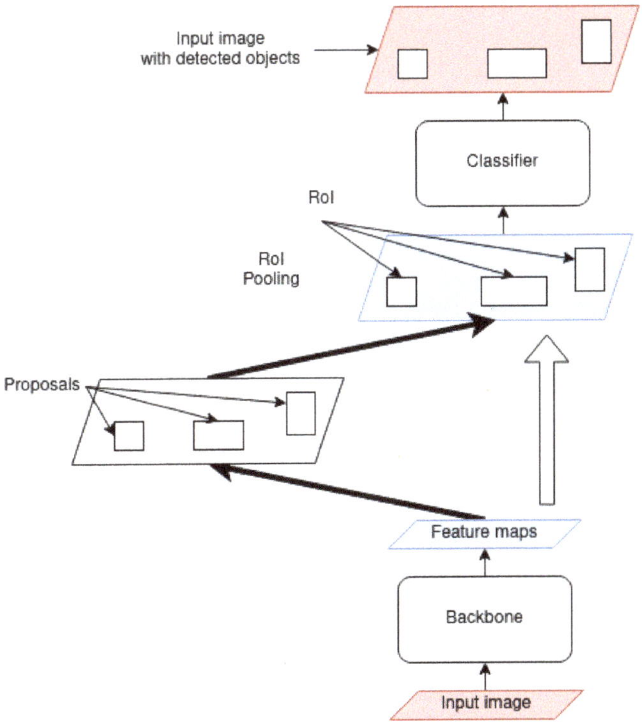

Figure 14. Faster R-CNN architecture.

4.3. Defect Length Measurement

Semi-autonomous length measurement is carried out by human controllers using a specifically developed software tool upon parts of acquired images from the PARADES system. Using a drawing tool coupled with depth information, the operator draws lines or splines on a merged 2D/3D image from a depth camera as shown in Figure 15. The 2D image from the depth camera is relatively close to the high-definition camera image used for detection: the field of view is quite similar and both are positioned close to each other and oriented the same way.

The defect length measurement tool provides two ways of measuring to the operator:

- The bird's eye distance between two points;
- the path length measurement (using pencil drawing).

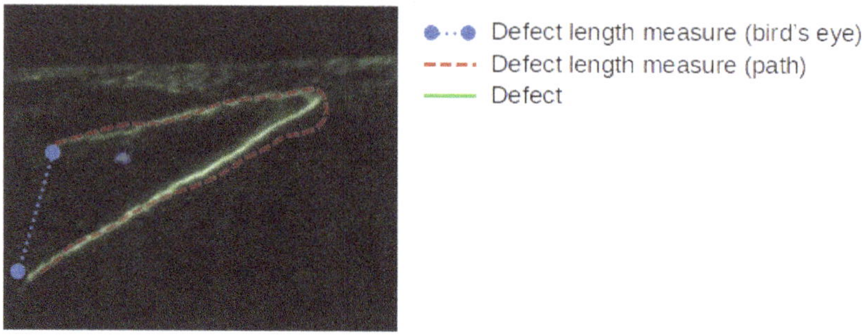

Figure 15. Two types of defect length measurement: straight line (bird's eye) and path.

Thanks to the depth camera system, pixels drawn by operators are converted into 3D voxels, using the manufacturer's SDK. The 3D coordinates of voxels in real space are extracted from depth array using SDK. To compute the path length (red line in Figure 15) we discretize the path drawn by the operator using the 3D voxels. The bird's eye distance (blue line in Figure 15) is computed using only the first and the last point. Both lengths are computed using euclidean distance.

5. Results

The goal of the PARADES machine is to help human operators detect defects between 3 and 25 mm on MPI processed parts. The first measuring performance criterion is PoD (same as recall in classification), with a requirement of >90%. Indeed, the PARADES system is used as a semi-autonomous machine to provide help for operators detecting defects. In this case, the machine shall not miss any defect but false detections will be easy and quicker for an operator to manually delete. So, precision is less important and is converted into average false detection per image to be more understandable by customers and human controllers. Average false detection per image shall not be more than 1. For defect length measurement, the Mean Absolute Accuracy (MAE) is chosen and must be less than 3 mm.

5.1. Defect Detection

A common performance metric in object detection is mean Average Precision (mAP). This represents the area under the recall/precision curve. In this study, according to the customer's requirements, the Intersection over Union (IoU) threshold is set to 0.02 for mAP computation. This can be seen as very loose but it makes sense for this use case. PARADES detection is then analyzed by a human operator who will adjust the surrounding box of detections to perfectly fit the defect.

The recall/precision curve of the PARADES machine, at IoU threshold = 0.02 is given in Figure 16. The mAP is 0.9043 with f1_max equal to 0.8434. To reach the minimal requirement of recall = 0.9, the chosen confidence score threshold is 0.31579, leading to:

- Recall = 0.9012 (>90%);
- Mean false positive per image = 0.9 (<1).

An example of defect detection on a test customer's part is given in Figure 17. As we can see, a defect is not detected (FN), there is a detection that does not match an actual defect (FP). Three other present defects (GT) are successfully detected (TP).

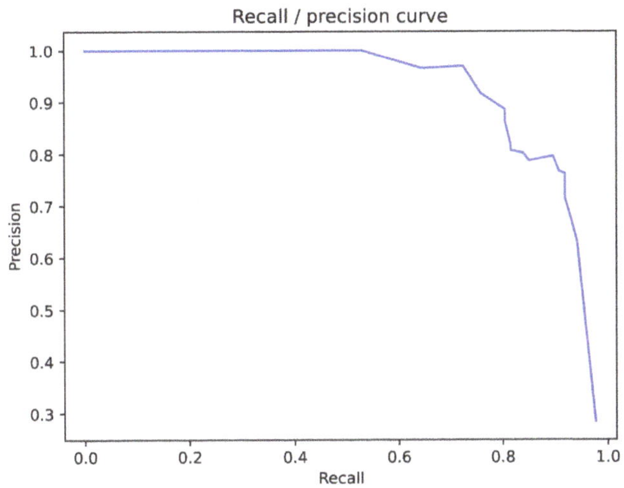

Figure 16. mAP@0.02 of the PARADES defect detection AI model.

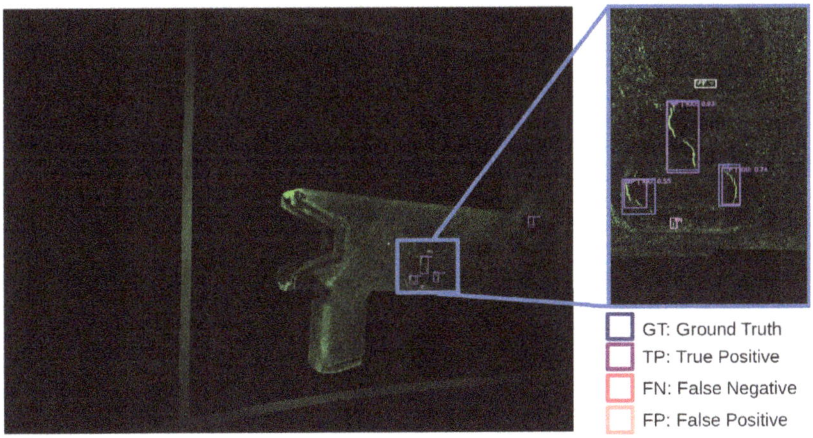

Figure 17. Example of detections on testset with correct detection (TP), missing detection (FN) and false detection (FP).

At this stage, the defect-detection model reaches the minimal recall requirement (i.e., 90%) without exceeding the average false detection per image limit (i.e., 1 FP per image). Increasing the number of parts in our dataset may increase those performances, as well as ongoing fine-tuning operations, enabling a possible fully autonomous machine.

5.2. Defect Length Measurement

Length measurement accuracy is defined as MAE on four artificially created ground truth straight paths representing defects. Those paths are straight lines ranging from 20 to 60 mm drawn on stickers which are applied on the curved surface of a real customer's part, as depicted in Figure 18. The results are summarized in Table 5. The computed MAE is 0.925 mm. We can see that MAE is not proportional to defect length. Indeed, the operator draws the defect line on an unzoomable image. This leads to operator-dependant position errors.

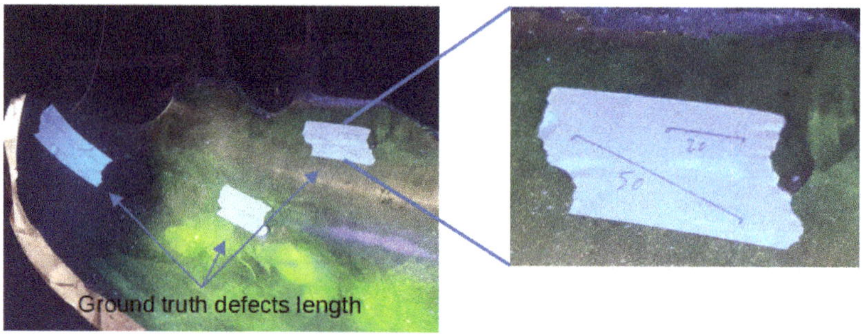

Figure 18. An example part with ground truth defect length (20 to 60 mm).

Table 5. Defect length measurement performance.

Ground Truth (mm)	Measured Value (mm)	Relative Error (%)	Absolute Error (mm)
50.0	51.8	3.6	1.8
20.0	20.0	0.0	0.0
60.0	60.1	0.2	0.1
30.0	28.2	6.0	1.8

Table 6 summarized the PARADES machine performances compared to previous work.

The paper [7] shows an industrial machine used to detect flaws longer than 10 mm and wider than 0.3 mm on steel wire rod and bar. The system is designed for a specific production line at Kobe Work, with a dedicated eight camera system. The work carried out in [8] shows an experimental setup to detect flaws of 15 mm length with widths of 0.2 mm, 0.4 mm, 0.6 mm and 0.8 mm. The machine is dedicated for small parts, for example a bearing of size 20 mm × 20 mm × 20 mm. Authors in [9] present a robotic MPI defect-detection system with external PC as computing unit. The machine is used to detect small defects on big mechanical parts (a defect represents 1/4000 image size). The paper [10] presents an automatic device for MPI defect detection dedicated to crankshaft. An experimental setup is presented with one UV light bulb and a camera, plugged into an external PC as a computing unit. In the paper [11], the authors present another way of detecting defects on crankshaft using the MPI technique. Authors in [13] present a software block to detect defects with MPI on bearing rings. The hardware side is not detailed.

As we explained in the previous paragraph, only some of the related works are industrial-grade products [9,11]; the others are experimental setups that cannot be used in industry. In those works, the detection performance (i.e recall) is either not given or is below our expectations. Only [9,11,13] perform better than our needs. In addition, some presented machines are dedicated to a chosen mechanical part; the system cannot be used for another part. Only [10,11] can be used with several mechanical parts. None of the presented related works offer a tool to measure defect length nor allow for wide mechanical part image acquisition (600 × 400 mm).

Based on this comparative study, we show that none of the previous works had success building a versatile and automatic defect-detection machine in MPI which can be used in production lines. Some works present advanced mechanical construction, but specifically dedicated to a kind of mechanical part. Other works show improvement in terms of defect-detection performance, but with an experimental and non-industrial machine that cannot be used in production lines. The PARADES machine gathers the following two important features: a versatile production-ready mechanical machine and high-performance defect-detection software. In addition, the PARADES machine provides a non-existing tool to accurately measure the defect length whatever the distance between the part and the machine.

Table 6. Comparison of automatic defect detection machines for MPI.

Machine	Industrial Grade	Defect Length Measure	Versatile Applications	>90% Recall	Field View 600 × 400 mm
[7]	?	No	No	?	No
[8]	?	No	No	No	No
[9]	Yes	No	No	Yes	No
[10]	No	No	Yes	?	No
[11]	Yes	No	Yes	Yes	No
[13]	?	No	?	Yes	No
PARADES	Yes	Yes	Yes	Yes	Yes

?: This property is not provided in the cited paper.

6. Conclusions

In conclusion, we have proposed the first semi-autonomous defect detection for MPI with a novel hardware block, AI-ready, standalone, performing UV-A lighting, image acquisition and detection for several kinds of visual NDT tasks. The versatile hardware built around a lighting system, a camera and the computation module in an industrial way should be used for any tasks requiring vision computer. This hardware is perfectly characterized, in particular the lighting system, enabling customization for smaller or wider part checking. Electronics is designed to be able to use human-spectrum UV spots for usual visual inspection. Features and the purpose of the machine can be easily extended with dedicated software development. Indeed, the Arm-based, AI-ready computer runs Ubuntu to make new application developments easy.

The PARADES machine is operated from outside the darkroom, so controllers are able to work in a clean environment with natural lighting, without UV light only, thus improving operators' working conditions and health. The PARADES machine reaches the minimal required specifications in terms of detection performance (PoD) and defect length measurement, so it is ready for production.

In the future, we will implement a zoomable depth image to lower the position error during defect length measurement. Ongoing works are focused on improving PoD by increasing dataset size and testing the SOTA transformer-based model. The PARADES machine is going to be integrated into a customer production line. Then, finetuning and using TensorRT will be implemented. An FPT images dataset is under construction for detecting FPT defects. An upcoming stitching feature will allow one to cover an area of 1200 × 800 mm using four 600 × 400 mm image acquisitions.

On the other hand, a new smaller handheld PARADES machine is being designed, in order to fulfill small parts manufacturers' needs as well as bigger complex and inaccessible parts such as cumbersome nuclear or hydraulic installation parts. A human controller handles the machine or fixes it with a dedicated articulated mechanical arm.

Author Contributions: J.B.: conceptualization, software, methodology and manuscript drafting, H.C.: supervision, A.P.: resources and supervision, P.M.: funding acquisition. All authors have read and agreed to the published version of the manuscript.

Funding: This research was funded by agence Nationale de la Recherche (ANR-21-PRRD-0047 01).

Data Availability Statement: The data presented in this study are available on request from the corresponding author. Dataset and AI software are not yet publicly available due to customer's restrictions.

Acknowledgments: This work was performed using HPC resources from GENCI–IDRIS (Grant 2023-AD011014417). Calculations were performed using HPC resources from DNUM CCUB (Centre de Calcul de l'Université de Bourgogne).

Conflicts of Interest: Authors Julien Baumeyer and Arnaud Pelletier were employed by the company CMPhy. The remaining authors declare that the research was conducted in the absence of any commercial or financial relationships that could be construed as a potential conflict of interest.

Abbreviations

The following abbreviations are used in this manuscript:

CCUB	Centre de Calcul de l'Université de Bourgogne
FP	False Positive
FPT	Fluorescent Penetrant Testing
IoU	Intersection over Union
MAE	Mean Absolute Error
mAP	mean Average Precision
MPI	Magnetic Particle Inspection
NDT	Non Destructive Testing
PARADES	Projet d'Automatisation de Reconnaissance et d'Analyse de DEfauts Surfaciques
PoD	Percentage of Detection
RMSE	Root Mean Square Error
SDK	Software Development Kit

References

1. Dwivedi, S.K.; Vishwakarma, M.; Soni, P.A. Advances and Researches on Non Destructive Testing: A Review. *Mater. Today Proc.* **2018**, *5*, 3690–3698. [CrossRef]
2. Smallbone, C. Welding and NDT—Enabling technologies to improve the global quality of life. In Proceedings of the 18th World Conference on Nondestructive Testing, Durban, South Africa, 16–20 April 2012.
3. Pelletier, A.; Crepeau, A.; Lemond, J.; Marquié, P.; Baumeyer, J.; Chatoux, H. Robotisation et Numérisation pour la Magnétoscopie et le Ressuage. *e-J. Nondestruct. Test.* **2023**, *28*. [CrossRef]
4. Ma, T.; Sun, Z.; Zhang, W.; Chen, Q. A machine vision assisted system for fluorescent magnetic particle inspection of railway wheelsets. *Proc. AIP Conf. Proc.* **2016**, *1706*, 150003. [CrossRef]
5. Willcox, M. *Automatic Defect Recognition in Magnetic Particle Inspection Applications*; Technical Report; Insight NDT Equipment Ltd.: Bedford, VA, USA, 2000.
6. Liu, J.; Yu, H.; Mei, L.; Han, B. Development of a new type of automatic magnetic particle inspection wall-climbing robot. *Adv. Mech. Eng.* **2021**, *13*, 16878140211047339. [CrossRef]
7. Nakata, H.; Hirata, M.; Tada, T. Fully automatic magnetic-particle inspection system for square billets. *IFAC Proc. Vol.* **1993**, *26*, 1–4. [CrossRef]
8. Chen, Y.; Kang, Y.; Feng, B.; Li, Y.; Cai, X.; Wang, S. Automatic defect identification in magnetic particle testing using a digital model aided De-noising method. *Measurement* **2022**, *198*, 111427. [CrossRef]
9. Wu, Q.; Qin, X.; Dong, K.; Shi, A.; Hu, Z. A learning-based crack defect detection and 3D localization framework for automated fluorescent magnetic particle inspection. *Expert Syst. Appl.* **2023**, *214*, 118966. [CrossRef]
10. Luo, J.; Tian, Z.; Yang, J. Fluorescent magnetic particle inspection device based on digital image processing. In Proceedings of the 11th World Congress on Intelligent Control and Automation, Shenyang, China, 29 June–4 July 2014; pp. 5677–5681. [CrossRef]
11. Tout, K.; Meguenani, A.; Urban, J.P.; Cudel, C. Automated vision system for magnetic particle inspection of crankshafts using convolutional neural networks. *Int. J. Adv. Manuf. Technol.* **2021**, *112*, 3307–3326. [CrossRef]
12. Ueda, A.; Lu, H.; Kamiya, T. Deep-Learning Based Segmentation Algorithm for Defect Detection in Magnetic Particle Testing Images*. *Proc. Int. Conf. Artif. Life Robot.* **2021**, *26*, 235–238. [CrossRef]
13. Yang, Y.; Yang, Y.; Li, L.; Chen, C.; Min, Z. Automatic Defect Identification Method for Magnetic Particle Inspection of Bearing Rings Based on Visual Characteristics and High-Level Features. *Appl. Sci.* **2022**, *12*, 1293. [CrossRef]
14. *ISO 9934-1:2015*; Non-Destructive Testing—Magnetic Particle Testing—Part 1: General Principles. ISO: Geneva, Switzerland, 2015.
15. *ISO 3059:2012*; Non-Destructive Testing, Penetrant Testing and Magnetic Particle Testing. ISO: Geneva, Switzerland, 2012.
16. Ozgenel, C.F. *Concrete Crack Images for Classification*; Mendeley Data: Burlingame, CA, USA, 2019; Volume 2. [CrossRef]
17. Silva, W.R.L.d.; Lucena, D.S.d. Concrete Cracks Detection Based on Deep Learning Image Classification. *Proceedings* **2018**, *2*, 489. [CrossRef]
18. Akgul, I. Mobile-DenseNet: Detection of building concrete surface cracks using a new fusion technique based on deep learning. *Heliyon* **2023**, *9*, e21097. [CrossRef] [PubMed]
19. Zhao, W.; Chen, F.; Huang, H.; Li, D.; Cheng, W. A New Steel Defect Detection Algorithm Based on Deep Learning. *Comput. Intell. Neurosci.* **2021**, *2021*, e5592878. [CrossRef] [PubMed]
20. Zhou, C.; Lu, Z.; Lv, Z.; Meng, M.; Tan, Y.; Xia, K.; Liu, K.; Zuo, H. Metal surface defect detection based on improved YOLOv5. *Sci. Rep.* **2023**, *13*, 20803. [CrossRef] [PubMed]
21. Redmon, J.; Divvala, S.; Girshick, R.; Farhadi, A. You Only Look Once: Unified, Real-Time Object Detection. *arXiv* **2016**, arXiv:1506.02640. [CrossRef]
22. Ren, S.; He, K.; Girshick, R.; Sun, J. Faster R-CNN: Towards Real-Time Object Detection with Region Proposal Networks. *arXiv* **2016**, arXiv:1506.01497.

23. Zhu, X.; Su, W.; Lu, L.; Li, B.; Wang, X.; Dai, J. Deformable DETR: Deformable Transformers for End-to-End Object Detection. *arXiv* **2021**, arXiv:2010.04159. [CrossRef]
24. Fang, Z.; Ren, J.; Sun, H.; Marshall, S.; Han, J.; Zhao, H. SAFDet: A Semi-Anchor-Free Detector for Effective Detection of Oriented Objects in Aerial Images. *Remote Sens.* **2020**, *12*, 3225. [CrossRef]
25. Mery, D. Aluminum Casting Inspection using Deep Object Detection Methods and Simulated Ellipsoidal Defects. *Mach. Vis. Appl.* **2021**, *32*, 72. [CrossRef]

applied sciences

MDPI

Article

A Dataset of Pulsed Thermography for Automated Defect Depth Estimation

Ziang Wei [1], Ahmad Osman [1,2,3,*], Bernd Valeske [3,4] and Xavier Maldague [1]

1 Faculty of Science and Engineering, Department of Electrical and Computer Engineering, University Laval, Quebec City, QC G1V 0A6, Canada; ziang.wei.1@ulaval.ca (Z.W.); xavier.maldague@gel.ulaval.ca (X.M.)
2 School of Engineering, University of Applied Sciences in Saarbrücken, 66117 Saarbrücken, Germany
3 Fraunhofer Institute for Nondestructive Testing IZFP, 66123 Saarbrücken, Germany; bernd.valeske@izfp.fraunhofer.de
4 Chair of Cognitive Sensor Systems, Saarland University, 66123 Saarbrücken, Germany
* Correspondence: ahmad.osman@izfp.fraunhofer.de

Abstract: Pulsed thermography is an established nondestructive evaluation technology that excels at detecting and characterizing subsurface defects within specimens. A critical challenge in this domain is the accurate estimation of defect depth. In this paper, a new publicly accessible pulsed infrared dataset for PVC specimens is introduced. It was enriched with 3D positional information to advance research in this area. To ensure the labeling quality, a comparative analysis of two distinct data labeling methods was conducted. The first method is based on human domain expertise, while the second method relies on 3D CAD images. The analysis showed that the CAD-based labeling method noticeably enhanced the precision of defect dimension quantification. Additionally, a sophisticated deep learning model was employed on the data, which were preprocessed by different methods to predict both the two-dimensional coordinates and the depth of the identified defects.

Keywords: pulsed thermography; deep learning; defect detection; depth estimation

1. Introduction

The critical role of Nondestructive-Evaluation (NDE)-based quality assurance within modern society cannot be overstated. The NDE technique allows for the assessment of components without causing any damage, making it indispensable in a wide range of industries, including aerospace [1–3] and civil engineering [4,5]. Over the past century, a number of NDE methods have been developed, such as ultrasound [2,6,7], eddy current [8,9], guided wave [10], and infrared thermography [11], among others.

Among these methods, Infrared Thermography (IRT) has gained widespread popularity for inspecting industrial components. Compared to other methods such as Computed Tomography (CT) or ultrasound, IRT offers a swift and cost-effective solution [11]. In practice, IRT experiments utilize an infrared camera to detect and record the transient thermal patterns on the surface of the testing specimen, with which valuable diagnostic information can be obtained.

An important variant of IRT experimentation is Pulsed Thermography (PT). In PT experiments, the testing specimen is exposed to a heat source, such as a high-power lamp, resulting in surface heating. Subsequently, the thermal wave propagates into the subsurface of the component where the potential defects could appear. This method exploits the distinct thermal properties and characteristics of sound areas versus defective areas, resulting in disparate thermal behaviors. As a result, PT effectively distinguishes defects from sound regions of the test component.

PT finds its greatest utility in inspecting materials characterized by relatively high thermal conductivity, including metals, Carbon-Fiber-Reinforced Plastic (CFRP), Glass-Fiber-Reinforced Plastics (GFRP), or Polyvinyl Chloride (PVC) materials.

In recent years, significant efforts have been dedicated to applying Artificial Intelligence (AI) techniques to the NDE data analysis. This includes the characterization of defects within ultrasound data [12,13], as well as advancements in CT data processing [14,15]. For IRT data analysis, current research primarily targets defect detection and the improvement of infrared imagery quality [16,17]. However, it is worth noting the significance of defect characterization in various industrial contexts. In particular, the depth estimation of defects is crucial for repair planning. Nonetheless, a major impediment to this advancement for IRT testing is the lack of publicly accessible datasets suitable for training deep learning models. Furthermore, accurately labeling of the 3D spatial location of defects poses a substantial challenge, as these anomalies are typically concealed beneath the surface of the test specimen.

In our previous work, a thermographic dataset for defect detection using deep learning methods was proposed [18]. However, this initial effort was limited by its focus on a single defect shape, reliance on Principal Component Thermography (PCT) images for labeling, and the absence of defect depth information. Building on this foundation, this study introduces a dataset with significant enhancements, including detailed defect depth information and refined labeling techniques, to further advance the field. Besides, to ensure accurate ground truth data, two distinct data labeling methodologies were implemented: a conventional manual approach and a semi-automated technique. Furthermore, we utilized the YOLOv5 deep learning model to train for the identification of defect locations and their depths.

The contributions of this study are threefold. First, a publicly accessible PT dataset specifically designed for defect detection and depth estimation is presented. This dataset, comprising 38 PT measurements, includes defects in two distinct shapes: rectangles and rounds, offering a standardized benchmark for testing various relevant algorithms. Second, a novel labeling methodology based on 3D-CAD files, subjecting it to qualitative and quantitative evaluation, is introduced, which showed a marked improvement over traditional manual labeling. Lastly, a YOLOv5-based defect-detection and depth-estimation method for PT data is proposed, delivering a comprehensive approach for PT data analysis.

2. Related Work

During a pulsed thermographic experiment, the specimen is subject to a controlled heat pulse followed by a cooling phase. This process is meticulously monitored using an infrared camera. The distinct thermal behaviors exhibited by defective regions at specific depths, as opposed to sound regions, facilitates the accurate determination of defect positions and depths. The temperature evolution on the specimen's surface is described by the following equation:

$$\Delta T = \frac{Q}{e\sqrt{\pi t}} \tag{1}$$

where ΔT represents the temperature increase and Q represents the input energy. t is time, and e is the thermal effusivity, which can be calculated using the following formula:

$$e = \sqrt{k\rho c_p} \tag{2}$$

In this formula, k, ρ, and c_p represent the sample's thermal conductivity, the sample's density, and the heat capacity of the sample, respectively. In homogeneous material areas within the specimen, abnormalities are not anticipated in the PT data. This is attributed to the uniformity in thermal effusivity within these areas. However, the presence of defects, such as foreign objects or structural anomalies, causes variations in thermal effusivity relative to the surrounding intact areas. Such variations appear as the contrast in thermographic imagery. This highlights the efficacy of pulsed thermography as an NDE technique.

Raw PT images often lack sufficient contrast to accurately identify the location of defects and, thus, require preprocessing for detailed analysis. Techniques such as PCT [19],

Pulsed Phase Thermography (PPT) [20], and Thermographic Data Reconstruction (TSR) [21] are prevalent in PT image analysis. These techniques are instrumental to enhance the contrast of defects and to reduce the dataset dimension. With these methods, the crucial features of the raw dataset can be effectively extracted.

PCT involves processing image sequences using principal component analysis along the time axis, typically utilizing the initial few Empirical Orthogonal Functions (EOFs) for in-depth analysis.

Conversely, PPT applies a Fourier transform along the time axis of the PT sequence, generating amplitude and phase coefficients. In general, the phase coefficients are observed to provide enhanced contrast in defect visualization.

TSR utilizes logarithmic polynomial fitting to model thermal profiles, as described by the following equation:

$$ln(\Delta T) = a_0 + a_1 ln(t) + a_2[ln(t)]^2 + a_3[ln(t)]^3 + ... + a_n[ln(t)]^n \quad (3)$$

where ΔT denotes the temperature increase and t is time. In infrared sequences, the frames index can be seen as a time representation. The coefficients of these polynomials are then used to reconstruct images, resulting in $n + 1$ frames containing TSR coefficients.

For manual labeling, PCT is a suitable choice, as it generates images with sufficient contrast for defect identification. Nonetheless, deducing the exact defect depth by directly inspecting the images produced by these methods is challenging. To overcome this, various approaches have been investigated, for example a method based on mathematical modeling. In [22], a mathematical modeling method for defect depth predication was reported. This approach is based on the analysis of a one-dimensional solution of pulsed thermography, where the temperature decay is multiplied by the corresponding time to derive a function, denoted as f. It is subsequently demonstrated that f exhibits linearity with respect to the square of the defect depth. Additionally, another defect depth estimation method utilizing peak contrast information from thermographic data was proposed in [23]. This method aids in quantitatively determining the defects' depth within stainless steel.

Ever since deep neural networks have emerged as state-of-the-art methods for data classification, efforts have been made to employ deep learning techniques to tackle the issue of depth estimation. For instance, in [24], the authors proposed a Long Short-Term Memory-Recurrent Neural Network (LSTM-RNN) to predict the depth information in a CFRP material inspected using laser infrared thermography. The proposed method notably outperformed the other two neural-network-based methods, CNN and RNN. Another study, presented in [25], employed a Gated Recurrent Unit (GRU) neural network to estimate the defects' depth in CFRP specimens. This research used a pulsed thermographic synthetic dataset, simulated by the Finite Element Modeling (FEM) software Comsol. The proposed methods demonstrated high performance, particularly noteworthy in terms of accuracy. Additionally, an artificial neural network was applied in [26] for defect depth prediction within CFRP coupons using PT. The model used in this study was trained on synthetic data and tested on synthetic datasets, as well as on experimental datasets. The network demonstrated remarkable accuracy, achieving 97% on synthetic datasets and 90% on experimental datasets.

3. Data Preparation and Pulsed Thermography (PT)

The material used for the specimens in this study was PVC. This is consistent with our previous research [18]. This set included an additional 19 specimens featuring flat rectangular indentations at the bottom. Figure 1 illustrates all potential defect locations within the dataset. The experimental setup remained identical to that described in our earlier work, with a schematic representation provided in Figure 2.

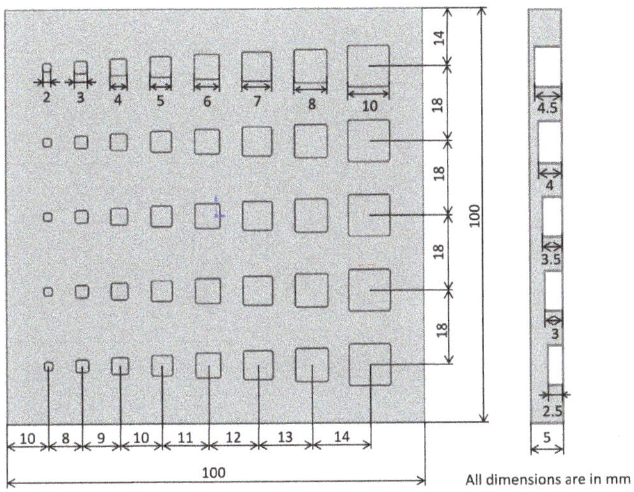

Figure 1. CAD image for all the possible defects' locations, sizes, and depths.

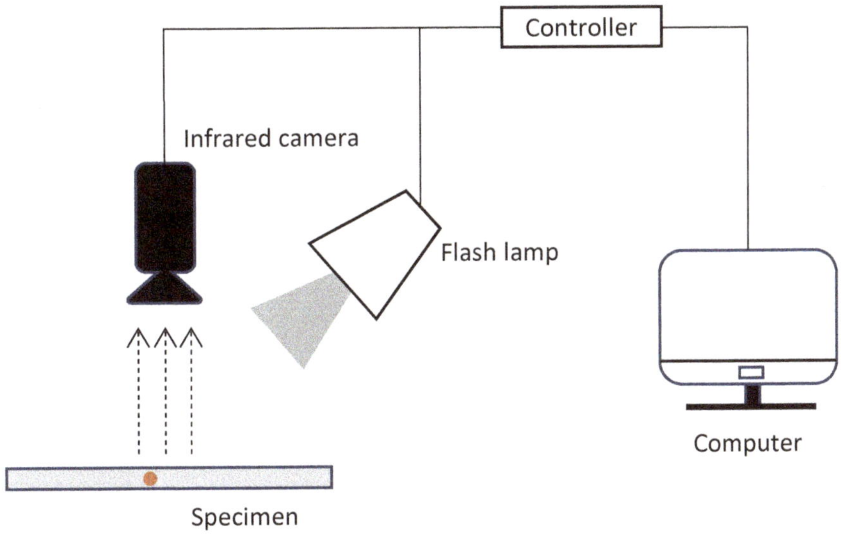

Figure 2. Experiment set up of PT inspection.

During the experiment, an infrared camera (FLIR SC5000) was employed to capture infrared images. Two high-power lamps, each with a power output of 4 KJ, were used to heat the specimens. The operation of the camera and the lamps was coordinated through a computer-controlled unit, ensuring precise timing for triggering events in predefined intervals. Initially, a signal is sent to activate the camera to start recording. After a certain time interval, a second trigger signal is sent to activates the flash. The camera, with a resolution of 320 × 256, records infrared video at a frequency of 10 Hz for 181 s. This process generated a dataset with dimensions of 256 × 320 × 1810 for each measurement.

4. Data Labeling

4.1. Principal Component Thermography Image-Based Method (Manual Method)

Figure 3 provides a visual comparison between an original infrared image and the result images generated by the three methods. In the original infrared frame, only four defects are discernible. However, after applying these methods, nine defects become visible, demonstrating a notable enhancement in defect contrast and detectability. Among these methods, the PCT images exhibited sufficient contrast for defects, making PCT the preferred choice for labeling tasks. Following the PCT process, one of the EOF images that contained the most-relevant information about the defects was selected for further analysis. The labeling of the defects' position can, therefore, be facilitated by examining this selected EOF image. Once the defect locations had been identified, the subsequent step involved integrating the depth information from the defect plan, as illustrated in Figure 4.

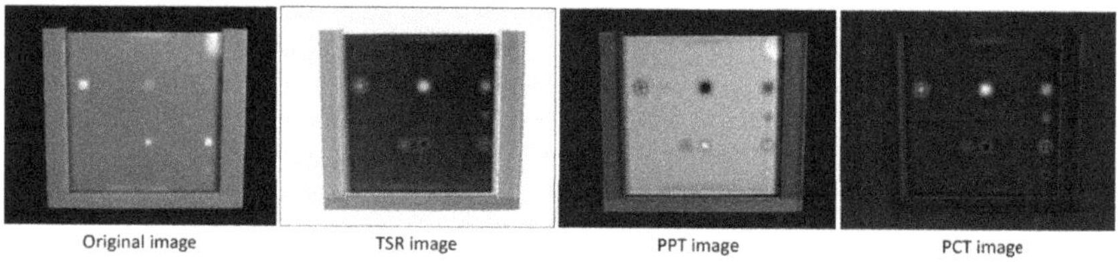

Figure 3. Comparison of original infrared frame and resulting images generated by TSR, PPT, and PCT, respectively.

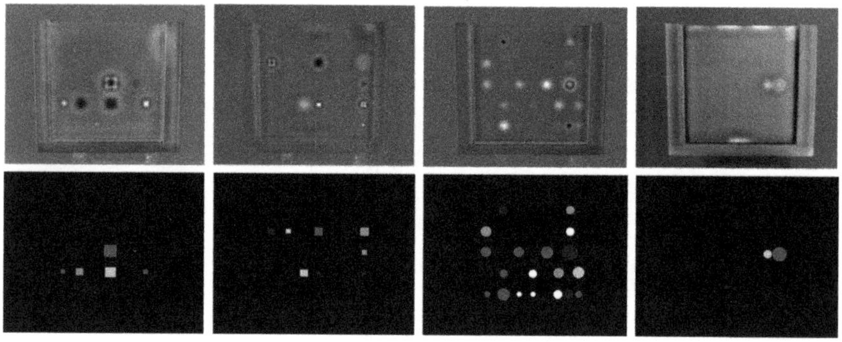

Figure 4. PCT images and labeled images, where different grayscale values represent different depths.

Specifically, one or more than one PCT images with the highest contrast indicating potential defects were selected for further processing. Subsequently, red masks were then manually drawn over these images to highlight abnormal regions. If the process involved multiple images, they were combined to create the final composite mask.

While this manual method enables precise defect localization through the careful adjustment of the masks to match abnormal regions, it is not without its drawbacks. Firstly, manually placing masks for each defect is a time-intensive task, especially when a large number of defects are present on a specimen. Moreover, accurately defining defect boundaries can be challenging, as the boundaries of abnormalities on the PCT image may not align perfectly with the actual defect boundaries. Secondly, some defects may not be visible in a single EOF image, often due to depth variations. In such cases, labeling must be performed on multiple EOF images, each of which is then merged to form the final composite mask. This additional step introduces complexity and potential inaccuracies in the labeling process.

4.2. Three-Dimensional CAD-Based Method

To overcome the limitations of the manual labeling method and effectively utilize the 3D information from the CAD files, a semi-automated labeling method was introduced. The first step of this method involved transforming the 3D model of each component into a two-dimensional depth image. The transformation was executed by simulating a virtual camera positioned above the specimen, focusing vertically on the specimen's bottom surface.

The next step involved aligning the transformed depth images with the PCT EOF images, as illustrated in Figure 5. To achieve this, four corresponding points were manually identified on both the depth image and PCT image to compute the perspective transform matrix. This matrix is described by the following equation:

$$\begin{bmatrix} Sx' \\ Sy' \\ 1 \end{bmatrix} = \begin{bmatrix} a_1 & a_2 & b_1 \\ a_3 & a_4 & b_2 \\ c_1 & c_2 & 1 \end{bmatrix} \begin{bmatrix} x \\ y \\ 1 \end{bmatrix} \tag{4}$$

where x, y denote the pixel locations from the CAD-generated depth image. The parameters a_1, a_2, a_3, and a_4 control the scaling, rotation, and shear mapping. b_1 and b_2 represent the translation of the image. c_1 and c_2 are projection parameters. Using this matrix, the depth image can be precisely overlaid onto the PCT image, enabling concurrent determination of the defects' depth and surface location.

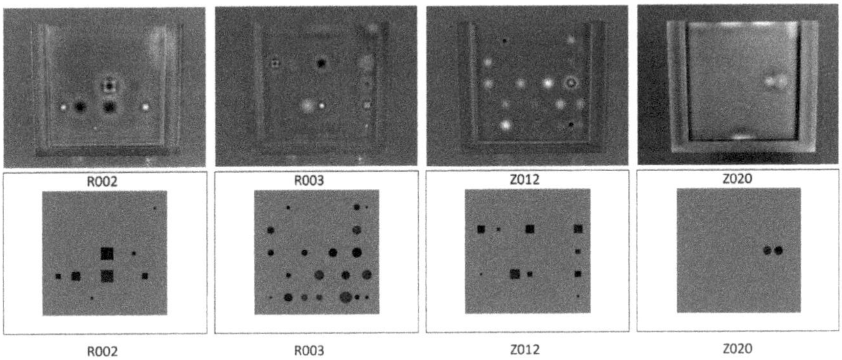

Figure 5. PCT images (**top**) generated from 3D files and the depth images (**bottom**).

The thermal contrast resulting from the boundaries of the subsurface defects extends into their neighboring areas, creating difficulty for manual labeling in accurately delineating the precise boundaries of these defects in PCT images. The semi-automated approach, which is based on a 3D-CAD file, was, therefore, anticipated to generate higher accuracy in defining defect boundaries compared to manual labeling.

4.3. Evaluation of the Labeling Quality

Qualitative evaluation: Figure 6 presents the overlaid images for a qualitative comparison of the labeling results obtained through both methods. For rectangular defects, the labeled masks were largely similar, with minor discrepancies observed in the manually labeled images. However, for round defects, a significant difference was noted between the masks generated by the two methods. The manual approach tended to generate larger masks, likely due to the difficulty in identifying precise reference points for round defects, as opposed to the more-straightforward four-corner identification in rectangles. This often resulted in the inclusion of adjacent regions in the mask, enlarging it beyond the actual size of the defect.

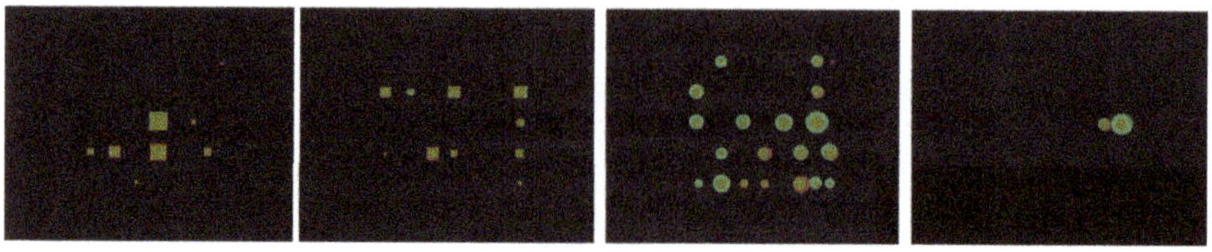

Figure 6. Comparison of mask images labeled by the manual method and the 3D-CAD-based approach, where the green masks are labeled by the manual method and the red masks are labeled by the 3D CAD-based method. Yellow areas highlight the overlap between these two masks.

While qualitative comparisons can be intuitively made, achieving an objective evaluation of the labeling quality is not straightforward. The actual defects lie beneath the surface of the specimen, making direct observation of their precise locations in the thermal imagery challenging. To overcome this, a reference distance was established to correlate the image dimensions with real-world measurements.

The dimension of the holder of the measurement can be derived directly from the CAD file, as shown in Figure 7. The measured distance on the upper part is 96 mm, corresponding to a pixel distance of 185 in the image. Equation (6) defines a linear transformation. Therefore, the distance between points A and B in the image can be used as a reference for evaluating the labeled sizes. To quantify the accuracy of defect size labeling, the Mean Absolute Error (MAE) and Mean Squared Error (MSE) were used. The MAE can be calculated using the following formula:

$$MAE = \frac{\sum_1^n |x_i - \hat{x}_i|}{n} \tag{5}$$

Here, x_i represents the size of the region in the labeled image and $\hat{x}_i$ represents the reference size of the same region derived from the reference distance and the 3D-CAD file. The MAE is a widely used metric for evaluating the model's performance, which is robust to outliers, particularly effective in scenarios with numerous outliers. This metric exhibits less sensitivity to large errors. Nevertheless, large errors are critically informative in this context. The MSE becomes relevant due to its characteristic of penalizing larger errors strongly. It can be calculated using the following formula:

$$MSE = \frac{\sum_1^n (x_i - \hat{x}_i)^2}{n} \tag{6}$$

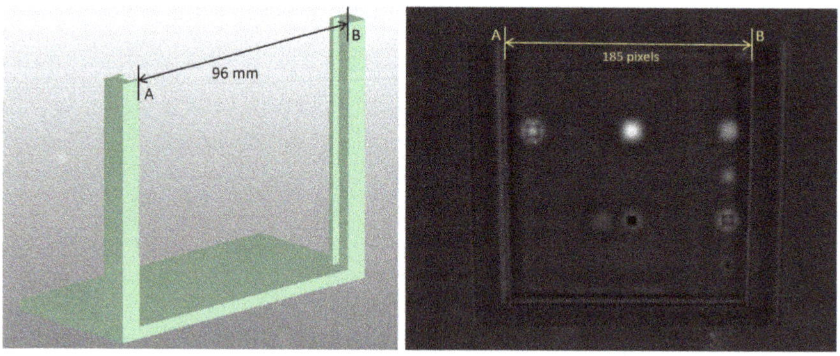

Figure 7. Reference distance from point A to point B. The left image shows this distance on the CAD image. The right image illustrates the corresponding distance on the PCT image.

Table 1 provides the results of the objective labeling quality evaluation for both the manual labeling method and the CAD-based labeling method. To be specific, labels for defects in varying dimensions were evaluated. The dimensions of the defects ranged from 2 mm to 10 mm for round defects and equivalent side lengths for square defects. On the one hand, the area for each defect was calculated in pixels based on the labeling data. On the other hand, each area was also calculated by employing the relative diameter or side length, which was derived using the reference length measured between points A and B. The results, obtained using the MAE and MSE, clearly demonstrated that the CAD-based method significantly outperformed the manual labeling method, as evidenced by the reduced magnitude of the errors observed. This suggests that relying solely on labeling methods based on PCT images, which is currently the most-popular approach, may frequently result in inadequate levels of accuracy. Furthermore, it is important to note that the availability of 3D-CAD models in this case is directly linked to the fabrication of the PVC specimens from these models. In other scenarios where such models are not used, manual labeling is typically employed.

Table 1. Objective labeling quality evaluation.

Labeling Method \ Defect Size	10 mm	8 mm	7 mm	6 mm	5 mm	4 mm	3 mm	2 mm
MAE with manual labeling	1.420	1.195	0.950	0.709	0.795	0.519	0.470	0.402
MAE with CAD-based labeling	0.126	0.115	0.086	0.088	0.075	0.068	0.065	0.055
MSE with manual labeling	3.6105	3.1602	1.9718	1.1343	1.5564	0.4883	0.5482	0.2947
MSE with CAD-based labeling	0.0252	0.0182	0.0099	0.0090	0.0072	0.0059	0.0053	0.0037

5. Experiments with Deep Models for Defect Detection and Depth Estimation

5.1. Network Architecture

YOLOv5, introduced in [27], is a single-stage object-detection network known for its high efficiency, making it particularly suitable for industrial applications [28–30]. Figure 8 illustrates the architecture of YOLOv5. It is composed of three main components: the backbone, neck, and head.

The backbone of the YOLOv5 model, CSPDarknet [31], efficiently extracts features from the input image. The neck of the YOLOv5 model is designed to generate feature pyramids, enhancing the model's robustness in detecting objects of varying scales. The specific network employed here was the Path Aggregation Network (PANet) [32]. Finally, the head of the model is responsible for generating the final detections, employing the features from the previous layers to produce outputs indicating class probabilities, object scores, and bounding box coordinates. The activation functions used in the model included Leaky ReLU [33] in the hidden layers and Sigmoid in the output layer.

It is worth noting that YOLOv5 offers several variants, including nano, small, middle, large, and XL. The small variant was chosen for its optimal balance between model size and performance.

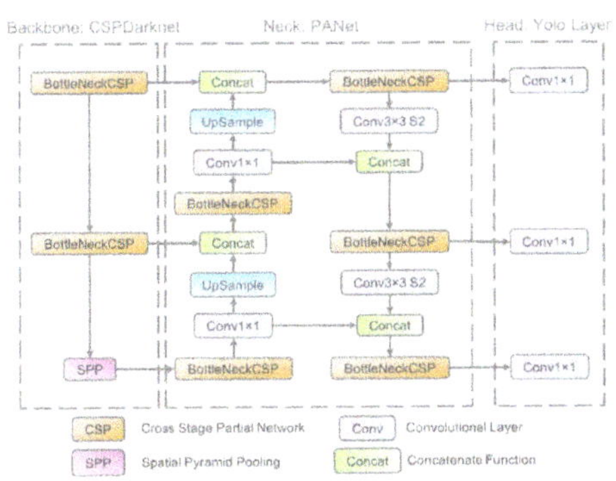

Figure 8. Network architecture of YOLOv5 model from [34], which is licensed under CC BY 4.0.

5.2. Model Training

The dataset consisted of 38 measurements, with 26 allocated randomly for training, 6 for validation, and the remaining 6 for testing. To optimize the model's performance, a linear learning rate scheduler was implemented during the training process.

All training data underwent preprocessing using the TSR, PCT, and PPT methods. Specifically, TSR was executed with a 5° decomposition. This preprocessing resulted in multiple frames of images generated by each method. The coefficients generated by TSR and PCT and the phase coefficient generated by PPT were selected for further processing. For integration with the ImageNet pretrained model, three specific frames were selected to construct RGB-like images, as is shown in Figure 9. For the PPT, PCT, and TSR coefficients, frames from the second to fourth sequences were chosen, excluding the first frame. The rationale for this exclusion was that the first frame represents only the average temperature of all thermal sequences, which is less relevant in this context, which focuses on defects with significant thermal contrast.

Figure 9. Comparison of training images generated by TSR, PPT, and PCT, respectively.

The model was trained on all three types of images: PPT, PCT, and TSR. Table 2 provides an overview of the hyperparameters employed in the experiments. Data augmentation techniques, including perspective transformations like rotation, scaling, and shearing, were dynamically applied to enhance the robustness of the training images.

Table 2. Hyperparameters to train the YOLOv5 model.

Hyperparameter	Value
Batch size	32
Optimizer	Stochastic gradient descent
Learning rate	0.005
Learning rate scheduler	Linear
Training epochs	500

The results are shown in Tables 3–5 and in Figure 10. Performance on various validation datasets was assessed using precision, recall, the mean average precision [35] at 50% overlap (mAP50), and the confusion matrix [36]. Precision quantifies the accuracy of the positive predictions, while recall measures the ability of the model to detect all the relevant instances. The mAP50 is a widely used metric for object detection. It considers a predicted bounding box as valid when it has more than 50% overlap with an actual target area. The mAP provides a comprehensive evaluation by integrating the precisions of the model at different recall levels. The confusion matrix extends the insights from recall and precision by offering detailed evaluation results for each class.

The results indicated that the PPT image preprocessing method achieved the highest mAP, signifying its superior efficacy. This can be attributed to its advanced image enhancement capability. Therefore, it is essential to recognize the significant impact of preprocessing methods in the hyperparameter selection process.

Table 3. Results on validation dataset of PPT data.

Defect Depth	Precision	Recall	mAP50
All	0.798	0.742	0.811
0.5 cm	0.648	0.889	0.887
1.0 cm	0.671	0.8	0.844
1.5 cm	0.81	0.846	0.807
2.0 cm	1	0.507	0.836
2.5 cm	0.863	0.667	0.681

Table 4. Results on validation dataset of PCT data.

Defect Depth	Precision	Recall	mAP50
All	0.403	0.557	0.478
0.5 cm	0.336	0.778	0.516
1.0 cm	0.347	0.467	0.487
1.5 cm	0.292	0.462	0.442
2.0 cm	0.248	0.3	0.206
2.5 cm	0.791	0.778	0.736

Table 5. Results on validation dataset of TSR data.

Defect Depth	Precision	Recall	mAP50
All	0.423	0.561	0.577
0.5 cm	0.625	0.889	0.802
1.0 cm	0.584	0.8	0.801
1.5 cm	0.403	0.573	0.654
2.0 cm	0.184	0.1	0.16
2.5 cm	0.319	0.444	0.47

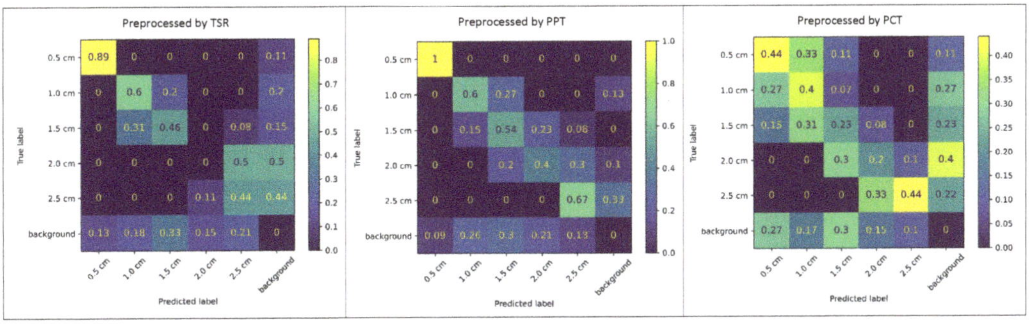

Figure 10. Confusion matrix of the results on the datasets preprocessed through TSR, PPT, and PCT.

6. Discussion

This study introduced a PT dataset comprising both defect location and depth labeling information, establishing a valuable resource for comparing different automated defect characterization methods. Initially, a conventional manual labeling method was applied to identify defect locations. Furthermore, considering the availability of 3D CAD files for each specimen, a CAD-based labeling approach was proposed to enhance the labeling accuracy and provide additional defect depth information.

The labeling results were evaluated through two methods: qualitative evaluation for the visual comparison of label masks and quantitative evaluation using a reference distance to calculate the regions. Qualitative evaluation revealed that the manual labeling method was accurate for defects in rectangular shapes. However, it may introduce bias in labeling round defects. In contrast, the quantitative evaluation demonstrated that the 3D CAD-based approach excelled in accurately determining the size of defect regions.

For automated defect depth and location prediction, YOLOv5 object-detection algorithm was applied to datasets preprocessed using the PPT, PCT, and TSR methods. The dataset that generated by PPT method enabled the model to attain an mAP of 0.811 across all depth classes, likely due to the rich temporal information in the phase images, which was more effective than the other preprocessing methods in emphasizing temporal details.

The YOLOv5 model, being lightweight, is well-suited for industrial applications. In contrast to one-dimensional thermal-profile-based defect depth estimation approaches, the proposed method leverages spatial–temporal information from thermal datasets to concurrently determine defect location and size, yielding higher accuracy in depth estimation. Nevertheless, this study is not without limitations. The YOLOv5 model provides only classification results, which are discrete values. Exploring a regression model capable of predicting continuous values could be particularly interesting for estimating defect depth, especially in scenarios where the depth of defects is uncertain.

7. Conclusions

In this study, a pulsed infrared thermography dataset with two distinct defect shapes and comprehensive defect depth information as the ground truth was introduced. This dataset makes a significant advancement over its predecessor, offering double the number of measurements and the advantage of using an innovative CAD-based labeling approach. This approach significantly enhanced the accuracy in defining the boundaries of the defect masks. The effectiveness of this novel labeling approach was further validated through two methods of labeling quality evaluation.

Additionally, the study successfully trained a deep learning model based on the YOLOv5 architecture, capable of detecting defect locations and predicting their depths. The results underscored the potential of the YOLOv5 model, particularly when utilizing phase images generated through the PPT method. The combination of high efficiency and accuracy in this model makes it an invaluable tool in industrial settings for defect analysis.

Looking ahead, it is worth exploring deep-learning-based recognition of defects with more-complex shapes, such as pyramids, wedges, or tapered cylinders. These shapes introduce novel challenges in thermal behavior analysis, presenting an opportunity to push the boundaries of current methodologies. Exploring these complex shapes could lead to significant advancements in the field of infrared thermography and defect characterization.

Author Contributions: Z.W.: methodology and manuscript drafting, A.O.: methodology and supervision. B.V.: supervision, X.M.: supervision. All authors have read and agreed to the published version of the manuscript.

Funding: This research was supported by the European Regional Development Fund (ERDF) and LUFO-Projekt RapidEF-Enablers from Fraunhofer IZFP institute.

Institutional Review Board Statement: Not applicable.

Informed Consent Statement: Not applicable.

Data Availability Statement: The dataset can be found on the website of Kaggle with the URL: https://www.kaggle.com/datasets/ziangwei/irt-pvc-depth (accessed on 21 November 2023). The dataset is available for academic and research use only. Papers using this dataset are kindly requested to refer to this paper.

Acknowledgments: Support from Fraunhofer IZFP is gratefully acknowledged. David Müller is acknowledged for the data acquisition. Yiting Yun is acknowledged for support in the method's implementation. Udo Nezelmann is acknowledged for supporting the publication of the dataset.

Conflicts of Interest: The authors declare no conflict of interest.

Abbreviations

The following abbreviations are used in this manuscript:

AI	Artificial Intelligence
IRT	Infrared Thermography
PT	Pulsed Thermography
PPT	Pulsed Phase Thermography
PCT	Principal Component Thermography
TSR	Thermographic Signal Reconstruction
NDE	Nondestructive Evaluation
IoU	Intersection over Union
EOF	Empirical Orthogonal Function
CFRP	Carbon-Fiber-Reinforced Plastic
GFRP	Glass-Fiber-Reinforced Plastics
PVC	Polyvinyl Chloride

References

1. Ciampa, F.; Mahmoodi, P.; Pinto, F.; Meo, M. Recent advances in active infrared thermography for non-destructive testing of aerospace components. *Sensors* **2018**, *18*, 609. [CrossRef] [PubMed]
2. Smith, R.A.; Mukhopadhyay, S.; Lawrie, A.; Hallett, S.R. Applications of ultrasonic NDT to aerospace composites. In Proceedings of the 5th International Symposium on Aerospace NDT, Singapore, 13–15 November 2013.
3. Ley, O.; Godinez-Azcuaga, V. Line scanning thermography and its application inspecting aerospace composites. In Proceedings of the 5th International Symposium on NDT in Aerospace, Singapore, 13–15 November 2013; pp. 13–15.
4. Hiasa, S.; Birgul, R.; Catbas, F.N. Infrared thermography for civil structural assessment: demonstrations with laboratory and field studies. *J. Civ. Struct. Health Monit.* **2016**, *6*, 619–636. [CrossRef]
5. Pozzer, S.; Rezazadeh Azar, E.; Dalla Rosa, F.; Chamberlain Pravia, Z.M. Semantic segmentation of defects in infrared thermographic images of highly damaged concrete structures. *J. Perform. Constr. Facil.* **2021**, *35*, 04020131. [CrossRef]
6. Blitz, J.; Simpson, G. *Ultrasonic Methods of Non-Destructive Testing*; Springer Science & Business Media: Berlin/Heidelberg, Germany, 1995; Volume 2.
7. Kurz, J.H.; Jüngert, A.; Dugan, S.; Dobmann, G.; Boller, C. Reliability considerations of NDT by probability of detection (POD) determination using ultrasound phased array. *Eng. Fail. Anal.* **2013**, *35*, 609–617. [CrossRef]
8. García-Martín, J.; Gómez-Gil, J.; Vázquez-Sánchez, E. Non-destructive techniques based on eddy current testing. *Sensors* **2011**, *11*, 2525–2565. [CrossRef] [PubMed]
9. Heuer, H.; Schulze, M.; Meyendorf, N. Non-destructive evaluation (NDE) of composites: Eddy current techniques. In *Non-Destructive Evaluation (NDE) of Polymer Matrix Composites*; Elsevier: Amsterdam, The Netherlands 2013; pp. 33–55.
10. Wilcox, P.; Evans, M.; Pavlakovic, B.; Alleyne, D.; Vine, K.; Cawley, P.; Lowe, M. Guided wave testing of rail. *Insight-Non-Destr. Test. Cond. Monit.* **2003**, *45*, 413–420. [CrossRef]
11. Ibarra-Castanedo, C.; Genest, M.; Piau, J.M.; Guibert, S.; Bendada, A.; Maldague, X.P. Active infrared thermography techniques for the nondestructive testing of materials. In *Ultrasonic and Advanced Methods for Nondestructive Testing and Material Characterization*; World Scientific: Singapore, 2007; pp. 325–348.
12. Budyn, N.; Croxford, A.J.; Bevan, R.L.; Zhang, J.; Wilcox, P.D. Characterisation of small embedded two-dimensional defects using multi-view Total Focusing Method imaging algorithm. *NDT E Int.* **2021**, *119*, 102413. [CrossRef]
13. Fradkin, L.; Uskuplu Altinbasak, S.; Darmon, M. Towards explainable augmented intelligence (AI) for crack characterization. *Appl. Sci.* **2021**, *11*, 10867. [CrossRef]
14. Tyystjärvi, T.; Virkkunen, I.; Fridolf, P.; Rosell, A.; Barsoum, Z. Automated defect detection in digital radiography of aerospace welds using deep learning. *Weld. World* **2022**, *66*, 643–671. [CrossRef]
15. Fuchs, P.; Gondrom-Linke, S. Information Extraction from Industrial CT Scans Using 3D Deep Learning. In International Conference on NDE 4.0, Berlin, Germany, 24–27 October 2022.
16. Liu, K.; Li, Y.; Yang, J.; Liu, Y.; Yao, Y. Generative principal component thermography for enhanced defect detection and analysis. *IEEE Trans. Instrum. Meas.* **2020**, *69*, 8261–8269. [CrossRef]
17. Wei, Z.; Fernandes, H.; Tarpani, J.R.; Osman, A.; Maldague, X. Stacked denoising autoencoder for infrared thermography image enhancement. In Proceedings of the 2021 IEEE 19th International Conference on Industrial Informatics (INDIN), Palma de Mallorca, Spain, 21–23 July 2021; pp. 1–7.
18. Wei, Z.; Osman, A.; Valeske, B.; Maldague, X. Pulsed Thermography Dataset for Training Deep Learning Models. *Appl. Sci.* **2023**, *13*, 2901. [CrossRef]
19. Rajic, N. Principal component thermography for flaw contrast enhancement and flaw depth characterisation in composite structures. *Compos. Struct.* **2002**, *58*, 521–528. [CrossRef]
20. Maldague, X.; Marinetti, S. Pulse phase infrared thermography. *J. Appl. Phys.* **1996**, *79*, 2694–2698. [CrossRef]

21. Shepard, S.M. Temporal Noise Reduction, Compression and Analysis of Thermographic Image Data Sequences. U.S. Patent 6,516,084, 4 February 2003.
22. Zeng, Z.; Tao, N.; Feng, L.; Zhang, C. Specified value based defect depth prediction using pulsed thermography. *J. Appl. Phys.* **2012**, *112*, 023112. [CrossRef]
23. Manohar, A.; Lanza di Scalea, F. Determination of defect depth and size using virtual heat sources in pulsed infrared thermography. *Exp. Mech.* **2013**, *53*, 661–671. [CrossRef]
24. Wang, Q.; Liu, Q.; Xia, R.; Li, G.; Gao, J.; Zhou, H.; Zhao, B. Defect depth determination in laser infrared thermography based on LSTM-RNN. *IEEE Access* **2020**, *8*, 153385–153393. [CrossRef]
25. Fang, Q.; Maldague, X. A method of defect depth estimation for simulated infrared thermography data with deep learning. *Appl. Sci.* **2020**, *10*, 6819. [CrossRef]
26. Saeed, N.; Abdulrahman, Y.; Amer, S.; Omar, M.A. Experimentally validated defect depth estimation using artificial neural network in pulsed thermography. *Infrared Phys. Technol.* **2019**, *98*, 192–200. [CrossRef]
27. Jocher, G. YOLOv5 by Ultralytics. 2020. Available online: https://doi.org/10.5281/zenodo.3908559 (accessed on 21 November 2023).
28. Zhao, Z.; Yang, X.; Zhou, Y.; Sun, Q.; Ge, Z.; Liu, D. Real-time detection of particleboard surface defects based on improved YOLOV5 target detection. *Sci. Rep.* **2021**, *11*, 21777. [CrossRef]
29. Li, C.; Yan, H.; Qian, X.; Zhu, S.; Zhu, P.; Liao, C.; Tian, H.; Li, X.; Wang, X.; Li, X. A domain adaptation YOLOv5 model for industrial defect inspection. *Measurement* **2023**, *213*, 112725. [CrossRef]
30. Xiang, X.; Wang, Z.; Qiao, Y. An improved YOLOv5 crack detection method combined with transformer. *IEEE Sens. J.* **2022**, *22*, 14328–14335. [CrossRef]
31. Wang, C.Y.; Liao, H.Y.M.; Wu, Y.H.; Chen, P.Y.; Hsieh, J.W.; Yeh, I.H. CSPNet: A new backbone that can enhance learning capability of CNN. In Proceedings of the IEEE/CVF Conference on Computer Vision and Pattern Recognition Workshops, Seattle, WA, USA, 14–19 June 2020; pp. 390–391.
32. Liu, S.; Qi, L.; Qin, H.; Shi, J.; Jia, J. Path aggregation network for instance segmentation. In Proceedings of the IEEE Conference on Computer Vision and Pattern Recognition, Salt Lake City, UT, USA, 18–23 June 2018; pp. 8759–8768.
33. Xu, B.; Wang, N.; Chen, T.; Li, M. Empirical evaluation of rectified activations in convolutional network. *arXiv* **2015**, arXiv:1505.00853.
34. Xu, R.; Lin, H.; Lu, K.; Cao, L.; Liu, Y. A forest fire detection system based on ensemble learning. *Forests* **2021**, *12*, 217. [CrossRef]
35. Russakovsky, O.; Deng, J.; Su, H.; Krause, J.; Satheesh, S.; Ma, S.; Huang, Z.; Karpathy, A.; Khosla, A.; Bernstein, M.; et al. Imagenet large scale visual recognition challenge. *Int. J. Comput. Vis.* **2015**, *115*, 211–252. [CrossRef]
36. Davis, J.; Goadrich, M. The relationship between Precision-Recall and ROC curves. In Proceedings of the 23rd International Conference on Machine Learning, Pittsburgh, PA, USA, 25–29 June 2006; pp. 233–240.

applied sciences

MDPI

Article

Pulsed Thermography Dataset for Training Deep Learning Models

Ziang Wei [1,2,3,*], Ahmad Osman [1,2,3], Bernd Valeske [3,4] and Xavier Maldague [2]

1 School of Engineering, University of Applied Sciences in Saarbrücken, 66117 Saarbrücken, Germany
2 Faculty of Science and Engineering, Department of Electrical and Computer Engineering, University Laval, Quebec City, QC G1V 0A6, Canada
3 Fraunhofer Institute for Nondestructive Testing IZFP, 66123 Saarbrücken, Germany
4 Chair of Cognitive Sensor Systems, Saarland University, 66123 Saarbrücken, Germany
* Correspondence: ziang.wei@izfp.fraunhofer.de

Abstract: Pulsed thermography is an indispensable tool in the field of non-destructive evaluation. However, the data generated by this technique can be challenging to analyze and require expertise to interpret. With the rapid progress in deep learning, image segmentation has become a well-established area of research. This has motivated efforts to apply deep learning methods to non-destructive evaluation data processing, including pulsed thermography. Despite this trend, there has been a lack of public pulsed thermography datasets available for the evaluation of various spatial-temporal deep learning models for segmentation tasks. This paper aims to address this gap by presenting the PVC-Infrared dataset for deep learning. In addition, we evaluated the performance of popular deep-learning-based instance segmentation models on this dataset. Furthermore, we examined the effect of the number of frames and data transformations on the performance of these models. The results of this study suggest that appropriate preprocessing techniques can significantly reduce the size of the data while maintaining the performance of deep learning models, thereby speeding up the data processing process. This highlights the potential for using deep learning methods to make non-destructive evaluation data analysis more efficient and accessible to a wider range of practitioners.

Keywords: pulsed thermographic dataset; deep learning; defect detection; non-destructive evaluation

1. Introduction

Pulsed thermography (PT) has gained widespread popularity as a Non-Destructive Evaluation (NDE) method in the past few years, due to its high accuracy and speed. It is especially suitable for detecting near-surface defects in materials such as metal, Carbon Fiber Reinforced Polymers (CFRP), and Polyvinyl Chloride (PVC). Its application spans several industries, including aerospace and automotive. However, the high complexity of PT data, which often contains noise, artifacts, and inconsistencies, makes its inspection challenging. Therefore, the application of deep learning methods in processing PT data is an important research topic. By using deep learning models, the efficiency and accuracy of defect detection within the PT data can be significantly improved. This is crucial for ensuring the reliability and safety of various engineering systems. Thus, the development of accurate and efficient deep-learning-based techniques for PT data processing can greatly enhance the capabilities of NDE for various industrial applications.

In recent years, deep learning algorithms have made remarkable progress in various image processing tasks such as image classification [1–3], object detection [4–6], image segmentation [7–9], and video understanding [10,11]. To evaluate the performance of these algorithms, commonly used publicly available datasets, such as ImageNet [12], PASCAL [13], COCO [14] for image classification and Cityscapes [15], ADE20K [16] for image segmentation, play a crucial role. These datasets serve as benchmarks for different deep learning methods. However, collecting such datasets in the NDE field presents several challenges. One of the main challenges is dealing with imbalanced datasets [17], where the majority of components being tested are free of defects or damage, while only a small proportion of them are defective. This can render an artificial intelligence (AI) model ineffective if proper data preprocessing is not conducted. In addition to the imbalanced dataset challenge, the limited availability of data is another challenge in the NDE field. In particular, obtaining PT data is more difficult than obtaining normal images, as it requires specialized equipment and expertise to perform experiments in a dedicated laboratory. Furthermore, the labeling of PT data presents a multi-faceted challenge. Accurately analyzing PT data and labeling it requires skilled expertise. Additionally, validating the absence or presence of defects can be a challenging task, as defects are often hidden beneath the surface of the object. Achieving accurate validation may require destructive testing or CT scanning, both of which can be costly or damaging to the testing object.

In the field of NDE, deep learning methods for data segmentation have been applied to defects defection [18–20]. However, access to publicly available datasets is often limited due to internal regulations. This paper addresses this issue by providing a well-designed yet relatively simple PT dataset that can facilitate future research in evaluating AI algorithms. The dataset can be found at https://www.kaggle.com/datasets/ziangwei/irtpvc (accessed on 23 February 2023). Additionally, we have experimented with several popular deep learning models for instance segmentation on this dataset. In summary, this work makes several contributions. Firstly, a publicly available and carefully designed PT dataset for PVC components with defects of various sizes and depths is developed. This dataset can serve as a benchmark for training deep learning models and further research in the field. Secondly, the study includes experiments and results comparison using several popular encoder-decoder based defect segmentation models on the developed dataset. Finally, the efficiency of the models has been improved through the use of appropriate data transform methods and employing a frames selection method on the PT data. Overall, this work provides a valuable resource for researchers and engineers seeking to improve the accuracy and efficiency of deep learning based methods for defects detection within PT data.

2. Related Work

2.1. Pulsed Thermography

Pulsed thermography is an active thermographic method that involves the use of a heat source to stimulate an object with a heat pulse. An infrared camera then records a video of the object to measure the cooling process on the object's surface, which can be described, using a 1D formulation, by the following Equation [21–23]:

$$y(z,t) = T_0 + \frac{Q}{e\sqrt{\pi t}} exp(-\frac{z^2}{4\alpha t}) \tag{1}$$

$$e = \sqrt{k\rho c_p} \tag{2}$$

where α is the thermal diffusivity of the sample. k is the sample's thermal conductivity. c_p is the heat capacity of the sample. e denotes the material's thermal effusivity. It describes the materials' capability of exchanging thermal energy with their surroundings. z is the depth. At the surface of the sample where z is zero, the equation can be simplified to:

$$y(t) = T_0 + \frac{Q}{e\sqrt{\pi t}} \tag{3}$$

Since the sound region has homogeneous materials, its thermal diffusivity and conductivity are similar. However, the material of the defective area has different parameters, resulting in a distinct thermal behavior. This difference forms the theoretical basis for PT as a method for non-destructive testing. PT has been applied to a range of materials, including

CFRP and glass fiber reinforced materials, as well as used for the inspection of spot welds and in the protection of cultural heritage [19,24–26].

2.2. Deep-Learning-Based Methods Used in Thermography for NDE

In recent years, the use of deep-learning-based methods has shown significant improvement in data classification and segmentation for PT data. In [20], Mask-RCNN has been applied to a synthetic experimental PT dataset to detect abnormal regions. In [19], authors have employed an U-Net model to segment the defect regions on curved CFRP samples using PT technique. In [27], Faster-RCNN models with Inception V2 and Inception Resnet V2 architectures are trained to identify defects within composite materials using thermographic images. Average precision is used to assess the models' performance. The results show that the model with Inception V2 outperforms the one with Inception Resnet V2. A hydrogen deep neural network that incorporates temporal and spatial information on the infrared data to detect the defects in composite and coating material is proposed in [28]. In [29], the authors use PT to detect the impact damage within the composite samples. They employ a support vector machine to classify the defective areas from the healthy area. They report the classification results ranging from 78.7 % to 93.5 %. In [30], recurrent neural networks and feed-forward neural networks are investigated to detect defects within non-planar CFRP specimens. The results indicate that a long short-term memory neural network outperforms a feed-forward neural network when dealing with time-dependent information. A feed-forward neural network to classify the defects into different categories is proposed [31]. Two models are trained on raw data and Thermographic Signal Reconstruction (TSR) coefficients, respectively. The model using TSR coefficients outperforms the other using raw data.

As the use of generative adversarial networks (GANs) has become increasingly popular, they have also been applied to processing thermographic data. A GAN-based thermal image argumentation method is applied to thermographic data to enhance the visibility of defects in CFRP components in [32]. In [33], the authors introduce IRT-GAN, which is trained on a large augmented numerical thermographic dataset. The dataset is fully simulated for composite materials with different defect types, sizes, and depths. The model takes TSR coefficients as input and is able to produce images for defect segmentation.

2.3. Public Available Thermographic Dataset

Infrared imaging and video datasets have been widely used in various applications, such as PTB-TIR [34] for pedestrian tracking and LLVIP [35] for low-light vision. However, most of the existing datasets are not designed specifically for NDE tasks. In a recent study, a PT dataset was introduced that contains composite material samples made of carbon and glass fiber-reinforced plastic [36]. However, this dataset is not publicly accessible and lacks annotations for machine learning applications. In [37], the authors describe the infrared measurement databases from Laval University, which include various materials such as composite materials, plexiglass, steel, aluminum, etc. The focus of this study is on defect segmentation in single infrared images and does not take into account the temporal information between thermal images. To the best of our knowledge, there is currently a lack of well-documented, publicly available PT datasets for defect segmentation that consider both spatial and temporal information within the thermal image sequences.

3. Data Preparation and Preliminaries

3.1. Specimens' Information and Data Acquisition

In this study, we introduce the PVC-Infrared dataset, which is a deep learning dataset that consists of 19 thermal image sequences for 19 PVC specimens. Each specimen has a dimension of 100 mm × 100 mm × 5 mm, with cylindrical holes of varying sizes and depths created on the bottom side to simulate subsurface defects, as depicted in Figure 1. The diameter of the cylindrical holes ranges from 2 mm to 10 mm, and the depth ranges from 2.5 mm to 4.5 mm from the bottom surface. Figure 2 provides CAD images of eight specimens viewed from the bottom, while Figure 3 illustrates a comparison of a CAD image and an optical image of a single specimen.

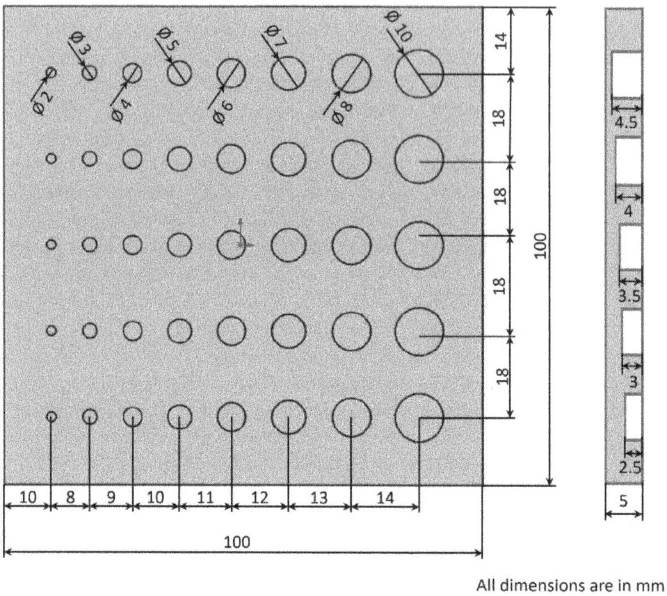

Figure 1. An example of specimens with defects in all possible sizes in all possible locations. The right part of the image shows all the possible depths of the defects, ranging from 2.5 mm to 4.5 mm from the right surface (bottom side).

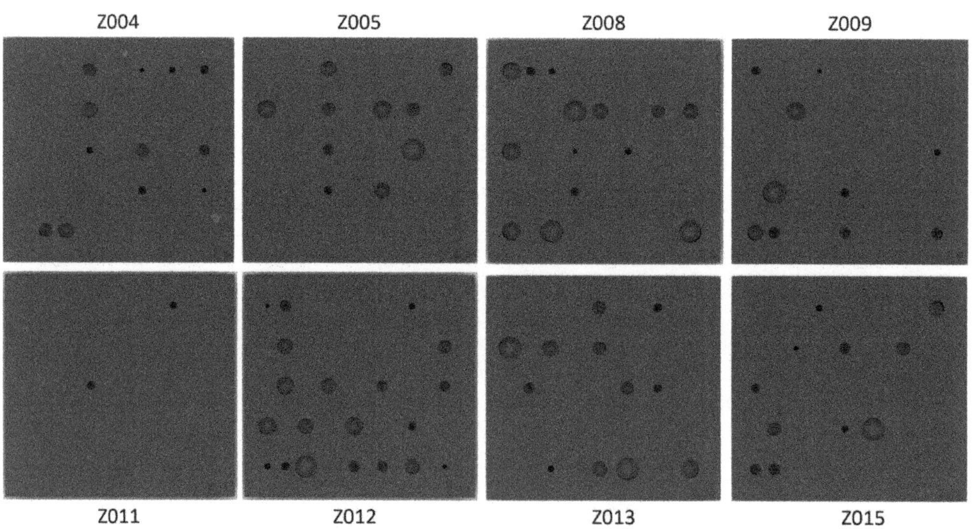

Figure 2. Visualization of eight specimens from the bottom side.

Figure 3. Images for a specimen, the **left** image is the CAD visualization, the **right** image is an optical image for it.

The experimental setup is demonstrated in Figure 4. The FLIR SC5000 infrared camera and lamps were controlled by a computer through a control unit to ensure synchronization. The recording process began with the trigger signal sent to the camera, which started recording after a pre-defined overall recording time. The lamps were then triggered to flash. In this experiment, two Broncolor lamps with a power of 6.4 KJ were used as heat sources for warming the specimens. The camera captured the infrared video with a resolution of 320 × 256 and recorded it for 181 seconds at 10 Hz, resulting in a video with a dimension of 320 × 256 × 1810 for each measurement.

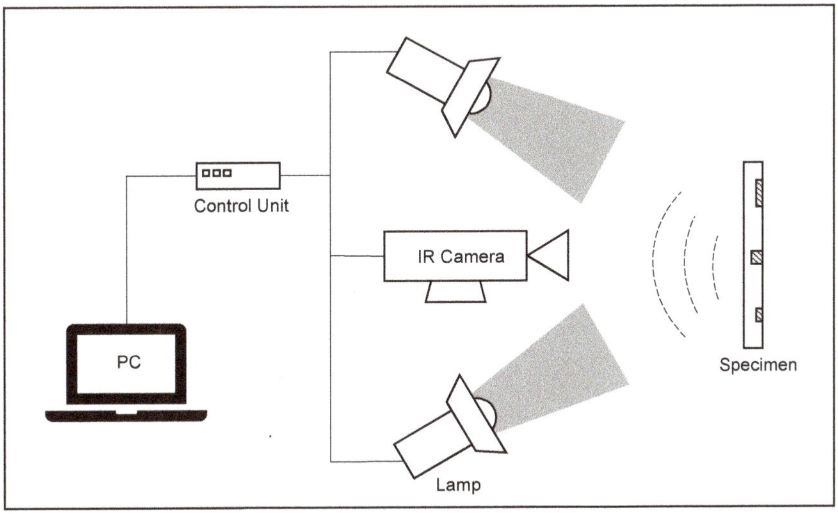

Figure 4. Setup of the PT experiment.

3.2. Pulsed Phase Thermography

Pulsed Phase Thermography (PPT) is a thermal image sequence processing that is proposed in [38,39]. The method involves the calculation of the phase of a thermal image sequence using a Discrete Fourier Transform (DFT). The DFT of a pixel value x_n at image frame n is calculated as:

$$y_k = \frac{1}{N} \sum_{n=0}^{N-1} x_n e^{-\frac{i2\pi kn}{N}} = Re_k + iIm_k \tag{4}$$

where Re_k and Im_k denote the real and imaginary parts of the DFT, respectively. The phase Φ_k and amplitude A_k of the DFT can be calculated using:

$$\Phi_k = atan(\frac{Im_k}{Re_k}) \tag{5}$$

$$A_k = \sqrt{Re_k^2 + Im_k^2} \tag{6}$$

Phase images of DFT results typically exhibit better contrast of abnormal regions, making them a preferred choice for further analysis. Usually, the initial 5 to 10 phase images provide an adequate representation of the raw thermal image sequence.

3.3. Principal Component Thermography

Principal Component Thermography (PCT) is introduced in [40]. This method applies Singular Value Decomposition (SVD) to the thermal image sequences. The 3D thermal video, with dimensions (N, W, H), is first reshaped to (N, WH) by a raster-like operation. The SVD of the reshaped raw data matrix X is then performed using:

$$X = U\Gamma V^T \tag{7}$$

where X is the $N \times WH$ raw data matrix. U is a $WH \times N$ matrix. Γ is a $N \times N$ matrix, and V^T is the transpose of an $N \times N$ matrix. When matrix A is arranged so that its columns contain time variations and its rows contain spatial variations, the columns of matrix U contain the Empirical Orthogonal Functions (EOFs). Typically, only the first few EOFs are used for further analysis.

3.4. PPT and PCT Images

Figure 5 shows the PPT images and PCT images for four specimens. The CAD images of these specimens are presented in Figure 2. However, the images in Figure 2 are horizontally flipped in comparison to the images in Figure 5, as the former shows a bottom view while the latter exhibits a top view. It is noteworthy that, qualitatively, the majority of defects are clearly identifiable in both PPT and PCT images, with good contrast. However, certain defects exhibit better contrast in PCT images than in PPT images, as indicated by the green arrows in Figure 5. Thus, PCT images are utilized for data labeling.

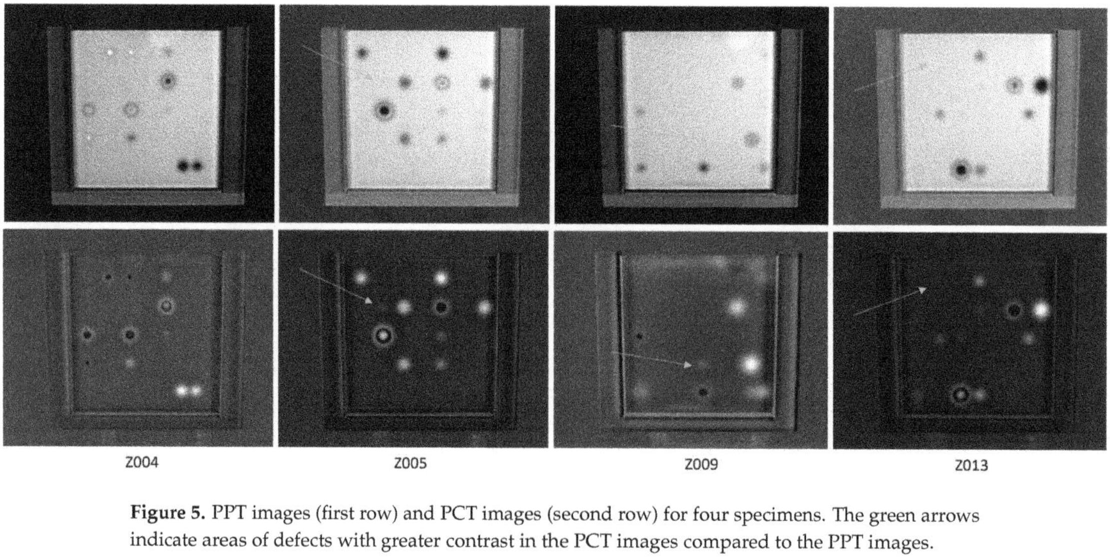

Figure 5. PPT images (first row) and PCT images (second row) for four specimens. The green arrows indicate areas of defects with greater contrast in the PCT images compared to the PPT images.

3.5. Data Labeling

The data labeling process is conducted manually on PCT images. The first 10 PCT images (EoFs) are visually inspected to identify the one with the best contrast of the defects. Manual labeling is then performed by placing circles on the selected PCT image. The labels are considered reliable as all defects are known and can be directly validated by viewing the bottom side of the specimens.

4. Experiments with Deep Models for Defect Segmentation

4.1. U-Net

Introduced in [7], U-Net is a popular and effective neural network approach for biomedical image segmentation. It is composed of two key components: the encoding stage and the decoding stage. The encoding stage involves several consecutive blocks of convolutional neural network operations, each followed by a max pooling operation. The decoding stage is comprised of an equal number of transposed convolution blocks to upsample the output back to the same dimensions as the input data. Importantly, U-Net also includes skip connections, which facilitate the transfer of high-resolution information from the encoding stage to the decoding stage by performing copy and concatenation of feature maps. In this work, we employ a lightweight version of U-Net, with a single convolutional layer used as an operation block, as shown in Figure 6.

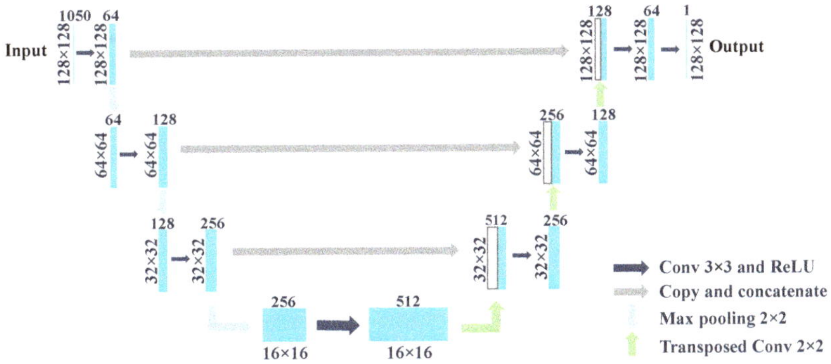

Figure 6. U-Net architecture presented in [19], which is licensed under CC BY 4.0.

4.2. SegNet

SegNet, proposed in [9], is similar to the U-Net architecture, which has encoding and decoding stages. The main difference is that, in SegNet, only the locations where max pooling is performed, which can also be referred to as pooling indices, are copied and used for upsampling to the decoding stage at the skip connection instead of copying the entire feature map from the encoding stage. Compared to the original SegNet, we have employed fewer steps both at the encoding and decoding stages. The model architecture used is illustrated in Figure 7.

4.3. Model Pre-Training

To improve the performance of the encoder-decoder model in processing thermal images, pre-trained models such as VGG Net [2] or ResNet [3], which have been trained on optical images, can be used as encoders due to their well-designed network architecture. However, as thermal images often have many more image channels than the standard three-channel RGB images typically used in these pre-trained models, they cannot be directly applied to thermal imaging data.

To address this issue, we modified the parameters in the first layer of the pre-trained models to adapt them to the thermal imaging data by calculating new weights using the following formula:

$$w_j^{'} = \frac{\sum_1^C w_i}{C'} \quad (8)$$

Here, w_i represents the weights of the neurons in the ith channel of the network's first layer, with C being the original channel number (3 for RGB images). $C^{'}$ represents the new channel number, which in our case is 1000. The new weights, $w_j^{'}$, are calculated by summing the weights of the original channels and dividing by the new channel number. This allowed us to utilize pre-trained VGG11 models on the ImageNet dataset to process our thermal imaging data.

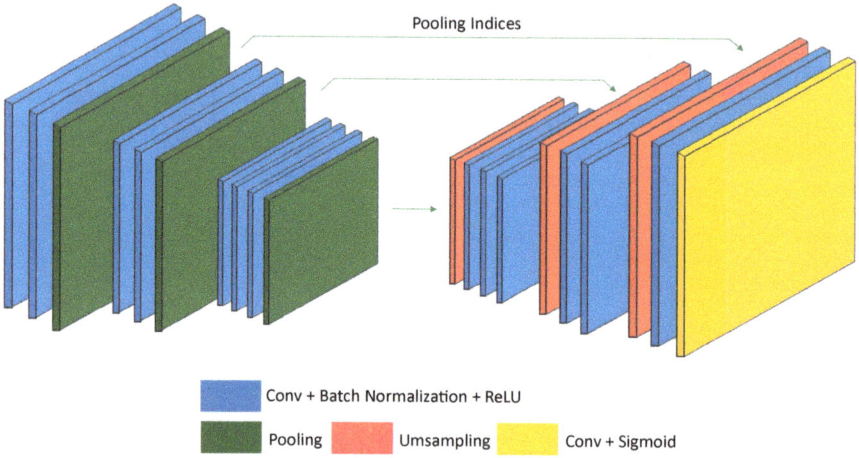

Figure 7. SegNet architecture adapted from a figure in [9], which is licensed under CC BY 3.0.

4.4. Experiments on the Original Dataset

In this study, the data are divided into three distinct sets: training, validation, and test. The training set consisted of 13 measurements, while the validation and test sets comprised of three measurements each. To evaluate the impact of pre-training on performance, several models are employed, including modified versions of U-Net and SegNet, as well as U-Net and SegNet models that utilized a VGG11 encoder backbone [41].

Stochastic gradient descent (SGD) is employed as the optimizer for training the models, which are trained for 3000 epochs. The loss function used in this study is binary cross-entropy, calculated according to the formula:

$$L = -\frac{1}{N}\sum_{i=1}^{N} y_i \cdot log(p(y_i)) + (1 - y_i) \cdot log(1 - p(y_i)) \quad (9)$$

where $y_i \in \{0, 1\}$ is the model's binary classification prediction, $p(y_i) \in [0, 1]$ is the model's prediction of the probability, and N represents the number of samples.

The quantitative results of the models' performance on the original test set are presented in Table 1. The evaluation metric used is Intersection over Union (IoU), and the average IoU is calculated for the three test samples. U-Net with a VGG11 encoder achieves the highest average IoU of 0.722. Interestingly, U-Net with a pre-trained VGG11 encoder produces comparable results, suggesting that pre-training does not significantly affect performance. To provide visual insight into the models' performance, Figure 8 displays sample segmentations obtained by U-Net and SegNet. Both models produce satisfactory segmentations of the defects, with SegNet tending to classify more pixels as foreground than U-Net. Finally, the processing speed of the models is evaluated. The results indicate that they are relatively slow, achieving 1–2 frames per second (FPS) on a machine with two Intel Xeon E5-2630 v4 @ 2.20GHz CPUs. This suggests that the models may not be suitable for industrial deployment where faster processing speeds are often required.

Table 1. Results on the original test dataset.

	Average IoU	Model Size
U-Net	0.683	18.4 MB
SegNet	0.668	17.2 MB
U-Net VGG11	0.722	91.4 MB
U-Net VGG11 pre-trained	0.712	91.4 MB
SegNet VGG11	0.588	95.3 MB
SegNet VGG11 pre-trained	0.599	95.3 MB

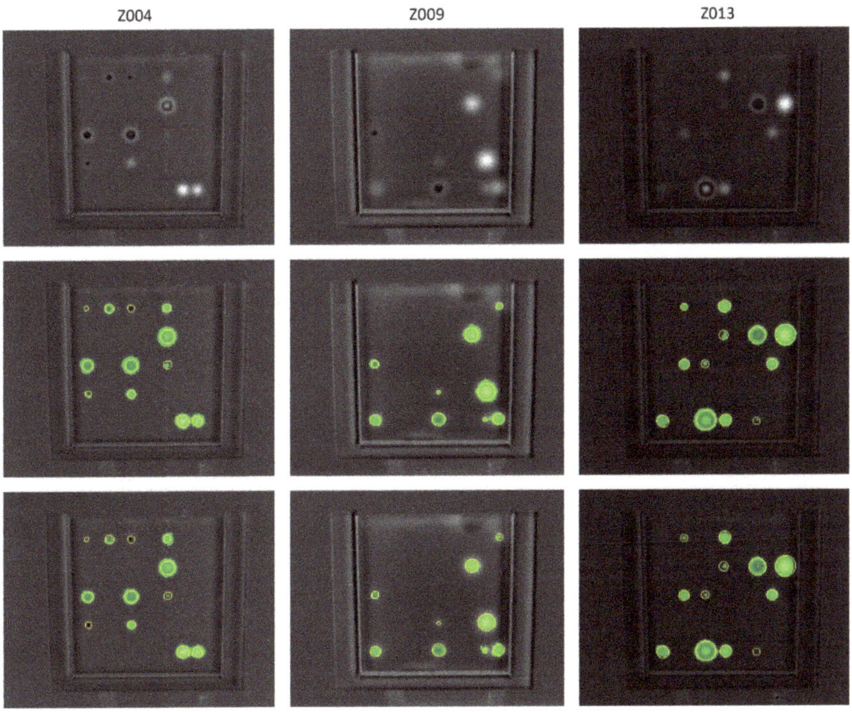

Figure 8. Visualization of segmentation results. The green masks represent the segmentations, and the yellow contours represent the ground truth areas. The first row shows PCT images of the PT measurements belonging to the test dataset. The second row displays the defect segmentation results obtained by applying SegNet on the original test dataset. The third row presents the corresponding results achieved by U-Net on the same test dataset of original data.

4.5. Efficient Data Segmentation Methods

The use of deep learning models is motivated by their ability to extract features effectively. Thus, it is reasonable to apply them directly to raw data. However, PT data are often substantial in size, consuming a significant amount of computational resources, and making industrial deployment challenging. To address this issue, we will explore two strategies for reducing the size of PT data in this section.

4.5.1. Experiment with PPT and PCT Images

Manually inspecting a large number of frames in each PT measurement can be labor-intensive and tedious. To mitigate this issue, PPT, PCT, and TSR are often applied to extract the most relevant information from the original PT data and generate a more concise representation, thus reducing the number of images that need to be inspected. This highlights the potential of these methods as effective data preprocessing techniques to decrease the size of the training data for deep learning models, which in turn reduces the computational cost required for training. In our study, the first 10 images generated by PCT and PPT are selected as the training data.

We evaluate the performance of two models on the PPT and PCT images, using the Adam optimizer [42] and consistent hyperparameters as described in Section 4.4. The results are summarized in Table 2. Additionally, the models processed the PPT and PCT images approximately three times faster than the models for the original PT data.

Table 2. Results on validation dataset of PPT and PCT images.

	Average IoU	Model Size
U-Net on original data	0.748	18.4 MB
SegNet on original data	0.729	17.2 MB
U-Net on PPT	0.732	14.5 MB
SegNet on PPT	0.717	13.3 MB
U-Net on PCT	0.638	14.5 MB
SegNet on PCT	0.625	13.3 MB

4.5.2. Experiment with Fewer Thermal Frames

The thermal profiles of six different areas, one sound and five defective, are compared in Figure 9. The results demonstrate that the majority of the thermal value difference occurs in the early stage. Hence, using only the initial portion of the thermal sequence as input data may prove to be an effective approach for deep learning models to detect defects.

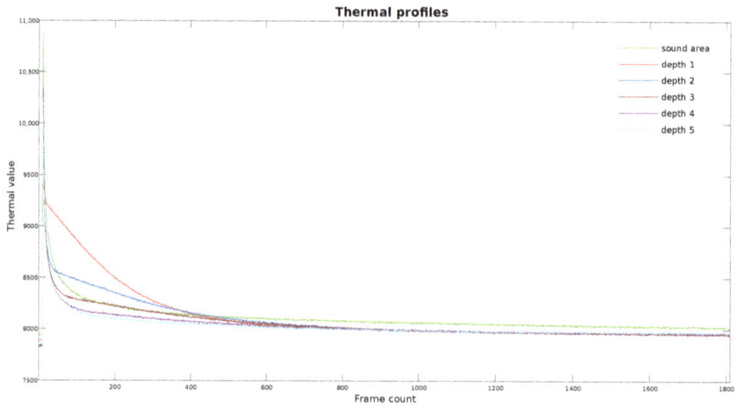

Figure 9. Thermal profiles comparison from sound area and defective areas with different defect depths.

To assess the impact of the number of frames on the model performance, the Average Intersection over Union (IoU) metric is used on the validation dataset. The results of the experiment, as shown in Table 3, highlight that the best performance is obtained with the use of the initial 200 frames of the thermal measurement. This suggests that the majority of the relevant information lies within the early stages, and that including additional trailing frames may not contribute significantly to the defect evaluation. The training hyperparameters are kept consistent with those used in Section 4.4. Additionally, using 140 or 200 frames resulted in faster processing times, with the models running two to three times faster compared to the original dataset.

Table 3. Results (Average IoU) with only leading frames of thermal sequence on the validation dataset. The original measurement contains 1810 frames.

	1810	200	140	100	50	25
U-Net	0.748	0.762	0.729	0.728	0.705	0.467
SegNet	0.729	0.769	0.730	0.727	0.723	0.478

5. Discussion

The advancement of applied AI heavily relies on the availability of quality datasets. Despite this, public datasets in the NDE field are scarce due to difficulties in data acquisition. To address this, we present a dataset for the training of deep learning models in the field of NDE. The dataset comprises of 19 measurements collected from 19 PVC specimens with artificially introduced defects.

Unlike common NDE datasets that contain completely hidden defects, the defects in our dataset are visible from the bottom side, allowing for accurate and easy labeling. Our dataset is larger than many NDE datasets used for AI applications, for example, the dataset used in [18], which contains fewer than 10 specimens. A larger dataset can help reduce the risk of overfitting in deep learning models.

Experiments are conducted using several deep encoder-decoder models for defect segmentation on the original dataset. The results indicate that the U-Net VGG11 model, without the use of pre-trained parameters, outperforms other models. The results also suggest that the use of a pre-trained model does not necessarily improve model performance, as observed in both U-Net and SegNet models. Incorporating VGG11 as the encoder backbone of the U-Net model improves its performance significantly, which is not the case for SegNet.

The results in Table 2 demonstrate that PT data transformation methods, such as PCT and PPT, can effectively reduce the PT data size without sacrificing the performance of the models. Furthermore, the thermal profiles reveal that the most significant difference in thermal values between sound and defective areas occurs early in the measurement process. This issue is further explored by using only the leading frames as training data, where the results show that the model's performance barely decreased. Models that use only 200 frames even performed better than those that use the entire dataset, suggesting that a significant portion of the raw thermal images contains redundant information.

The study has some limitations. Firstly, it does not consider the defects' depth information, which is essential for some applications. Secondly, the models are only tested on defects of a single shape, and it would be valuable to evaluate their ability to detect defects of various shapes. Lastly, the dataset only contains one type of material, PVC. However, the proposed methods have the potential to apply to other materials, such as CFRP and steel, provided that the defects are visible within PT data. Despite these limitations, the study provides a valuable contribution to the development of AI-based methods for defect detection in PT data. The dataset and the results can serve as a benchmark for future research in this area.

6. Conclusions

In this study, we introduce the PVC-Infrared dataset, a dataset designed specifically for the detection of defects using PT as an NDE method. The purpose of this dataset is to further advance the research in this field and contribute to the growth of deep learning technologies. We have performed experiments with various deep learning models, including U-Net, SegNet, and their variants with a VGG network as a backbone. These experiments are conducted on both the original dataset and on transformed datasets that are processed using PPT and PCT techniques. Additionally, we have explored the impact of using only the leading thermal frames of the PT measurement. The results show that utilizing fewer frames can lead to comparable or even better performance compared to using the entire dataset. It opens up opportunities for improving the efficiency of deep learning methods for PT data processing in industrial settings. In future research, it is important to consider the depth information of defects and evaluate the effectiveness of AI models on PT datasets with defects of varying shapes. Additionally, the exploration of different frame selection methods can lead to improvements in the efficiency of defect segmentation models. By considering these factors, further research can enhance the accuracy and robustness of AI models for defect detection in PT data.

Author Contributions: Z.W.: methodology, data labeling, training the models and original drafting of the manuscript. A.O.: supervision, expertise in AI, manuscript review and editing. B.V.: supervision, funding acquisition, manuscript review and editing. X.M.: supervision, manuscript review and editing. All authors have read and agreed to the published version of the manuscript.

Funding: This research was supported by the European Regional Development Fund (ERDF).

Data Availability Statement: The PT dataset used in the paper can be accessed via the URL https://www.kaggle.com/datasets/ziangwei/irtpvc (accessed on 23 February 2023). The dataset is available for academic and research use only. Papers using this dataset are kindly requested to refer to this paper.

Acknowledgments: Support from Fraunhofer IZFP is gratefully acknowledged. David Müller is acknowledged for the design and preparation of the test specimen used in the paper within their research on the evaluation of deep learning approaches for quantitative defect testing and reconstruction with thermography. Udo Netzelmann is acknowledged for providing advice for the experimental setup.

Conflicts of Interest: The authors declare no conflict of interest.

Abbreviations

The following abbreviations are used in this manuscript:

AI	Artificial Intelligence
PT	Pulsed Thermography
PPT	Pulsed Phase Thermography
PCT	Principal component Thermography
TSR	Thermographic Signal Reconstruction
NDE	Nondestructive Evaluation
IoU	Intersection over Union
CFRP	Carbon Fiber Reinforced Polymers

References

1. Krizhevsky, A.; Sutskever, I.; Hinton, G.E. Imagenet classification with deep convolutional neural networks. *Commun. ACM* **2017**, *60*, 84–90. [CrossRef]
2. Simonyan, K.; Zisserman, A. Very deep convolutional networks for large-scale image recognition. *arXiv* **2014**, arXiv:1409.1556.
3. He, K.; Zhang, X.; Ren, S.; Sun, J. Deep residual learning for image recognition. In Proceedings of the IEEE Conference on Computer Vision and Pattern Recognition, Las Vegas, NV, USA, 26 June–1 July 2016; pp. 770–778.
4. Girshick, R.; Donahue, J.; Darrell, T.; Malik, J. Rich feature hierarchies for accurate object detection and semantic segmentation. In Proceedings of the IEEE Conference on Computer Vision and Pattern Recognition, Columbus, OH, USA, 23–28 June 2014; pp. 580–587.
5. Girshick, R. Fast r-cnn. In Proceedings of the IEEE International Conference on Computer Vision, Santiago, Chile, 7–13 December 2015; pp. 1440–1448.
6. Bochkovskiy, A.; Wang, C.Y.; Liao, H.Y.M. Yolov4: Optimal speed and accuracy of object detection. *arXiv* **2020**, arXiv:2004.10934.
7. Ronneberger, O.; Fischer, P.; Brox, T. U-net: Convolutional networks for biomedical image segmentation. In Proceedings of the International Conference on Medical Image Computing and Computer-Assisted Intervention, Munich, Germany, 5–9 October 2015; pp. 234–241.

8. Zhao, H.; Shi, J.; Qi, X.; Wang, X.; Jia, J. Pyramid scene parsing network. In Proceedings of the IEEE Conference on Computer Vision and Pattern Recognition, Honolulu, HI, USA, 21–26 July 2017; pp. 2881–2890.
9. Badrinarayanan, V.; Kendall, A.; Cipolla, R. Segnet: A deep convolutional encoder-decoder architecture for image segmentation. *IEEE Trans. Pattern Anal. Mach. Intell.* **2017**, *39*, 2481–2495. [CrossRef]
10. Karpathy, A.; Toderici, G.; Shetty, S.; Leung, T.; Sukthankar, R.; Fei-Fei, L. Large-scale video classification with convolutional neural networks. In Proceedings of the IEEE conference on Computer Vision and Pattern Recognition, Columbus, OH, USA, 23–28 June 2014; pp. 1725–1732.
11. Wang, X.; Girshick, R.; Gupta, A.; He, K. Non-local neural networks. In Proceedings of the IEEE Conference on Computer Vision and Pattern Recognition, Salt Lake City, UT, USA, 18–23 June 2018; pp. 7794–7803.
12. Deng, J.; Dong, W.; Socher, R.; Li, L.J.; Li, K.; Fei-Fei, L. Imagenet: A large-scale hierarchical image database. In Proceedings of the 2009 IEEE Conference on Computer Vision and Pattern Recognition, Miami, FL, USA, 20–25 June 2009; pp. 248–255.
13. Everingham, M.; Van Gool, L.; Williams, C.K.; Winn, J.; Zisserman, A. The pascal visual object classes (voc) challenge. *Int. J. Comput. Vis.* **2010**, *88*, 303–338. [CrossRef]
14. Lin, T.Y.; Maire, M.; Belongie, S.; Hays, J.; Perona, P.; Ramanan, D.; Dollár, P.; Zitnick, C.L. Microsoft coco: Common objects in context. In Proceedings of the European Conference on Computer Vision, Zurich, Switzerland, 6–12 September 2014; pp. 740–755.
15. Cordts, M.; Omran, M.; Ramos, S.; Rehfeld, T.; Enzweiler, M.; Benenson, R.; Franke, U.; Roth, S.; Schiele, B. The cityscapes dataset for semantic urban scene understanding. In Proceedings of the IEEE Conference on Computer Vision and Pattern Recognition, Las Vegas, NV, USA, 27–30 June 2016; pp. 3213–3223.
16. Zhou, B.; Zhao, H.; Puig, X.; Fidler, S.; Barriuso, A.; Torralba, A. Scene parsing through ade20k dataset. In Proceedings of the IEEE Conference on Computer Vision and Pattern Recognition, Honolulu, HI, USA, 21–26 July 2017; pp. 633–641.
17. Chakravarthy, A.D.; Bonthu, S.; Chen, Z.; Zhu, Q. Predictive models with resampling: A comparative study of machine learning algorithms and their performances on handling imbalanced datasets. In Proceedings of the 2019 18th IEEE International Conference On Machine Learning Furthermore, Applications (ICMLA), Boca Raton, FL, USA, 16–19 September 2019; pp. 1492–1495.
18. Wei, Z.; Osman, A.; Gross, D.; Netzelmann, U. Artificial intelligence for defect detection in infrared images of solid oxide fuel cells. *Infrared Phys. Technol.* **2021**, *119*, 103815. [CrossRef]
19. Wei, Z.; Fernandes, H.; Herrmann, H.G.; Tarpani, J.R.; Osman, A. A deep learning method for the impact damage segmentation of curve-shaped cfrp specimens inspected by infrared thermography. *Sensors* **2021**, *21*, 395. [CrossRef] [PubMed]
20. Fang, Q.; Ibarra-Castanedo, C.; Maldague, X. Automatic defects segmentation and identification by deep learning algorithm with pulsed thermography: Synthetic and experimental data. *Big Data Cogn. Comput.* **2021**, *5*, 9. [CrossRef]
21. Chung, Y.; Lee, S.; Kim, W. Latest advances in common signal processing of pulsed thermography for enhanced detectability: A review. *Appl. Sci.* **2021**, *11*, 12168. [CrossRef]
22. Fleuret, J.R.; Ebrahimi, S.; Ibarra-Castanedo, C.; Maldague, X.P. Independent component analysis applied on pulsed thermographic data for carbon fiber reinforced plastic inspection: A comparative study. *Appl. Sci.* **2021**, *11*, 4377. [CrossRef]
23. Ibarra-Castanedo, C.; Genest, M.; Servais, P.; Maldague, X.P.; Bendada, A. Qualitative and quantitative assessment of aerospace structures by pulsed thermography. *Nondestruct. Test. Eval.* **2007**, *22*, 199–215. [CrossRef]
24. Jensen, F.; Terlau, M.; Sorg, M.; Fischer, A. Active Thermography for the Detection of Sub-Surface Defects on a Curved and Coated GFRP-Structure. *Appl. Sci.* **2021**, *11*, 9545. [CrossRef]
25. Verspeek, S.; Ribbens, B.; Maldague, X.; Steenackers, G. Spot Weld Inspections Using Active Thermography. *Appl. Sci.* **2022**, *12*, 5668. [CrossRef]
26. Yao, Y.; Sfarra, S.; Lagüela, S.; Ibarra-Castanedo, C.; Wu, J.Y.; Maldague, X.P.; Ambrosini, D. Active thermography testing and data analysis for the state of conservation of panel paintings. *Int. J. Therm. Sci.* **2018**, *126*, 143–151. [CrossRef]
27. Bang, H.T.; Park, S.; Jeon, H. Defect identification in composite materials via thermography and deep learning techniques. *Compos. Struct.* **2020**, *246*, 112405. [CrossRef]
28. Luo, Q.; Gao, B.; Woo, W.L.; Yang, Y. Temporal and spatial deep learning network for infrared thermal defect detection. *Ndt Int.* **2019**, *108*, 102164. [CrossRef]
29. Alhammad, M.; Avdelidis, N.; Ibarra-Castanedo, C.; Torbali, M.; Genest, M.; Zhang, H.; Zolotas, A.; Maldgue, X. Automated Impact Damage Detection Technique for Composites Based on Thermographic Image Processing and Machine Learning Classification. *Sensors* **2022**, *22*, 9031. . [CrossRef]
30. Tao, Y.; Hu, C.; Zhang, H.; Osman, A.; Ibarra-Castanedo, C.; Fang, Q.; Sfarra, S.; Dai, X.; Maldague, X.; Duan, Y. Automated Defect Detection in Non-planar Objects Using Deep Learning Algorithms. *J. Nondestruct. Eval.* **2022**, *41*, 1–11. [CrossRef]
31. Duan, Y.; Liu, S.; Hu, C.; Hu, J.; Zhang, H.; Yan, Y.; Tao, N.; Zhang, C.; Maldague, X.; Fang, Q.; et al. Automated defect classification in infrared thermography based on a neural network. *Ndt Int.* **2019**, *107*, 102147. [CrossRef]
32. Liu, K.; Li, Y.; Yang, J.; Liu, Y.; Yao, Y. Generative principal component thermography for enhanced defect detection and analysis. *IEEE Trans. Instrum. Meas.* **2020**, *69*, 8261–8269. [CrossRef]
33. Cheng, L.; Tong, Z.; Xie, S.; Kersemans, M. IRT-GAN: A generative adversarial network with a multi-headed fusion strategy for automated defect detection in composites using infrared thermography. *Compos. Struct.* **2022**, *290*, 115543. [CrossRef]
34. Liu, Q.; He, Z.; Li, X.; Zheng, Y. PTB-TIR: A thermal infrared pedestrian tracking benchmark. *IEEE Trans. Multimed.* **2019**, *22*, 666–675. [CrossRef]

35. Jia, X.; Zhu, C.; Li, M.; Tang, W.; Zhou, W. LLVIP: A visible-infrared paired dataset for low-light vision. In Proceedings of the IEEE/CVF International Conference on Computer Vision, Montreal, BC, Canada, 11–17 October 2021; pp. 3496–3504.
36. Erazo-Aux, J.; Loaiza-Correa, H.; Restrepo-Giron, A.D.; Ibarra-Castanedo, C.; Maldague, X. Thermal imaging dataset from composite material academic samples inspected by pulsed thermography. *Data Brief* **2020**, *32*, 106313. [CrossRef] [PubMed]
37. Fang, Q.; Ibarra-Castanedo, C.; Maldgue, X. University Laval Infrared Thermography Databases for Deep Learning Multiple Types of Defect Detections Training. *Eng. Proc.* **2021**, *2*, 8032.
38. Maldague, X.; Marinetti, S. Pulse phase infrared thermography. *J. Appl. Phys.* **1996**, *79*, 2694–2698. [CrossRef]
39. Maldague, X.; Largouët, Y.; Couturier, J.P. A study of defect depth using neural networks in pulsed phase thermography: Modelling, noise, experiments. *Rev. Générale de Therm.* **1998**, *37*, 704–717. [CrossRef]
40. Rajic, N. Principal component thermography for flaw contrast enhancement and flaw depth characterisation in composite structures. *Compos. Struct.* **2002**, *58*, 521–528. [CrossRef]
41. Iglovikov, V.; Shvets, A. Ternausnet: U-net with vgg11 encoder pre-trained on imagenet for image segmentation. *arXiv* **2018**, arXiv:1801.05746.
42. Kingma, D.P.; Ba, J. Adam: A Method for Stochastic Optimization. In Proceedings of the ICLR (Poster), San Diego, CA, USA, 7–9 May 2015.

MDPI AG
Grosspeteranlage 5
4052 Basel
Switzerland
Tel.: +41 61 683 77 34

Applied Sciences Editorial Office
E-mail: applsci@mdpi.com
www.mdpi.com/journal/applsci

www.ingramcontent.com/pod-product-compliance
Lightning Source LLC
Chambersburg PA
CBHW040805250926
51850CB00007BA/1088
9783725851270